THE MOSQUITO

THE MOSQUITO:

Its Life, Activities, and Impact on Human Affairs

J. D. Gillett

1972
Doubleday & Company, Inc., Garden City, New York

The author and publishers wish to thank Laurence Pollinger Ltd. and the estate of the late Mrs Frieda Lawrence for permission to reproduce the poem 'The Mosquito' on pages 191–3, from *The Complete Poems of D. H. Lawrence*, published by William Heinemann Ltd.

Originally published in England by Weindenfeld & Nicolson under the title *Mosquitos*.

Library of Congress Catalog Card Number 72–76160

Printed in the United States of America
First Edition in the United States

CONTENTS

FIGURES

PLATES

between pages 148 *and* 149

between pages 172 *and* 173

between pages 196 *and* 197

ACKNOWLEDGMENTS

Figures

All but one of the text figures (figure 4, included by permission of the Commonwealth Agricultural Bureaux) were drawn especially for this book by Dr Judith G. Smith under the author's supervision. Wherever possible these were made from living or freshly caught material. The sources of those which had to be adapted from other drawings are acknowledged (figures 7, 10, 11, 12). The sources of the data from which the graphs have been made are also acknowledged (figures 13–17).

Plates

Plates 11, 15, 16 and 18 were made especially for this book, plate 11 by the SEATO Laboratories, Bangkok (by kind permission of Dr J. D. Gould), while plates 15 and 18 were painted by Dr Judith G. Smith. The photograph of *Anopheles gambiae* feeding on the author was taken by Mr John Champion (Brunel University). Plates 9 and 28–30 were taken by Mr Gregory R. Ballmer of Riverside, California.

Acknowledgments for the other plates are as follows: Plates 1 and 8—Mr S. A. Smith (London School of Hygiene and Tropical Medicine); Plates 2, 3, 4 and 5—Professor H. E. Hinton, FRS (Bristol University); Plate 6—*American Journal of Tropical Medicine*, September, 1951 (The Williams and Wilkins Co., Baltimore); Plates 7 and 24—*Arctic;* Plate 10—Dr P. Galindo and the

Gorgas Memorial Laboratory, Panama; Plate 17–the author; Plate 12–Professor A. J. Haddow and the Commonwealth Agricultural Bureaux; Plate 13–The author and Mr L. E. Hewitt (East African Virus Research Institute); Plate 14–Mr J. Champion and Dr M. D. R. Jones (Brunel University); Plates 19–22–World Health Organisation (Geneva); Plate 23–Professor Chamlong Harinasuta, Mr Pratan Sarikavanija and Miss Vanida Kerdpibule (Mahidol University, Bangkok); Plate 25–Dr C. R. Twinn (Canada); Plate 26–Professor B. Hocking (University of Alberta); Plate 27–Dr H. Townson (Liverpool School of Tropical Medicine); Plates 31–33, 35–38–The Wellcome Institute of the History of Medicine (London); Plate 34–The Royal Society (London).

Specimens

Live and preserved specimens were provided by the following:

Dr C. B. Cuella (Ross Institute of Tropical Medicine); Mr Angus McCrae (East African Virus Research Institute); Dr P. F. Mattingly and the Trustees of the British Museum (Natural History); Dr T. H. G. Aitken (Virus Laboratory, Belem, Brazil).

To

Sir S. Rickard Christophers F.R.S.,

whose original work on mosquitos, extending over a period of more than sixty years, has withstood the severe test of time.

PREFACE

It is now some seventy years since mosquitos came to be generally recognised as playing a vital part in the affairs of men. For more than half this period it has been my lot to study them, for the most part in Africa but in other parts of the world too. During this time I have seen the emphasis change from mainly descriptive to experimental, from field observation to the discipline of the bench. Both approaches are, of course, necessary but it has long seemed to me that either approach by itself is not altogether satisfactory, for the one leads to theories that are difficult or impossible to put to the test, the other to oversimplification by a deliberate removal of most of the complexities of the real life situation. Nevertheless, it is the laboratory experiment that usually provides a clue to the mechanisms that underlie overt behaviour.

My own work on mosquitos has been more or less equally divided between field and laboratory, ecology and epidemiology on the one hand and physiology, behaviour and the capacity to transmit specific disease organisms on the other. During this work I think I can claim to have acquired more than a nodding acquaintance with these seemingly frail but formidable creatures. Indeed, in some respects I have, I suppose, come to know them rather intimately. Much of what I have been privileged to find out on the subject has been published in the scientific press. It seemed to me, however, that enough is now known about mosquitos to make possible a coherent and interesting story for the specialist and general reader alike.

It is, then, partly to the general reader that I have aimed this book, the reader who knows something about mosquitos and wants to know more, and the reader who is interested in all the fellow-creatures that manage to share our lonely blue, white and brown planet. Throughout each chapter I have kept in mind the questions that are so often put to me whenever the subject of mosquitos is raised. While I have tried to answer most of these, my real hope is that some of my own interest and I trust not misplaced enthusiasm will rub off on to the reader and perhaps stimulate further questions. I, for one, never fail to marvel at these tiny flying-machines with their problems of aerodynamics, navigation, refuelling and safety, with their computer-like control systems ensuring that some at least survive to bring us their deadly load. To the reader who asks, 'What can you possibly see to interest you in these miserable creatures?' or 'So you work on mosquitos, so what?' (according to which side of the Atlantic he is on) I would say that I find it sobering to reflect that mosquitos have at times played a more effective part in determining the course of human history than the generals and other leaders who set themselves up to guide our fortunes. But when the *inevitable* question is asked, I can only ask in return, 'What *use* is a man?'

Because of the economic and medical importance of mosquitos, they have been studied particularly thoroughly. Indeed, the subject is now so vast that few of us can hope to do more than specialise in certain aspects: structure, classification, flight, biting habits, endocrine control of reproduction, inborn or circadian rhythms, transmission of disease and so on. In this book I have for the most part drawn from my own experience, but naturally I have had to lean heavily on that of my colleagues where my own deficiencies would have left conspicuous gaps in the story. I have been able to acknowledge most of these in the body of the text. If I have managed to say anything of interest or of use to my colleagues, I shall be delighted. Many of them may well disagree with some of the interpretations and points of view which I have not hesitated to express throughout the book. If these, or any of the speculations in which I have indulged (particularly in Chapter 1) are unacceptable or annoy, I am sorry. But if any of them suggest further lines of inquiry, then they will have served a useful purpose.

I am particularly pleased to have an opportunity of thanking all those who have helped with the preparation of this book. Most of all I wish to place on record my gratitude to Mrs Carol James, who not only typed and retyped various revisions of the text with accuracy and interest but maintained this interest even when typing and checking the long and what must have been tedious lists of species in the appendices; these I have adapted from Stone, Knight and Starcke's invaluable catalogue of the mosquitos of the world.

I am also greatly indebted to Miss Arwa Waliji of Dar-es-Salaam, who willingly gave up her own time to type some of the earlier chapters under what were rather difficult conditions. To my cousin, Miss Gertrude Hopper and to my son Dr Richard Gillett, I owe my special thanks for their critical and extremely helpful reading of the early draft.

I have been fortunate in obtaining the co-operation of Dr Judith Smith, who is responsible for all but one of the twenty-two line-drawings in the text and made the painting of *Sabethes belisarioi* on the cover; she is also responsible for Plates 15 and 18. I am also greatly indebted to Dr Thomas Aitken in Brazil, Dr Pedro Galindo in Panama and Dr Harold Trapido in Colombia, for their quick and magnificent response when I sent out what amounted to an SOS for specimens and photographs of *Sabethes;* the beautiful collection of freshly caught specimens which Dr Aitken made especially, together with his descriptions and photographs of the resting attitude of these remarkable mosquitos, made Dr Smith's pictures possible.

The sources of the photographs in the other plates are acknowledged on page ix. I must, however, especially thank Professor H. E. Hinton, FRS, for allowing me to select any pictures I particularly liked from his outstanding collection of electroscan micrographs. I must also thank Professors B. Hocking and William

Reeves and Dr C. R. Twinn for sending me prints of some of their previously published but none the less valuable pictures. It gives me special pleasure too to thank Mr Sidney A. Smith of the London School of Hygiene and Tropical Medicine for his beautiful pictures of eggs and of a feeding adult of *Mansonia;* the latter has not been published before. I must also place on record the kindness and generosity of the following: Dr Harold E. Stark of the SEATO Laboratories, Bangkok, for arranging for the photograph of artificial insemination; Mr Gregory R. Ballmer for the fine pictures of *Anopheles balabacensis.* I should also like to thank Miss Jane Papworth and Mr M. J. Dorling of the Library, Brunel University, for their valuable assistance. Finally, I should like to record the pleasure I have had in all my dealings with Mr Richard Carrington, formerly Scientific Editor of the World Naturalist series and with Mr Julian Shuckburgh and Miss Mary O'Connell of Messrs Weidenfeld and Nicolson, the [British] publishers.

THE MOSQUITO

1. MOSQUITOS IN GENERAL

Mosquitos have been coming to man and making a nuisance of themselves for nearly a million years, yet it is only recently that man has begun to turn his serious attention towards them. Till the end of the nineteenth century mosquitos were regarded merely as unpleasant little creatures that had unfortunately to be tolerated, or if possible avoided. Only occasionally were voices raised to suggest a more important role, but these were little more than guesses and for the most part went unheeded. It is true that the amazing life-histories of some species had been worked out in considerable detail, but the accounts of these were not then generally available. Since the turn of the century, however, the situation has undergone a dramatic change and these insects have been receiving an ever-increasing amount of our attention. Indeed, they have now probably been as intensively studied as have any other single group of animals.

It was not the surprising life-histories of these small two-winged flying animals that attracted our attention: such complicated ways of going about things were already familiar in other partially aquatic insects. Nor was it merely that they paid us visits on warm summer evenings to drink our blood and, as if that was not enough, left us with unpleasant reminders in the shape of ugly red swellings often accompanied by intense irritation. Something much more sinister brought us round to studying mosquitos; the discovery that these flying visits could be the prelude to sickness and death; the realisation that this lowly creature, this insignificant gnat could be responsible for whole epidemics, epidemics that in the past had helped shape the course of human history.

Once the importance of mosquitos in human affairs was recognised, work on this fascinating group of animals started in earnest. But at the turn of the century most of the two thousand and more species of mosquitos still awaited discovery, and before anything else could be contemplated a taxonomic task of Herculean proportions lay ahead. Before we can proceed to piece together the astonishing events that make up the life of a mosquito it will help us if we attempt to summarise briefly the results of this early period of exploration, description and classification.

It is not possible to pay tribute to all those who have been responsible for unravelling this gigantic tangle. Much was achieved at the turn of the century when F. V. Theobald set out to discover the unknown mosquitos of the world and described them in a series of now classic monographs [18]. Much was added during the years that followed by L. O. Howard, H. G. Dyar and F. Knab [13]. But if one had to pick out any one name from the many I think most workers on mosquitos today would agree on F. W. Edwards, whose monumental contributions span the period from 1911 till his untimely death some thirty years later [8]. While it may be invidious to single out any of the more recent contributors, special mention must nevertheless be made of Alan Stone for his world-wide study of mosquito taxonomy; his great catalogue of the mosquitos of the world, published in 1959 in collaboration with Kenneth Knight and Helle Starcke sits on the shelves of every mosquito worker today, whether primarily interested in classification or not [17].

Mosquitos belong to the order of two-winged flies known as the Diptera, which include the house-flies, stable-flies, bluebottles, tsetse-flies and very many others. The hind wings in all the Diptera have been modified as the halteres, small organs that vibrate during flight and are believed to have a gyroscopic function providing a constant flight attitude [199]. Mosquitos themselves belong to a suborder of the Diptera known as the

Nematocera. This suborder includes the midges, sand-flies, black-flies, moth-flies and harlequin-flies, mosquitos belonging to the family Culicidae.

The Culicidae are divided into three subfamilies: the Anophelinae including the well-known genus *Anopheles,* many species of which are responsible for the transmission of malaria; the Toxorhynchitinae, the enormous larvae of which eat other mosquito larvae; the Culicinae which, with about 2000 species, are further divided into two tribes: the Culicini and the Sabethini. The culicine mosquitos include such well-known genera as *Culex, Aedes* and *Mansonia,* the sabethines include *Sabethes, Wyeomyia* and *Malaya.*

The 2700 or so species of mosquitos at present known are arranged altogether in thirty-four genera. As might be expected most of these occur in the tropics; Central and South America top the list with about 750 species between them, but tropical Asia has almost as many, while Africa south of the Sahara Desert, together with the huge island of Madagascar, has some 550 species. Thus about 2000, or three-quarters of the total number of species, come from the tropics and subtropics. None the less, this still leaves about 600 species for the other regions: Australia with the Pacific Islands, including New Zealand and New Guinea, has nearly 500 of these, while the great northern regions made up of the United States, Canada, Greenland, Europe and non-tropical Asia as far east as Japan together with North Africa and the Middle East share about 290 species. About 130 of these occur in North America (sixty of them in Canada), the rest in Europe, the Mediterranean and northern Asia, thirty-two of them occurring in the United Kingdom.

Order DIPTERA
- Suborder Nematocera
  - Family Culicidae
    - Subfamily Anophelinae
      - Genus *Chagasia*
        - *Bironella*
        - *Anopheles*

Subfamily Toxorhynchitinae
Genus *Toxorhynchites*
Subfamily Culicinae

Tribe Sabethini
Maorigoeldia
Genus *Tripteroides*
Trichoprosopon
Wyeomyia
Phoniomyia
Limatus
Sabethes
Malaya
Topomyia

Tribe Culicini
Genus *Ficalbia*
Coquillettidia
Mansonia
Uranotaenia
Hodgesia
Zeugnomyia
Orthopodomyia
Aedeomyia
Psorophora
Heizmannia
Mattinglyia
Udaya
Eretmapodites
Aedes
Ayurakitia
Armigeres
Haemagogus
Opifex
Culiseta
Culex
Deinocerites

If the total number of species appears to come to more than 2700, this is because some of them are found in more than one of these regions. Indeed, some species are very widespread indeed. *Aedes aegypti,* for example, is found in most tropical places, both in the New and Old Worlds, while *Ae. vexans* is found in most subtropical regions and in the north and south temperate zones as well. But the truly cosmopolitan species is *Culex pipiens,* which, with its subspecies and varieties, is found almost all over the world wherever there is land, except in the far north and south.

The word mosquito itself is of Spanish derivation meaning a little fly—a *muscato, muskitto* or *musqueto* to give a few examples of its earlier spelling. Its use has crept in during the past three hundred years and gradually ousted the Middle English word 'gnat'. In many ways I prefer this older word with its common derivation with and appropriate kinship to gnash and gnaw, but usage now dictates 'mosquito' as the accepted word and 'gnat' is rather inappropriately relegated to the non-blood-sucking relatives of mosquitos.

During the last thirty years increasing attention has been paid to the mosquito as a living animal. To be sure there were always a few who studied behaviour and other aspects of mosquito biology, but these were exceptions at the time. Now the picture has changed, and competent mosquito taxonomists are becoming hard to find; we have ecologists, ethologists, physiologists, endocrinologists and geneticists studying mosquitos, some in the field, others in the laboratory and a few in both. But the number of species receiving this attention is still relatively small—most of them being those that are known to be associated with the transmission of disease.

In this book we are going to take a look at mosquitos for themselves and then go on to consider their influence on human affairs. Much has been learned about these insects, not only of their fascinating ways but also of the complex workings of their bodies and, if you like, of their minds. It is easy and all too usual to think that because mosquitos are small they must also be simple. Nothing could be further from the truth. Smallness is a relative term; things are either large or small when measured against ourselves. But is an elephant the more complex for being eighty times our weight? In biomolecular terms a mosquito is immense. A single cell from the lining of the gut of a mosquito, or from any other insect is infinitely more complex than the most sophisticated piece of hardware devised by man. Even the modern electronic computer is a model of naïveté by comparison.

Each mosquito contains everything that is required not only for the complexities of life itself in an inhospitable world but for the continuity of the species; it is able to reproduce itself. As a larva it is able to swim in water, as an adult to walk on land and to fly in the air. In flight it has control over speed and direction and can make split-second adjustments to its aerodynamics in order to cope with the ever-changing situation. It is able to hover over a potential landing site or to take sudden evasive action. To a mosquito flying through rain each drop must appear as a gigantic missile several times its own weight, yet the insect can fly safely through and land still dry and ready for a meal. Indeed, mosquito activity often increases dramatically during rain of moderate intensity. In other words, a mosquito is capable of recognising or otherwise perceiving when it is threatened and in no sense is its flight merely a haphazard wind-dependent affair.

Much of this was put more succinctly by the younger Pliny during the first century in his references to the smallness of mosquitos. 'The *culices*', he says, 'are one of the wonders of nature providing on such a small scale all the five senses and other requirements of life.' While it is clear that an insect's behaviour does not exhibit intelligence in the usually accepted sense it is nevertheless a marvel of precision and co-ordination that cannot ever be dismissed as simple.

If we find the behaviour of mosquitos difficult to put into every day terms it is partly because it differs in kind from anything in our own experience. The mosquito's brain and nervous system is not a scaled-down version of our own, it is altogether differently organised. A mosquito's behaviour owes much to fixed patterns of response which it inherits along with its colouring, its shape and its general ornamentation. Just as there are programmed patterns of development, a phenomenon common to all living creatures, so there are also programmed patterns of response. These are probably found to some extent in all animals, but mosquitos and indeed all insects have a particularly large share of them. Such behaviour used to go under the convenient name of instinct, but this is no longer fashionable and instinct tends now to be treated as a dirty word. If I should use it in the course of writing this book it will be as a convenient short-hand for patterns of behaviour that appear to us to be predetermined, whether at the neural level or through the agency of hormones.

If the life of a single mosquito seems complex, what of the large number of interacting individuals that make up a whole population or of the collection of populations that make up a species? It is the species that goes on not the individual. It is the species that in its sum total of individuals must be perfectly fitted to the environmental conditions in which it lives. Deviation of form or of behaviour of the individual may mean death to it or at least less success in the production of viable offspring, but the species goes on. In each generation, however, a few such individuals may be carrying genes that fit them or their offspring to take over when gross changes in the outside world occur, changes originating from natural causes and changes brought about by the intentional and unintentional interference of man. At such times the roles of normal top-dog and deviant under-dog tend to become reversed. The normal individuals, closely adapted to a particular environment and making up the bulk of the population, find themselves no longer able to cope; they may fail to survive or their reproductive potential may be so reduced as to bring about their relative decline. Meanwhile the deviants, those reserve variants previously making up only an insignificant proportion in each generation, no longer at a disadvantage, find themselves as the favoured few; the line will go on although the species may change.

Mosquitos are of course really no more complex or more interesting than any other kind of insect. They merely happen to have been studied particularly thoroughly; thoroughly enough, I dare say, to make us realise what little we really know about them and how much there remains to find out, but thoroughly enough for us to have accumulated many things of interest. It is from this vast body of information that I propose to draw. I make no claim to be complete; new developments, new things and new interpretations of old things are constantly being added, and at an ever-increasing rate; I shall not attempt to include all of these. Whatever the gaps, there is already ample material to fill out our story.

We shall begin at the beginning, remembering that mosquitos spend their early days in water. Broadly speaking there are two types of aquatic animal: those that live in water but obtain their oxygen from the atmospheric air, and those that live in water and take up oxygen from the water itself. Mosquito larvae and whales belong to the former type; both are perfectly adapted for an aquatic life but must make periodic visits to the surface in order to get rid of accumulated carbon dioxide and to take in a fresh supply of oxygen. Otherwise both live and feed wholly in the water and neither survives long on land. Each is shaped to provide efficient locomotion through the water by sinuous movements of the whole or of a portion of the body. Apart from respiratory exchange at the surface, commitment towards life in water is absolute. Yet sooner or later the larva must undergo the amazing transformation into a flying mosquito. It is hard to imagine any creature less promising. During its aquatic life it undergoes growth and development through four successive larval stages or stadia, moulting at each stadial change. But on reaching the fourth and final stadium the creature appears no more promising as a future flier than before; there is no obvious progressive change that gives any indication of the dramatic events to follow.

How is this astonishing transformation achieved? To say that there is a transitionary or pupal stage interposed between the fourth stage larva and the flying adult mosquito does not get us very far. The pupal stage usually lasts for two or three days only and the pupa itself is also highly adapted for rapid movement through an aquatic medium; superficially it does not appear any more promising than the larva. But during this incredibly short period old tissues are broken down and resorbed to be replaced by new ones; muscles that by alternate contraction and relaxation brought about the characteristic sine-wave motion of the larva are, so to speak, melted down. New sets of muscles that will be needed for flight, for walking and for mating and oviposition, are laid down in their place. Yet during this fantastic melting-pot period the pupa remains fully mobile and can move with astonishing agility, though the mode of locomotion is entirely different from that of its old larval days.

The adult mosquito that emerges from the pupa is faced with problems that are far removed from those of its aquatic youth. Added to the more obvious differences associated with swimming and flying, there is the violent change from the friendly watery medium to the hostile environment of air, in which the threat of death by desiccation is never far removed. The adaptations that are necessary for life in air are, of course, common to terrestrial insects in general and are, therefore, largely outside the scope of this book. When, however, there is an abrupt change from a wholly aquatic to a terrestrial or aerial life these adaptations appear all the more impressive. But it is flight more than anything else that impresses most. A tiny *Hodgesia sanguinae* preparing to alight on a suitable host for a drink is a marvel to watch, even when the drink of choice happens to be one's own blood. With a total wing span of less than four millimetres this mini-mosquito hovers, dips and dives first over one shoulder or arm then over the other, each dip bringing it nearer to the surface of the skin as if testing it in some way. It may keep up these antics for five minutes or more before finally alighting, usually by then out of sight. A single pair of tiny wings vibrating rapidly one on either side of its elegant little black and silver body carry it under perfect control even when it departs with a shining red pay-load weighing several times its own unladen weight.

The whole question of flight in insects, whether the slow whining flight of a mosquito or the fast buzzing flight of a Sphingid moth raises many interesting questions and I trust I shall be forgiven if I digress from the main theme for a minute or two. Flight in birds, mammals and reptiles has been made possible by the modification of a pre-existing arm, hand and finger respectively to form a wing. The wing of a flying fish is merely a modified pectoral fin. But the wing of an insect is different in origin. It is something added to an already full complement of functional limbs, a distinction it shares with fairies and angels. One may be forgiven for wondering how a wing that is not a modified limb or other organ could ever arise by a process of natural selection. Could it be that insects' wings have evolved from organs that were primarily associated with the control of water- or heat-loss or both, and later perhaps with respiratory ventilation? In other words, these organs may have functioned not as wings but as adjustable covers protecting the spiracles, like the opercula protecting the gills of fish, or perhaps as tiltable radiators one on each side of the thorax and extending along the abdomen or even as a pair of fans controlling air supply to the spiracles.

This view would demand either that insects were primarily aquatic animals and then left the water for the land or that at some time after their early evolution there was a general drying up of the Earth. The former is hardly tenable since everything about an insect's organisation points towards primary adaptation to a life on land. Also, if insects were primarily aquatic one would surely expect to find some at least of the hundreds of thousands of species inhabiting the world today still living an entirely aquatic existence throughout all stages of their life cycle. We are left, then, with the alternative that the world became drier and perhaps hotter for a spell.

We have evidence from a study of vertebrate palaeontology that there was indeed a time during the Devonian period when the earth or large portions of it did become much drier than it had been. I am suggesting that it was perhaps then that these structures first developed, only to become redundant millions of years later when more humid conditions returned as we know they did during the Carboniferous. Thus these already tiltable and perhaps vibratable structures, freed from their original function, could be put to an entirely new use—flight. It would be out of place to pursue this line of thought any further here, except to add the necessary corollary that the pre-adult phases were spent in sheltered places. Far-fetched or not, it is no more so, in my opinion, than some of the other views that have been put forward on the origin of flight in insects.

Thus far we have taken the mosquito out of the water and launched it in the air. We have already touched upon some of the typical activities of the adult mosquito; in all these and in others as well, flight is the key. It dominates every phase of adult life. It was flight that removed the mosquito from the surface of the water in the first place. It is flight that now enables it to join in the swarming dance that plays such a characteristic part in the life of the young adult, and again for mating, for many species of mosquito copulate in flight or at any rate a pair may begin the act on the wing even though it finishes it off with at least four of the twelve feet on the ground. Then, flight enables the young adults of both sexes to reach a suitable flower or fruit as a source of carbohydrate to replenish the fuel that is used up in all this flight activity. And in the female it is flight that brings the mosquito to its target host for a blood meal and safely back to a place of shelter and rest while the eggs within its abdomen are growing in size. Finally it is flight that brings the gravid female back again to the water where it can safely deposit its eggs in preparation for the aquatic upbringing of the next generation.

So far we have avoided saying anything about the eggs of mosquitos. Aristotle, in the fourth century BC, recognised that *empis* (his word for a mosquito) was an aquatic creature during its early life and only later changed into the familiar flying form. He believed, however, that *empis* did not copulate or lay eggs, and that the aquatic stage was a result of spontaneous generation in putrifying waters, a view which, as Sir Rickard Christophers has emphasised, was still largely held some 2000 years later [3]. As late as 1562 one reads, '–some moysture (of the Elm Tree) after yt is dried up, is resolved into litle flies like Ganattes'. In fact all mosquitos lay eggs and in many ways these are no less interesting subjects for study than the other stages. Mosquito eggs are laid on the water surface singly or in clusters, lying flat or standing on end. Some of those laid singly are deposited one by one on the surface, others are dropped on to it while the mother is in flight. Others are laid below the surface and yet others are actually laid on a suitable substrate just above the surface and must await a rise in water-level before they can hatch. Some even are merely scattered over more or less dry land in areas that will later become submerged. We shall have more to say about all this in the next chapter.

It is safe to say that little would be known about mosquitos if they did not make such a serious impact on man. The misery that they can cause by their persistent attentions is not confined to the warmer parts of the world. Indeed, some of the worst attacks on man occur in the far north above the arctic circle—in Lapland and in Canada. In tropical regions one finds it necessary to protect oneself with a mosquito net at night, whereas in the far north these pests pounce on one in such hordes that one is obliged to seek some sort of protection even in the open. It is not unusual for a man to be seen walking in the open with netting draped over his head and neck region. Fortunately the mosquito season in these places is not long, owing to the shortness of the warm period. Further south, the mosquitos do not, so to speak, find it necessary to concentrate their attack in this way; they can spread it over a longer period. It would almost seem that there is within wide limits an optimum size for an adult population, or perhaps a minimum size below which the species cannot go without endangering its future; that absolute number is an important factor in maintaining a viable gene pool in a species, whether its individual insects are concentrated in space and time as in the far north, or whether they are spread over a vast area and are active for most of the year, as in the equatorial regions further south.

If the population density of a species occupying the warmer parts of the world is comparatively low, this is more than compensated for by the great variety of species that occurs in these regions. One sees the same thing in other animals. Whereas there are far more species of fish found in a given area of tropical waters than elsewhere, no part in the tropics can compare with the gigantic population sizes to be found in a single species occupying northern waters. There are other ways of interpreting these phenomena; whether for a mosquito or a fish, life is comparatively easy in the warm equable conditions of the tropics and scope for speciation is enormous since in the fullness of time at least some variants are likely to find an unexploited niche for themselves and others of their kind. In the harsh cold regions the individuals that make up a species must conform or perish. Thus interspecific competition is likely to be less acute and vast populations of a single species may result.

It is one thing to note facts and quite another to interpret them.

The careful and accurate observer of these things records them once and for all. But few interpretations or explanations of natural phenomena survive for long. Probably the explanation for these recorded facts embraces a little of both interpretations, each far more complex than we ever imagined, the one interacting with the other and both with other factors yet unsuspected. It is dangerous to make such interpretations, yet these are the very stuff of which science is made. Unfortunately when it comes to ecology most interpretations are little removed from speculation. Nevertheless, I said in the preface to this book that I would allow myself the privilege of digression and speculation in this first chapter and keep more to the recording of facts in the chapters that follow.

Between the intolerable but happily short-lived mosquito clouds of the north and the ever-present disease carrying tormenters of the tropics there is hardly a land surface that does not support its own populations of active mosquitos for at least part of the year. It matters little whether these lands are built up into cities and vast conurbations as, for example, London or New York, or whether they are wild and untouched by man. 'What would summer be,' said one lady I knew, 'without a few mosquito bites to scratch?' Most of us notice only those species that come to us for blood. There are, however, many other species that hardly ever or never even attack man. There are bird-feeders and rodent-feeders. Some even restrict their feeding activities to ants; they are unable to feed on vertebrates. Others feed exclusively on fruit or flowers, and a recent report indicates that some may even attack caterpillars.

In the chapters that follow we shall look in greater detail at all these aspects of mosquitos and their ways of life. We shall not only record their habits and ways but look into the mechanisms that determine some of these. We shall take a look at the so-called internal clock mechanisms that regulate the activity of mosquitos and at the other factors that in general make them tick. Later we shall look in some detail at the part played by mosquitos in the transmission of disease from man to man, from animal to animal and from animal to man: filaria and its hideous sequel, elephantiasis; malaria, still the most important incapacitator and killer of man; yellow fever, of former terror and present threat. The mosquito alone is responsible for the maintenance and spread of these major diseases of the world. But there are others, some of them also spectacular but less well known, others comparatively mild and hardly known at all. It is this aspect of mosquitos, their role as transmitters of disease, that has forced our attention on them and made this book possible.

2. EGGS

Ask most people about the eggs of mosquitos and the chances are they will be unable to say a thing about them. If, however, you address your question to a field naturalist or perhaps a teacher of school biology, they will as likely as not state unhesitatingly that mosquito eggs are laid in rafts; that is, many eggs are attached together to form a raft-like structure that floats on the surface of the water. In fact of the thirty-four genera of mosquitos it is only in *Culex, Culiseta, Coquillettidia* and *Uranotaenia* that egg-rafts are commonly found. The others, far and away the majority, mostly lay their eggs singly or, as in the case of *Mansonia,* some *Ficalbia* and even some *Culex,* lay them in clusters attached to floating vegetation.

In the genus *Anopheles,* which includes the malaria mosquitos, the females of the 300-odd species lay their eggs singly either directly on the surface or by scattering them over the water. Each egg is boat-shaped with pointed and often upturned ends, giving it the appearance of a miniature canoe. Along each side is a float structure resembling the ballast tanks that flank each side of a submarine. These floats ensure that the egg maintains a correct attitude on the water even when the surface is being constantly agitated by wind or, as in the case of *Anopheles funestus* and other species that sometimes lay their eggs along the margins of comparatively fast-flowing rivers and streams, by the eddies and other disturbances set up by the current. Sometimes the surface of the water is violently disturbed as, for example, when a two-ton hippopotamus steps into an otherwise peaceful puddle on which are floating some eggs of *An. gambiae*. An egg that is not washed out of the puddle altogether and stranded on dry land, and escapes physical injury or even crushing under the intruding foot, is none the worse for its experience. The float ensures that it remains at or returns to the surface, and the general lifeboat-like shape with its upturned ends increases its chances of remaining right way up. In other words, the eggs are reasonably unsinkable and self-righting. It is interesting to note that the eggs of *Aedes* mosquitos are occasionally also laid singly on the water surface. These eggs are cigar- or spindle-shaped, according to species. They do not have upturned ends and they do not possess floats; if the surface of the water is agitated they sink and remain at the bottom.

Although *Anopheles* eggs are laid directly or are dropped independently on the surface of the water, it sometimes happens that several eggs find themselves in close proximity. These may be drawn to one another by surface forces to form little groups of usually not more than five or six eggs in each. Such groups tend to

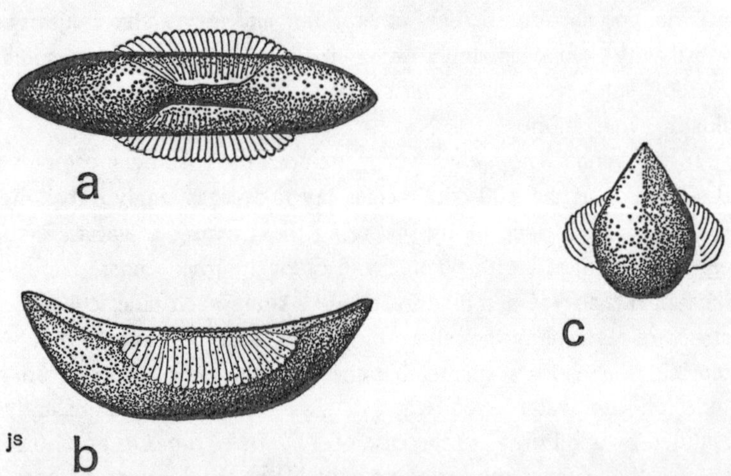

Figure 1. Single eggs of *Anopheles: a.* top view, *b.* side view, *c.* end view. Each egg floats on the water surface, resembling a miniature lifeboat with its pointed upturned ends and float chambers.

adopt geometric formations, often arranging themselves in beautiful stellate patterns resembling the spokes of a wheel or the petals of a flower. One group joins with another to form a delicate network over the surface of the water (Fig. 2). It is not known whether this tendency confers on them any advantage.

When one of the surface-laying *Anopheles* mosquitos is ovipositing it stands on the water, the long tarsal segments of the legs distributing its weight over a comparatively large area. An egg is laid here, another there, the mosquito walking on all six legs from place to place. Some of these eggs are laid in open, fully exposed pools, as for example with *An. maculipennis* and *An. gambiae,* others in among thick vegetation or under overhanging banks, as for example with *An. claviger* and *An. funestus.* There are actually very few observations of eggs being laid in the wild. One makes the assumption that the aquatic stages of *An. gambiae,* one of the two important malaria transmitters in Africa, are found in puddles and pools fully exposed to the sun in daytime, because

Figure 2. Cluster of *Anopheles* eggs floating on the surface of the water in typical stellate and geometric patterns.

the mother chose such places for egg-laying in preference to more shady places. Similarly, with *An. funestus,* the other important malaria transmitter in Africa, one makes the assumption that the mother actively selects shady pools, streams and rivers with emergent or floating vegetation. It seems more than likely that this is indeed the case. Yet there is no real evidence with these two species that selection of the breeding site takes place at the time of egg-laying. It could be argued that the eggs of both species are laid indiscriminately in any water, and that selection operates at a later stage, allowing *An. gambiae* to survive in the open pools and *An. funestus* in the shady places. One knows that the light tolerance of the larvae of these two species is very different (sunlight acting either directly or perhaps indirectly by heating the water) but one cannot be certain about the eggs. On the other hand, laboratory observations on *An. maculipennis* and *An. claviger* in Europe show quite clearly that selection of the breeding site is made by the egg-laying female; *An. maculipennis,* the larvae of which are found in exposed places, selects water at 22–29°C; *An. claviger,* the larvae of which are found in shady places, selects water at 12–20°C [1, 23, 46–8].

The eggs of *An. culicifacies,* an important malaria transmitter in India and Ceylon, are laid in paddy fields at the time of early growth of the rice plants. Later, however, when the plants become taller, breeding in the paddy fields ceases. One might easily jump to the conclusion that light again is important and that it operates after the eggs have hatched. Paul Russell and Ramachandra Rao, who made these observations on rice and *An. culicifacies* breeding, found that they could simulate these conditions by placing vertical glass rods at regular intervals through water. Provided the rods projected less than thirty centimetres above the water surface the mosquitos flew to and fro above the rods, dropping eggs on the water surface below. When the rods were taller egg-laying ceased [53]. Thus *An. culicifacies* does not venture more than thirty centimetres above the water surface for its egg-dropping operation and when the rods or the rice stems exceed this height they act like barrage balloons and force the mosquito to fly at a height at which for some reason egg-dropping is no longer carried out. It is clear, therefore, that in this well-documented case it is indeed the selection by the egg-laying mother that determines that young rice is dangerous for man and old rice is safe. These remarks apply to *An. culicifacies* only. *An. funestus* and *An. hyrcanus* will also breed in rice, but since their eggs are apparently laid directly on the surface of the water tall rice is no obstacle and larvae may be found in the water supporting rice of almost any age.

We may appear to have been side-tracked into describing egg-laying behaviour of the adults rather than dealing with the eggs themselves. But it is not possible to separate the two entirely. Even the intricate sculpturing on the surface of mosquito eggs is determined by the cells lining the oviduct of the mother, and the formation of the beautiful egg-rafts is partly a function of the mother's behaviour at the time of egg-laying.

The typical egg-raft of *Culex* consists of several hundred eggs laid at a single 'sitting', each egg standing on its end and cohering to its neighbour, giving the whole an inverted honey-comb appearance, that is, with protuberances in place of the cells. We noted before with the *Anopheles* egg that it floated on the water with its convex side downwards and that its general boat-shape often with pointed upturned ends gave it stability on the water even under disturbed conditions. If these are important features necessary for the floating eggs of *Anopheles,* how can we reconcile the complete absence of these features in the upright-floating eggs that we are now considering. In *Culex* it is the whole raft, that is the complete cluster of a hundred or so eggs, that takes over this function. The underside of the raft tends to be convex and the upper surface concave, the ends of the raft being pointed. Thus the whole structure, which may be 0.3 cm long and 0.2 cm wide, is perfectly adapted for floating on the surface even though each individual egg comprising it is virtually standing on its head.

The egg-laying *Culex* mosquito places the first egg with its future head end downwards on the surface of the water. The second egg is laid immediately adjacent to the first and thus comes to push the first egg backwards away from the mother by a distance equal

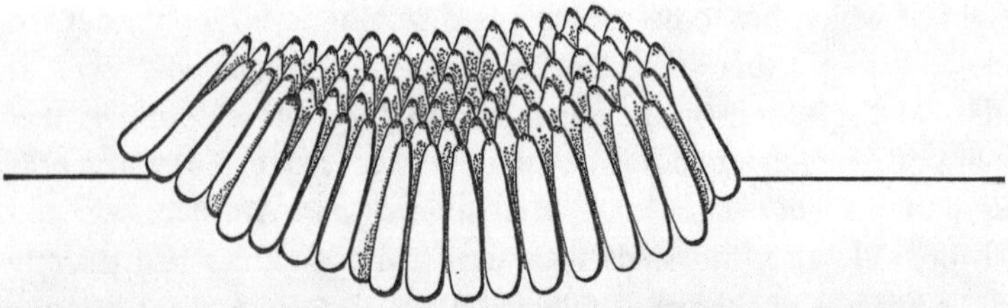

Figure 3. Collection of eggs of *Culex* forming a composite floating raft of many eggs, each with its head end downwards. The whole egg-raft is typically convex below and concave above with upturned ends.

to its own diameter. This process is repeated over and over many times until a row of eggs is laid, each egg standing on end but cohering to the previous egg for most of its length. The egg-laying mosquito then moves itself or its abdomen backward and starts a second row immediately alongside the first row, each egg in the new row being staggered in position with respect to those in the previous row. Other rows are then added, the middle rows tending to be longer than those along each side, and the eggs in each row being staggered in position with those in the next row. Thus the rows come to be straight from end to end but zig-zag from side to side, like the arrangement of bottles in a divisionless crate.

We see, then, that it is primarily the behaviour of the mother that determines the general shape of the raft, that is, the arrangement of the upturned eggs in staggered rows. But so far we have omitted to say how the completed raft comes to acquire its characteristic boat shape, which in some species is its most striking feature. In fact the convexity of the underside of the whole composite structure is ensured by the shape of the individual eggs comprising it. Each egg tapers slightly towards its upper end and since it coheres closely with its neighbour throughout much of its length, the result must be to give the whole raft when completed this characteristic outline. The degree of upturning will naturally depend on the number of eggs in a row, hence in an elongated raft the ends will be more upturned than the edges or sides, thus adding the same self-righting property as we found in the individual eggs of some *Anopheles*. This is essential, since the eggs open on hatching at the lower end, allowing the young larvae to escape directly into the water below. Hatching would not be possible from an overturned egg-raft.

Some of the most beautiful egg-rafts are those laid by species of *Coquillettidia,* a genus of some forty-five species represented in most parts of the world outside the arctic regions. The rafts themselves are often much longer and narrower than the more familiar rafts of *Culex,* and the shape of the raft can sometimes be used to identify the species. Among the African *Coquillettidia* there is a complete range of raft shapes from the *Culex*-like wide raft of *Coq. metallica,* through the longer more boat-like rafts of *Coq. pseudoconopas* and *Coq. maculipennis* to the extremely long, elegant, slender canoe-like rafts of *Coq. aurites* and *Coq. fuscopennata* in which there may be as few as three or four rows of eggs only. Indeed, the rafts of *Coq. fuscopennata* usually consist of only two full-length rows of about thirty to fifty eggs, with two much shorter rows, one on either side making the whole structure resemble a long narrow racing skiff [33]. Except in *Coq. metallica* the eggs themselves do not taper and there is thus usually no convexity to the under-surface of the whole structure. Perhaps this difference is connected with the very special habits of the genus, the eggs always being laid on water associated with floating or emergent vegetation. We shall have more to say about the habits of this remarkable genus when we come to consider larvae and pupae in the next two chapters.

It is clear from what we have already said that in the characteristics of the egg-rafts of the various species of African *Coquillettidia,* we are really dealing with differences in the behaviour of the mother. Whether there are ten to fourteen comparatively short rows of eggs as in *Coq. metallica* or three to five rows only, with two or three of them very long, as in *Coq. aurites,* can only be determined by the mother at the time of laying [33]. To lay many short rows of eggs or a few long ones is as characteristic of these two species as is, respectively, their typical metallic blue or brilliant shining gold coloration. We are justified then in stating that egg-raft shape is an inherited quality, even if only indirectly. It is difficult to see what advantage there can be in this complete range of raft shapes, but we must not suppose that this means there is none. Too often, in my opinion, biologists in their arrogance are apt to pronounce that there can be no possible selective advantage in this or that when what they really mean is that they themselves cannot at present see one.

The egg-raft itself, whether of *Culex* or one of the other genera, holds together as a single entity. It is possible, and in fact it is quite easy, to separate a single egg or a group of eggs from the remainder of the raft, but under normal conditions the raft remains intact. The eggs are not cemented together in the ordinary sense,

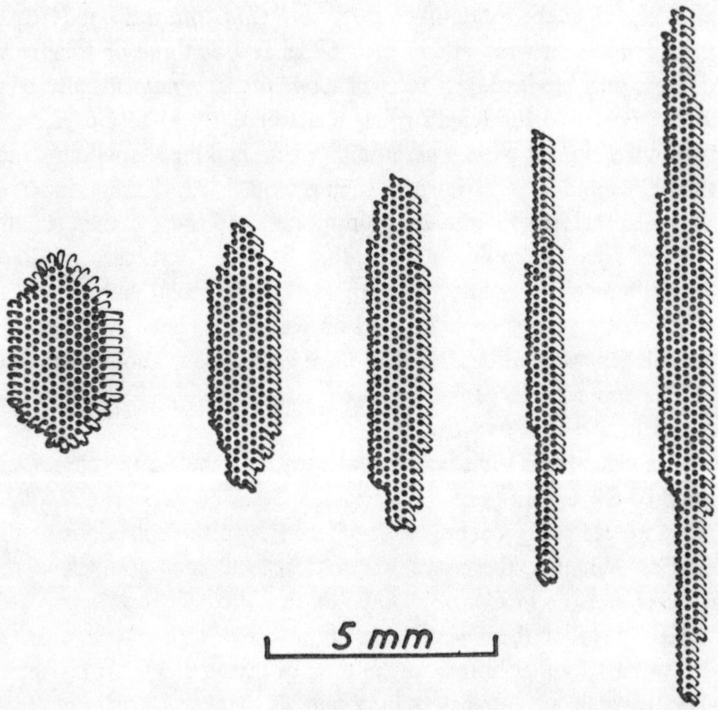

a b c d e

Figure 4. Egg-rafts of *Coquillettidia* spp., showing characteristic shapes (after Gillett): *a. C. metallica, b. C. maculipennis, c. C. fraseri, d. C. fuscopennata, e. C. aurites.*

they merely cohere to one another as a result of surface forces set up between neighbouring eggs [27]. It is the remarkable sculpturing of the egg surface that brings about this cohesion. Each egg being circular in cross-section and not hexagonal, a small space is left between adjacent eggs, as there would be between coins arranged in adjacent rows. The surface of the eggs is endowed with unwettable properties and the spaces remain filled with air. These spaces thus act as air tanks and give added buoyancy to the whole raft. If for any reason the water-repellent surface is destroyed, water enters these spaces, floods the convex top and the raft sinks.

Closely related to the egg-raft-forming *Coquillettidia* is the equally remarkable *Mansonia,* a genus comprising some twenty species found throughout the warmer parts of the world. The eggs of *Mansonia uniformis* are also laid in clusters, but these are attached to the underside of the leaves of aquatic vegetation and are thus laid underwater. The egg-laying mother stands on the edge of a leaf of some floating or emergent plant and by extending the end segments of her abdomen reaches down through the water to find the underside of the leaf. The eggs are laid standing on end and touching one another to form a single cluster. Since each egg is wider at the end away from the leaf, the whole cluster tends to radiate outwards to form a circular rosette (Plate 1). While the *Culex* and *Coquillettidia* eggs stand on their heads on the surface of the water and release their young larvae into the water below, the *Mansonia* egg stands on its 'tail', but being upside-down on the underside of a leaf, it is also in a position to release its young larva directly into the water. The underside of a single leaf may come to bear several clusters laid by more than one female. Indeed, egg-masses of this sort may consist of thousands of eggs all tucked away clear of the surface with its attendant risks. We shall see later on that the eggs, larvae and pupae of these extraordinary mosquitos all spend their lives independent of the surface, even for the needs of respiration. The same is true of *Coquillettidia,* except for its eggs which as we have seen remain on the surface like those of any ordinary *Culex.*

The genus *Culiseta* is particularly interesting since the egg-laying behaviour, even of the same species, differs according to geographic region. In Europe the eggs have been recorded as being laid singly on the water surface or attached to the moist substrate a little distance above the surface [14]. In America, on the other hand, the eggs have usually been found in rafts floating on the surface of the water. Recently Robert Wallis and Loring Whitman have observed that *Culiseta morsitans* not only lays its eggs in rafts but attaches the rafts to the moist substrate some three to five centimetres above the surface of the water.

Some mosquitos lay their eggs inside damaged bamboos and reeds. Some of these have had to adopt various dodges for getting the eggs inside, since the entrance is often too small to allow normal entry. This problem has been solved in *Armigeres dolichocephalus* and *Ae. angustus* by the evolution of an abnormally long and narrow thorax (usually the bulkiest region of the adult), which allows the former actually to enter beetle-larva holes only 1.5 mm in diameter. *Ar. flavus,* on the other hand, solves the problem in an entirely different manner: it lays its eggs on its own hind legs and later uses these as an auxiliary ovipositor [45]. But the most accomplished is *Sabethes chloropterus.* The egg-laying female hovers in flight in front of a suitable hole in a bamboo and with a sudden jerk of her abdomen flicks the eggs in one by one [391]. Also extraordinary is the egg-laying habit of *Ae. woodi,* which deposits long almost needle-like eggs in the narrow space within the axils of sedge stems.

We now come to the problem-eggs of *Aedes, Haemagogus* and *Psorophora,* which by their resistance to desiccation and delayed and erratic hatching have been the perennial subjects for discovery and rediscovery for fifty years and more. In the large and important subgenus of *Aedes* known as *Stegomyia,* which includes the well-known yellow fever or tiger mosquito, *Aedes* (*Stegomyia*) *aegypti,* the eggs are laid singly but not on the surface of the water. The egg-laying female attaches each egg just at the water's edge or in the wet or damp zone just above the water's edge. These eggs, which are laid flat like those of *Anopheles,* are usually placed around the edges of small containers of one sort or another, even in such small and unlikely places as pineapple tops.

When laying her eggs, the female *Aedes* does not lay the whole batch in one spot; she lays one egg, walks on round the edge, lays another and so on round and round the edge of the water or even a centimetre or two above the water's edge. Thus a single batch of some seventy eggs may be spread thinly right round the container at varying distances from the water's edge. Several females may lay in the same container either at the same time or add their quota on later occasions. Since the eggs cannot hatch until they become submerged by a subsequent rise in the water-level, the edge of the water may become surrounded by a continuous mass of eggs, like pebbles along the margin of the sea-shore. Thus a hundred eggs lying close together in one small area may have been laid by a dozen or more different mothers [32].

The eggs themselves are mostly cigar-shaped and, not unexpectedly, devoid of floats. They are white and soft when first laid, and somewhat smaller than mature eggs. But they soon swell up to reach their full size and after about thirty to fifty minutes begin to darken, the process beginning at the future head end. The period of darkening is complete within one to two hours, depending on temperature and on the species concerned [121]. The eggs are not actually cemented to the substrate, but when laid on moist filter paper in the laboratory they remain attached even when the paper is dipped in water. Nor does one have to be particularly careful in carrying out this operation: the paper bearing the eggs may be thrust forcibly into the water many times or even waved about quite violently from side to side without disturbing the eggs. Yet an egg may be moved gently from one part of the paper to another without losing its adhesion even when the paper is later submerged, provided that sliding contact is maintained throughout this manoeuvre. If it is lifted even once from the surface, or the motion is by rolling instead of sliding, then the eggs will become detached when the paper is dipped in water [121]. As in the cohesion of *Culex* eggs it seems that surface forces are responsible, and the typical deep sculpturing of the egg surface is almost certainly involved. The eggs of *Ae. aegypti,* by far the most thoroughly studied of all mosquitos, are sometimes laid on the water itself. This trait is seen more often in certain strains of this species than in others, but it occurs from time to time in most strains, particularly when laying conditions round the edge of the water are unfavourable, or when the mosquito laying the eggs is injured or otherwise infirm. Such eggs are perfectly viable, but do not usually hatch unless submerged.

Other species of this large genus, notably some of those belonging to the subgenera *Aedimorphus, Neomelaniconion* and *Ochlerotatus,* lay their eggs on or over more or less dry ground. Or, perhaps, it would be truer to say, on ground subject to inundation by flooding, by melting snow or even by tidal action. The areas thus sown with eggs may be very extensive indeed, species of *Ochlerotatus* being typical of the large salt-marshes that occur in many regions of the north temperate zone. How these places come to be selected is far from being fully understood. Many of these eggs tend to be more spindle-shaped than the familiar sausage-shaped eggs of the tree-hole-breeding *Aedes,* and do not adhere to the substrate. Indeed, those of the African *Ae.* (*N.*) *circumluteolus* and the European *Ae.* (*O.*) *rusticus* are almost diamond-shaped when viewed from above. It is interesting to note, however, that the tree-hole eggs of *Ae.* (*Finlaya*) *ingrami* also have this curious shape [121]. These too do not adhere to the substrate, and thus differ markedly from those of the better-known species of *Finlaya;* perhaps the taxonomic position of *Ae. ingrami* needs to be looked at again.

The eggs of mosquitos, like those of other insects, are not fertilised until a moment before they are actually laid. Sperm, stored in the little sacs of the female known as the spermathecae, enters each fully developed egg or oöcyte as it should strictly be called as it passes the openings of the spermathecal duct on its way out. Once the eggs have been laid, development of the embryos proceeds apace. The eggs of *An. gambiae* hatch within two or three days of being laid, the actual period depending on the ambient temperature. Within the usual bounds of variation found in all biological material, and with the notable exception of *Aedes, Haemagogus* and *Psorophora,* the eggs in a batch all hatch together or over a short period of time. The mature embryo bears a hardened protuberance or spine on the top of its head known as the hatching spine or egg-tooth. Before hatching the embryo swells up, presumably by the imbibition of water, and exerts pressure on the shell of the egg, most of which is concentrated on the point of the hatching spine. In some eggs there is a line of weakness round the head end of the shell, and it is at this point that the spine exerts its pressure. Suddenly the shell splits open round this line of weakness and the first-stage larva works its way out. It is a remarkable sight to see a hundred or so tiny heads poking out at the lower end of all the eggs in a raft of *C. pipiens.*

Earlier we referred to the problem-eggs of *Aedes, Haemagogus* and *Psorophora,* and it is to these that we must now turn our attention, in particular to those of *Aedes.* Once the embryos have completed their period of development, the duration of which not only varies with temperature but with the species concerned (two to three days in *Ae. aegypti,* five to six days in *Ae. africanus* and as long as eleven days in *Ae. longipalpis*), these eggs may remain unsubmerged and almost entirely dry for comparatively long periods without loss of viability. There are records of the eggs of *Ae. aegypti* hatching after being kept dry for periods of six months or more. Such drought resistance is in fact extremely widespread in the eggs of all three genera; litter collected from dry tree-holes in parts of equatorial Africa will more often than not produce first-stage larvae of *Aedes* of various species when later placed in water, often more than one species from a single tree-hole; turves from dried-up swamplands or flood-plains in Canada will likewise produce first-stage larvae of *Ae. hexodontus,* even when submerged in water under laboratory conditions weeks or months after collection. Provided that the soil does not become completely dry, eggs of *Ae. squamiger* in California will survive for two years, while those of the North American flood-water species *Ae. sticticus* and the widely spread *Ae. vexans* have been known to survive for four years [28–30, 35, 40, 54].

It is this ability to withstand drying that enables these eggs to survive when laid above the level of the water or when scattered over dry ground subject to inundation at some future time. It is an adaptation to the utilisation of breeding waters of a precarious nature, either by virtue of their smallness of size and consequent risk of drying, or by their fleeting nature owing to fluctuations in water-level and general impermanence. It allows whole populations of certain species to tide over dry periods, only to burst into activity at the return of the rain or raising of the water-table. In tropical regions where subsequent development of the aquatic stages may take only a matter of days, it accounts for the sudden invasions of marauding mosquitos that may appear only a week after the rains have set in. With embryonic development already completed before the dry period started, hatching can now occur within minutes of inundation, and growth and development of the larvae can proceed at once. Other mosquitos that spend their aquatic youth in less precarious situations can, so to speak, afford first to wait for the gravid females to locate the water on which to lay their eggs, and then to wait for the embryos to develop. But when the breeding places may remain filled for a week only, then every day counts and the loss of these three or four days before larval development can begin, could well be disastrous. Drought resistance not only ensures survival through periods of low or irregular rains but allows maximum 'use' to be made of each breeding place once it is filled.

Nothing that we have said so far would warrant the epithet 'problem-eggs'. Quite the reverse, in fact, for they seem by their ability to withstand desiccation to be singularly problem-free, and if this were the whole story one would have to agree. But it is very far from being the whole story as we shall see. It is true that the eggs may wait in a state of more or less suspended animation ready, as it were, to take immediate advantage of the first rain or flooding. But while it is also true that on flooding hatching may occur within minutes, this is not always the case, for while some of the eggs may hatch readily enough, others may remain unhatched for varying periods, even though they remain continuously submerged. Indeed, hatching may go on in small instalments at irregular intervals over a very long period [21, 22, 30]. A single turf collected from a dry flood-plain and placed in water at intervals over a period of some months may produce a fresh instalment of first-stage larvae each time it is submerged. In 1926 Dr E. Hearle took two pieces of turf from a well-known breeding area of *Ae. vexans* in Canada and alternately flooded and dried them. In five floodings he obtained from one piece 1593, 1529, 72, 24 and 0 larvae and from the other, 1553, 12, 3, 0 and 0 larvae respectively. Marshall, who published his classic book on British mosquitos more than thirty years ago, working with *Ae. cantans* found that once hatching of submerged eggs begins, first-stage larvae may appear in instalments over a period of three months.

We have already seen that a mass of *Aedes* eggs lying a little above the water in a tree-hole will come to exhibit heterogeneity of age. The obvious explanation, then, is that hatching by instalments can be accounted for by differences in age, since the eggs in any given egg-mass were themselves laid in instalments. But obvious explanations are usually suspect, and more often than not they turn out to be wrong. Eggs of *Ae. africanus* laid all on the same day, or even within one hour, have been shown to exhibit exactly the same phenomenon: hatching goes on in instalments over a long period. Even the eggs laid by a single female following a single mating do not all hatch at once [30]. A single batch of seventy to one hundred or so eggs of *Ae. africanus* kept under water almost from the time of being laid, and thus being in a position to hatch at the earliest opportunity, typically shows the following pattern: five or six days after being laid the eggs reach embryonic maturity and a proportion of them hatch. The remainder, which is usually well over half of them, do not hatch then but do so at irregular intervals and in small instalments over the ensuing weeks and months. A very similar pattern occurs in *Ae. aegypti*, only embryonic maturity is reached in a much shorter period and the initial burst of hatching activity occurs on day two or three.

The scientist is always on the lookout for simple elegant explanations to account for the facts that he observes. But no simple explanation can satisfactorily account for all the known facts about these eggs. And, really, there is no reason why the explanation should be simple; life is far from being simple. What seems to be happening is this: on reaching maturity the eggs begin to hatch in the normal way, but shortly afterwards those eggs still unhatched pass into a state of dormancy or depression of varying duration and depth, which can only be broken by the application or presence of a suitable hatching stimulus. Eggs not only differ from one another in the duration and depth of this depression but the depth of depression itself varies with the passage of time in any one egg [30]. This state of depressed hatching ability resembles in many respects the widespread phenomenon known as diapause. It differs, however, in certain fundamental features and for this reason it might be better called simply depression. Both terms are, in fact, used here, the term diapause being used when it is felt that by so doing the meaning is made clearer.

Many different factors have been described as acting as hatching stimuli, and it is evident that some of these are more potent than others. Any given stimulus will not affect an egg until its depth of depression has risen above the threshold for that stimulus. Hatching, when it occurs, is usually followed by a small burst of hatching in companion eggs over the next two days, making each instalment cover a period of two to three days. It is not known whether the first to hatch stimulates others to hatch or whether there is a time-lag on the part of some of the eggs in responding to the stimulus. A succession of hatching factors or stimuli has been described over the past fifty years since the subject of erratic hatching first came to be noted. Indeed, hardly a year passes without more names being added to the list. It seems that many different physical changes can act as hatching stimuli but for the past thirty years emphasis has been placed on those factors which reduce the amount of dissolved oxygen in the water containing the eggs [34]. Certainly most of the factors described as hatching stimuli are accompanied by a reduction in the amount of dissolved oxygen. Nevertheless, there are a number of apparent exceptions. While hatching is undoubtedly usually accompanied by a decrease in the amount of dissolved oxygen in the water, I sometimes wonder myself if the two are really *causally* associated. The consensus of opinion, however, seems to be that they are [41].

By and large, *Aedes* utilise either small containers of water such as ground pools, rot-holes in trees, leaf-axils of plants and man-made articles of every description, or larger collections of water but of a temporary or precarious nature such as flood-plains, salt-marshes and arctic snow pools. Such breeding places are subject to violent fluctuations in level and even to sudden drying or to freezing. Survival in northern *Aedes* depends on all or most of the eggs going into a state of hatching-depression deep enough and long enough to outlast the winter. This is an inherited property; the eggs do not normally hatch during the winter months even if they are brought into a warm laboratory. In the tropical small-container species, on the other hand, survival depends on erratic rather than delayed hatching. The production of larvae by instalments ensures that all will not be lost if the container becomes dry before larval and pupal life is completed; there will be unhatched companion eggs ready to take advantage of any further flooding. Their ability to withstand desiccation will ensure their survival in the interim [30].

This interpretation has been queried by some workers, including no less an authority than Sir Rickard Christophers, who sees no place for this phenomenon in nature and attributes delayed and erratic hatching in *Ae. aegypti* to laboratory artifact [3]. I cannot, myself, see how such precarious places can be utilised for breeding unless there is some insurance mechanism of this sort. A scattering in time is as reasonable to expect as a scattering in space, and quite essential when space is severely restricted and all the eggs have to be placed in one 'basket'.

A variable dormancy is well known in the seeds of certain plants, particularly in those that are not widely dispersed. Instead of all the seedlings developing together in one restricted area and running the risk of total annihilation, there is a succession of seedlings over a period of time as dormancy is terminated by instalments. If the first crop is lost there will be others to follow. The two situations appear to me to be wholly analogous. Simultaneous hatching or simultaneous germination in a restricted area must surely lead to extinction. But in the eggs it is a combination of the two qualities, drought resistance and variable diapause, that between them provide a mechanism ensuring against loss by fluctuations in the water supply.

A factor that has not helped in reaching agreement on these matters is the extreme genetic plasticity of *Ae. aegypti,* the species which has received by far the most attention. Each colony of this species is apt to differ from the next as far as its behaviour is concerned. This and differences of experimental technique have led to conflicting claims and to failure of one worker to repeat the results of another. In a colony of *Ae. aegypti* originating from Lagos, Nigeria, the females laid each batch of eggs at a single 'sitting'. That is, while they walked round and round the edge of the water in the manner described, they continued until all the eggs in that batch were laid. In contrast to this, a colony of the same species from southern Tanzania divided each batch of eggs into three or four daily instalments: some were laid on one day, more on the next day and so on till the batch was completed. Now the eggs of this East Coast strain mostly hatched soon after reaching maturity, in marked contrast to those of the West African strain that hatched in the more usual small instalments over a long period of time. The eastern strain *laid* its eggs by instalments and thus gave them a *spatial* scatter, while the western strain laid its eggs in one place, but *hatching* was by instalments, and thus gave them a *temporal* scatter. Hatching by instalments was, so to speak, unnecessary in the eastern strain [31].

We have noted that there is a burst of hatching activity on reaching maturity. This is a very regular feature, but the significant thing is that rarely do all the eggs hatch during this initial pre-depression period. It seems that to begin with there is nothing to stop the eggs from hatching once they have reached maturity. Indeed, this is precisely what happens and in this respect they resemble what we might call ordinary eggs. Pre-depression hatching on reaching maturity could be disastrous, if it were not soon brought under control. Evidence has been produced which suggests that each egg that hatches releases into the water a diffusible substance that acts as an external messenger or pheromone [32]. As more and more eggs hatch the concentration of this pheromone builds up in the water until it reaches a concentration at which it inhibits hatching in the remaining unhatched eggs. A day later these unhatched eggs pass safely into the characteristic state of depression or diapause and will not now hatch unless subjected to a hatching stimulus. The mechanism ensures that at least some of the eggs pass on to the diapause state and thus become 'salted' away for a rainy day. We shall look at the whole question of erratic hatching again when we come to discuss strains and species, but then we shall be looking at it as a perfect example of balanced polymorphism.

3. LARVAE

The larva cannot be dismissed merely as the growth phase in the life of a mosquito. To be sure, it covers the period of growth—the only period of growth if we discount development of the embryo within the egg—for no further growth takes place in the pupa or in the adult. But it is surely something more than this; the larva lives its own independent life in a medium so unlike that of the adult flying mosquito that development and change in the evolutionary sense have gone off on paths of their own quite uninfluenced by future adult needs. Nor can the larva be said merely to be a transition phase, for the larvae of some species of mosquitos live for many months, some in fact living longer than the adults that follow. Indeed, it could be argued that the adult stage is interposed solely for the mixing of genes and dissemination, in preparation for the exploitation of water by the larvae. But here we are verging on the realms of philosophy and the argument can be, of course, and has been, carried further to include the reproductive phase of all organisms. Nevertheless, the argument becomes a little more easily assimilable when there is an independent larval life of comparable duration to that of the adult, and utterly unlike it in habit and form. Does the caterpillar produce the butterfly or the butterfly the caterpillar? A meaningless question perhaps, but one that is not entirely unrewarding to ponder.

It is often said that mosquitos breed in stagnant water, a term that denotes lack of movement and usually implies a certain degree of foulness as well. In fact almost every type of freshwater is exploited by one species or another. While the gulley trap and pit latrine may be suitable and in fact may be the preferred sites for some species, others are much more particular and will breed only at the edges of clear-running streams and rivers. There are forest-pool breeders, brackish-water breeders, tree-hole, plant-axil and even empty snail-shell breeders. There are others that breed almost exclusively in small quantities of water held up in fallen leaves. Yet others are found only in hollow reeds. Even swamps, the other all embracing breeding place of popular imagination, impose their restrictions and only a limited number of species actually breed in them, although the sheer size of some swamps may result in a prodigious output of those that do breed there. Nor is the type of breeding place and water in it the only factor concerned. Tree-hole breeders in the tropics, as might be expected, are different from those of the temperate regions. But this is not merely a result of distance and isolation—the conditions and requirements of each are very different. In the temperate regions provision must be made for overwintering. Even within the tropics tree-hole breeders in the lowland rain-forest are different from those of mountain forest where cold conditions and tropical heat alternate in a 24-hour cycle.

Mosquito larvae are essentially filter feeders. On either side of the mouth is a brush-like structure of about a thousand curved hairs or bristles. When the larva is feeding these brushes move in a rhythmic sweeping fashion, drawing a current of water towards itself, like two hands simultaneously beckoning towards the mouth. Large particles are turned away on reaching the mouth and there is a temporary cessation of the movements of the brushes. Somewhat smaller particles that enter the mouth may be forcibly ejected. Other particles that enter the mouth are subjected to a further screening process on reaching the pharyngeal hairs inside. The individual bristles that make up the brushes are simple in most species of *Culex* but a varying proportion of them are serrated or toothed in some mosquitos in which the feeding habits have departed from the basic pattern. Competition for food is reduced not only by the utilisation of different types of water but by a stratification of feeding levels in the same site, some species tapping the surface film, others different levels below the film.

The various types of feeding behaviour are closely associated with the mode of breathing. All mosquito larvae are dependent either directly or indirectly on atmospheric air. Two spiracular openings leading into the branched tracheal system serve the purpose of respiratory exchange. Both are situated at the posterior end of the larva, lying side by side on the dorsal surface. In the surface-feeders the pair of spiracles is situated directly on and flush with the dorsal surface of the eighth abdominal segment. In others that feed below the surface the spiracular openings are placed on the tip of a rigid turret or breathing siphon, allowing the larva to make respiratory contact with the surface while the rest of the larva's body remains submerged, like a submarine with its main bulk lying hidden below the surface while the top of the conning tower remains in contact with the air. Just as in the submarine water is prevented from entering the top of the conning tower when the craft dives by a trap-door that is held firmly into its seating by the pressure of the water above, so when the mosquito larva submerges the paired openings are protected by a number of flaps that automatically close and seal them off. These flaps or valves spread outwards and expose the openings when contact is again made with the surface. An oily secretion from the perispiracular glands keeps the surface of the valves and the surrounding area in a water-repellent or hydrophobe condition [69]. There are also muscles within the siphon that open or close the spiracles. An *Anopheles* larva, in which there is no respiratory siphon is like a man without a nose, or perhaps more like a gorilla with a non-protuberant nose; the respiratory openings lie directly on the body surface itself. Nevertheless, there are similar devices for sealing off the openings on submergence and for ensuring immediate contact with the air on resurfacing.

All mosquito larvae are fascinating to watch, none more so than those of *Anopheles,* the genus that includes the transmitters of human malaria. These siphonless larvae are surface-film feeders, each larva feeding and breathing simultaneously as it lies full length at the top of the water. The hydrophobe area surrounding the spiracles helps hold the posterior end of the elongated larva to the surface. The rest of the body is held up by a series of paired palmate hairs which spread out on surfacing like the spokes of a wheel and act as floats. But even with the whole larva lying parallel with the surface in this way it would not be possible for it to feed on the surface film itself were it not for the remarkable ability of the larva to swivel its head round 180° and so bring its mouth brushes, which normally point downwards, directly in contact with the surface film. It is as if a man could float in water back uppermost but be able to turn his head round so that he could look upwards at the sky. Why not surface, one may be forgiven for asking, with the ventral side uppermost and so avoid the necessity of these gymnastics. But the spiracular openings are dorsally placed in all mosquito larvae, and if the creature is to feed and breathe at the same time it presumably has been easier to evolve the head-swivelling mechanism on an already narrow and mobile neck than to remould the fundamental layout of the tracheal system.

Anopheles larvae have no need to leave the surface except in order to escape from danger. On coming to rest at the surface, the head remains for a moment or two facing downwards and the feeding brushes are motionless. Then if all is clear, the head suddenly swivels round 180° and the brushes start again their beckoning movements. When the brushes are in operation a steady current may be seen moving rapidly across the surface film from a point in front of the larva to its mouth. On reaching the mouth smaller particles pass directly in, but the main current divides into two and passes to either side of the larva's head, carrying the larger particles with them. When a particle too large to 'handle' repeatedly arrives at the region of the head as the current brings it round a second and third time in an endless cycle, the brushes may abruptly cease their motion and the head be returned through 180° to its normal downward-facing position. If the particle is still worrying the larva it may be spat out in a downward direction and so put an end to this cycle. The surface of the water may support so many larvae that it is thereby darkened with their presence. It is interesting to watch the play and interplay of currents set up in the crowded conditions. Sometimes heads collide as the larvae themselves are drawn towards one another. Again the heads will be rotated through 180° to resume the normal position. Only after a minute or two will the heads turn round again to take up the feeding attitude once more.

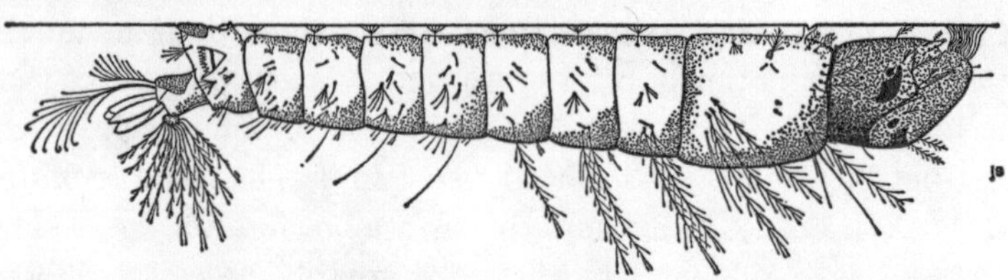

Figure 5. Larva of *Anopheles,* showing palmate or float hairs and upswivelled head enabling it to feed on the surface. Note absence of breathing siphon.

Most species of *Culex* also feed while breathing at the surface, and only leave it when disturbed or threatened. The presence of the siphon, however, holds the rest of the body away from the surface and thus, receiving no further support, it hangs down at an angle of about 45°. Feeding, therefore, occurs a little way below the surface, the actual distance depending on the age and size of the larva, the angle at which it hangs from the surface and on the length of the siphon. In *C. nebulosus* and in some other species the siphon is short, but in *C. horridus* and *C. albiventris* and in many other species of *Culex* the siphon is extraordinarily long and slender, being pulled out almost to needle-like sharpness and being nearly as long as the whole of the rest of the abdomen. Every intermediate is to be found between these extremes. In some, like *C. pipiens,* the siphon is barrel-shaped with convex sides, in others it is churn-shaped with concave sides.

While most species of *Culex* feed just below the surface with the tip of the siphon breaking the surface film, there are exceptions. *C. moucheti,* for example, appears to feed at the bottom and to make only very short but frequent visits to the surface. One cannot be quite certain about *C. moucheti* because the water in which these larvae were found when these particular observations were made was so foul and turbid that one could not see what they were doing each time they dived. At all events their behaviour was very different from the more commonly observed members

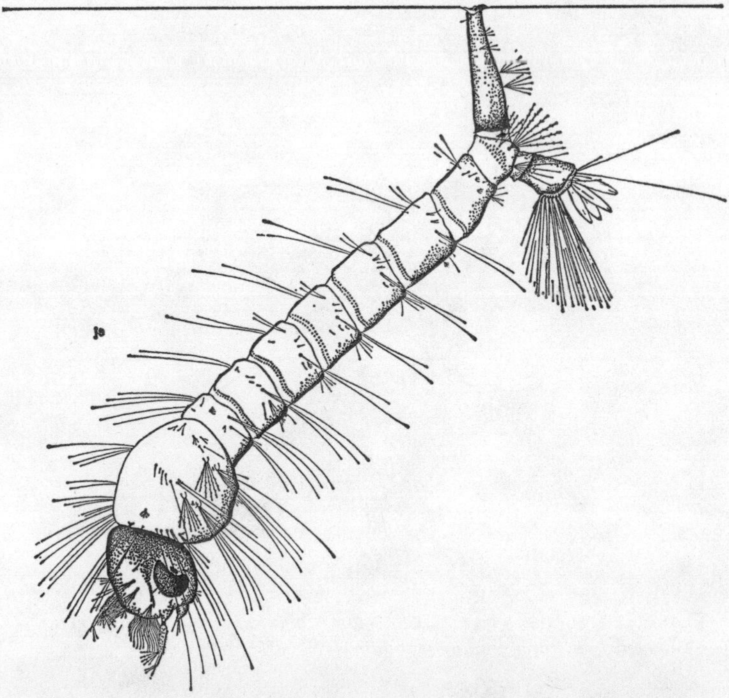

Figure 6. Larva of *Culex,* showing breathing siphon enabling it to feed below the surface of the water while still in respiratory contact with the air above.

of the genus. These particular specimens were captured in a very deep, dark and rather inaccessible and distinctly unsavoury pit. A weighted container fastened to the end of a succession of pieces of banana fibre knotted together was first lowered into the water, or liquid, and then hauled with its contents to the top. It may well be that the surface shyness which these larvae exhibited was

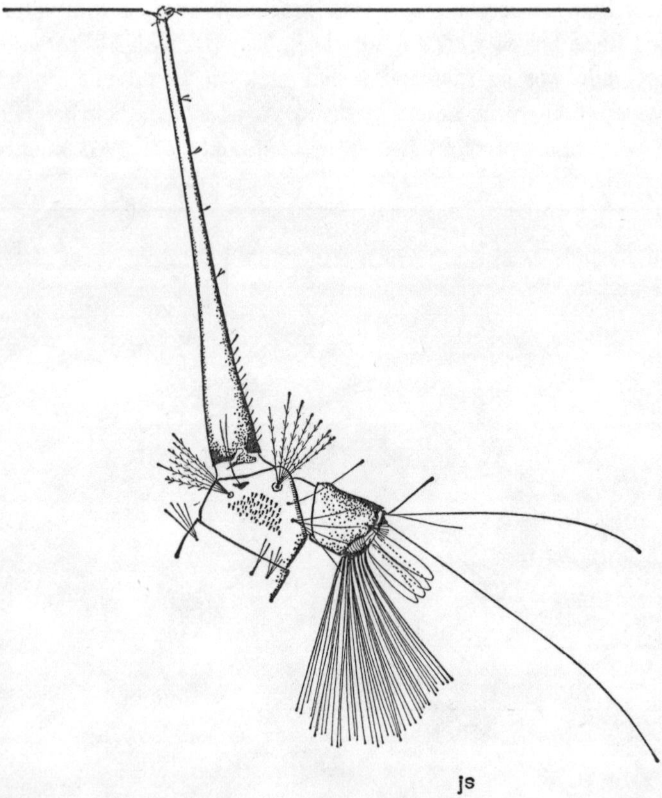

Figure 7. Terminal segments of *Culex rubinotus* larva showing greatly elongated breathing siphon (modified from Hopkins).

a reaction either to the unaccustomed light in their new environment or to the wholesomeness of the air which they now found each time they darted to the surface. At any rate no sooner did they arrive at the top of the tin than they darted back down again and out of sight. My attention was first drawn to the possibility of something interesting going on down there when a certain chief complained that his wives were being bitten on the bottom.

Mosquito larvae, then, do breed in stagnant water, even of the foulest kind, despite all that I have said earlier. If I overstated the case for the purity of their upbringing I did so deliberately in order to try to correct at the outset a widespread misconception about mosquito-breeding. Foul and stagnant water is indeed colonised by mosquito larvae, but the species inhabiting such places are few. Actually certain subspecies of the commonest, best-known and most widespread species in the world, *C. pipiens,* are often found in gulley traps, pit latrines and even in cess-pits. *C. pipiens* is the classic mosquito larva of popular imagination appearing in enormous numbers wherever there is standing water, whether in temperate, subtropical or tropical regions. It is the larva of the water-butt, the horse-trough and the schoolroom. In parts of the tropics the subspecies *fatigans* competes with *Ae. aegypti* as the domestic-breeding species, sharing old coconut shells and husks, tins and waterlogged canoes and even tending to replace it in the special environment provided by discarded motor-tyres in parts of Tanzania. Unlike *Ae. aegypti,* however, it also breeds happily enough in the foulest of places as indicated above.

The larvae of *Aedes* are usually much longer in the body than those of *Culex* and they lack what the enthusiast might call the smart appearance of *Culex*. Instead of sticking out stiffly at an angle from their point of suspension at the surface, like soldiers standing at attention, albeit upside-down, they hang almost vertically in a rather untidy and haphazard fashion. But these characteristics serve them well, for with their short barrel-like siphon attached to the surface they can either feed below the surface like any other culicine or with their elongated abdomen looped round in U-fashion they can feed on the surface film after the manner of *Anopheles*. They do not, however, 'need' to contort their heads because their looped posture automatically brings the mouth brushes into contact with the surface. Thus they can feed below or at the surface while breathing. Now the bristles comprising the brushes are made up of both simple and toothed types, the latter being stouter and more rigid than the others [3]. These modified mouth brushes enable the larvae to nibble at quite large items of food and they characteristically browse round the sides and bottom of their containers, particularly when suspended matter is in short supply. They will move round browsing on rotting leaves and even on dead animal matter, merely rising to the surface for an occasional 'breather'. Indeed, they will tackle small living animals and are not averse even to eating their younger brothers and sisters. The feeding versatility of these larvae adapts them well to living in small containers of water of precarious permanence where survival must often depend on a feeding race against time.

These trends and traits are carried to their logical conclusion in the larvae of *Eretmapodites,* a genus found only in the continent of Africa. These peculiar larvae tend to occupy the smaller kinds of small container: plant axils, old snail shells and fallen leaves and bracts of flowers. They also possess a somewhat truncated siphon, but the rest of the abdomen is so elongated that they can in their chosen shallow environment, make contact with the air while feeding on the bottom. The bristles of the mouth brushes show even greater modification than those of *Aedes* [90] and cannibalism is so common that rarely are more than a few larvae found together in any one container–a further adaptation to life in very small collections of water which clearly would be unsuitable to house large populations. They are facultative cannibals [66]. Normally they attack only their smaller and weaker companions, but if overcrowding is severe or if other sources of food are scarce they will take on larvae almost of their own size. By these severe means are the numbers adjusted to the optimum for the size and productivity of their territory. Thus, contrary to some popular writing, man is not alone in attacking his own kind when threatened with inadequacy of space and food. His methods may be somewhat less melodramatic, but this is open to question; his motivation may be wholly at the subconscious level and might once have been attributed to instinct, but the results are the same.

Somewhat less objectionable in the eyes of all but the most objective observers are those species whose larvae have become predators attacking larvae of other species, although even these will eat their own kind when given the opportunity. *C. tigripes,* a large and handsome *Culex* with a short straight-sided siphon, is wholly predacious, having rejected entirely the primitive filter feeding of its relatives and ancestors. It lies parallel with, but just below the surface in an attitude superficially resembling an *Anopheles*. Its head is thus in reach of *Anopheles* and with a little adjustment of its very long abdomen it can reach larvae of other genera too. The mobility of the abdomen also enables it to twist round on itself when it perceives a likely prey behind it. The larvae of *C. tigripes* are found in all sorts of breeding places from small artificial containers to large swamps. A single larva may devour as many as eight large mosquito larvae or pupae in a single day. Similar in habits but very different in appearance and relationships are the larvae belonging to the subdivision of the family Culicidae known as Toxorhynchitinae. There are some sixty species belonging to this tribe of mosquitos found in America and in the Old World tropics; all are obligatory predators, for the most part on mosquito larvae and pupae. These are the giants among mosquitos, a single larva being a centimetre or more in length and very wide across the thorax. They are mostly found in relatively small containers and rarely is more than one larva found in any one container. Whether this is because the female mosquito distributes her eggs in this manner or for more ominous reasons is not known, but judging from laboratory observations one rather suspects the latter. Like its unrelated partner, the other obligatory predator and part-time cannibal, it lies parallel with and just below the surface and has been observed in the laboratory to consume as many as sixteen large mosquito larvae per day. *Toxorhynchites* are long-lived and a single specimen may, in its time, consume as many as 200 mosquito larvae.

Another group of wholly predacious mosquito larvae is the subgenus of *Aedes* known as *Mucidus*. Like most predators, including the other predacious mosquito larvae, all are large and all have modified mouth parts. If I have dealt at some length with these carnivorous larvae it is on account of the surprising role they play as allies of man. The number of species that behave in this way, whether as facultative or obligatory carnivores is only a small proportion of the total. Indeed one would not expect it to be otherwise. We shall have more to say about them when we come to consider predators as agents of control in Chapter 11.

Before we leave the subject of feeding and breathing we must say something about the larvae of *Mansonia* and *Coquillettidia*, whose way of obtaining air while feeding is so strange as almost to tax one's confidence in the theory of evolution by natural selection. The larvae of both genera puncture the roots and stems of aquatic plants and draw their air supply from the air spaces within the plants. The flaps at the tip of the modified siphon no longer operate merely as valves to expose or close off the tracheal openings. Each of the anterior valves is extremely small but is provided with a stout, curved bristle, while the posterior valves are enormous, as long as the whole siphon. They are curved round on themselves to form a tube or sheath protecting a veritable tool-kit including a stout toothed saw [12]. This complicated apparatus is used in conjunction with six pairs of hooks, first to pierce and then for attachment to aquatic plants. Like all mosquito larvae, *Mansonia* and *Coquillettidia* swim backwards and thus approach a root or stem siphon first. To begin with, the saw is used to cut a hole in the plant and then with violent rhythmic movements of the whole body it works the narrowed tip of the siphon into the opening, anchoring itself by means of the six pairs of hooks which now spread outwards. The whole larval life and indeed the pupal life too, is spent attached in this way. They hang upside-down, that is not only head downwards but ventral surface uppermost, looking for all the world like small rootlets or extensions of the plants themselves. The pale, off-white colour of some species and the bright green and blue-green of others helps this illusion [98]. Only at the time of moulting do they let go, and then only just so long as to allow them to find another anchorage close by. The actual process of moulting is, in fact, carried out while the larva is still in position, and the old larval skin remains attached to the plant.

These incredible larvae are by no means rare; indeed, quite the reverse. *Coq. richiardii* can be found all over Europe, including parts of Britain and is a most fascinating and challenging subject for study. Although widely distributed over much of the world larvae of *Mansonia* and *Coquillettidia* are not always easy to find.

Coq. fuscopennata is one of the most widespread and indeed one of the most troublesome mosquitos over large areas of tropical Africa, appearing in unbelievable numbers wherever there are suitable swamps. Yet, to my knowledge, only four larvae have ever been found; these were identified by breeding them through to the adult stage. Other species are easier to come by, once one knows how to set about capturing them. Although they hang to the roots, stems and leaves of emergent and floating plants, it does not take much in the way of disturbance of the vegetation to induce them to drop off and disappear into the mud below. One way of capturing them then is to uproot or remove the plants very suddenly with one hand while simultaneously thrusting in a bowl or basin below the plants with the other. Larvae that have been alarmed will be caught in the bowl, where they will sink to the mud and debris at the bottom. You will not find them until you have decanted off most of the water. Just as you are about to give up and there is less than a centimetre depth of water left in the bowl they appear to take fright again and this time reveal themselves while actively swimming against the flow of the emptying bowl, as if trying to prevent themselves from being washed away.

Everything about these larvae is slow. They move extraordinarily slowly compared with other mosquito larvae, lumbering along as if with difficulty. It may be argued that they do not normally need to use their powers of swimming and that these are therefore poorly developed. But even the movement of the mouth brushes is very slow, those of *Coq. aurites* at 23°C making only 2.5 sweeps per second, compared with the 5.8 per second in *Ae. aegypti* at the same temperature [33]. Then, too, the life-cycle is long-drawn-out; *Coq. aurites* takes forty to fifty days from egg to adult at 23°C compared with about ten days for *Ae. aegypti; M. uniformis* and *M. africana* take twenty-five to forty days at 26° to 30°C [71]. It seems more likely that natural selection has produced a slow-geared creature that fits it to its special mode of life with the limitations imposed on it by its peculiar method of obtaining oxygen. An animal with a high rate of energy output would make too heavy a demand on the meagre oxygen supply. One is not the cause of the other, but the two are geared to act together. The evidence for this view will become more apparent when we come to consider the pupae and adults, both of which also exhibit this leisurely pace except when it comes to the adults' biting.

This highly specialised mode of obtaining oxygen by the harnessing of plants and the remarkable set of tools on the end of the siphon that makes this possible, have caused more than one orthodox biologist to scratch his head and begin to wonder at the theory of evolution. How could such a specialised habit and such a specialised means of achieving it be produced by natural selection? We have all heard the same arguments over the human eye often enough (although why the human eye more than, say, that of an eagle, I am never quite sure). But, like the eye there are intermediate states that help enormously in our understanding of the probable steps involved. In the genus *Ficalbia* there is a complete series of species showing differences in the degree of specialisation towards the tapping of air from plants: in *F. pallida* the tip of the siphon is fully adapted for the piercing of plant tissue, although the modifications are not carried to such a degree as those in *Coquillettidia* and *Mansonia;* in *F. perplexens* and *F. splendens* the modifications at the tip of the siphon allow it to tap the air trapped on the underside of floating leaves but do not actually permit of penetration of the plant tissue; in *F. plumosa* there is hardly any modification at all and the larva behaves in the normal manner and can be found in the complete absence of vascular plants. Some other genera too add their quota of intermediate forms. The larvae of the peculiar *Aedeomyia* show modifications similar to those shown by some of the *Ficalbia* with the lateral valves modified into large hooks. *Culex poicilipes* also has enormously enlarged valves and is able to obtain its oxygen from the air film on submerged plants. Similar modifications are also found on the siphons of the tiny *Hodgesia* larvae and it seems probable that they too make use of plants in this way [12].

The larvae of *Aedeomyia* are peculiar in other ways. They have enormously enlarged antennae which are sometimes used for clasping vegetation, particularly while the modified siphon is being jockeyed into position on the surface of a leaf. Evidence has been produced to show that the antennae are also used as auxiliary organs of respiration, picking up oxygen dissolved in the water. Two large air sacs, one at the base of each antenna, act as reservoirs and enable these larvae to remain submerged for forty minutes or more. Similar but smaller air reservoirs are found also in certain other mosquitos, including *Aedes argenteopunctatus,* which is unusual for an *Aedes* in that it breeds in swamps, and in certain species of *Culex,* all of which are also able to remain submerged for long periods.

We have already seen that there is hardly any type of water that does not have its own special fauna of mosquito larvae. This does not mean that only one species will be found in any particular habitat. On the contrary, it is quite usual to find several different species living together even in some of the smallest breeding places. Species living together in one type of water will be found all to possess specialised characters that fit them for the particular conditions found there. Different types of water vary in the amount of chloride available, an essential in the life of any mosquito larva. Chloride is taken up by four special organs known as the anal papillae. In general, larvae living in habitats such as tree-holes in which the concentration of chlorides is particularly low, are found to have excessively large anal papillae, whereas larvae from large swamps have exceptionally small ones, especially those from brackish waters [62, 70, 74, 79, 86, 88, 89, 93, 94]. The predacious larvae living in small containers form an exception, for the anal papillae of *C. tigripes* and *Toxorhynchites* are minute; presumably these larvae obtain their salt supply indirectly from the larvae which they eat. A few mosquito larvae actually live in pools containing sea-water, a rare medium for any insect. *Ae. australis* is found in sea-water containing over seven per cent salts, while *Ae. detritus* has been found in sea-water containing ten per cent salts. The anal papillae of these brackish and salt-water breeders are reduced to tiny knobs and no longer take part in the uptake of chloride [4].

These differences in the size of the anal papillae are not entirely genetically determined, for Sir Vincent Wigglesworth showed many years ago that larvae of *Ae. aegypti* raised in water rich in chloride have very much smaller papillae than those brought up in water of low chloride content. I once had to give evidence as an expert witness in a case brought against a man for allowing mosquito-breeding to occur in a drum of stored water kept on his premises. At that time it was an offence to store water in such a manner as to allow mosquito-breeding to take place and people who had no piped water supply and were therefore obliged to store water in drums and tanks were required to empty these once a week. The unfortunate fellow stubbornly maintained that he did not store water at all but had each day's supply brought up on the head of his wife (eight gallons at a time), who drew it from the nearby swamp some five or six miles away. One did not have to get down to all the business of identifying the actual species. A glance at the small size of the anal papillae was enough to show that the larvae were, in fact, swamp-dwellers, or at any rate had not spent their life in artificial containers on the premises, and the man was duly acquitted.

Mosquito larvae are found in ground pools on the floor of dense forest and woodland, pools filled with rotting leaves and general forest litter, pools half hidden by fallen logs and without living vegetation. The number of species that inhabit such places is too many to list here, but one may mention *Ae. rusticus* in Europe and *Ae. circumluteolus* in Africa. But one does not find any, say, of the British species of *Anopheles,* or any *Aedes* belonging to the subgenus *Stegomyia,* nor *Coquillettidia* or *Mansonia.* Ground pools can be very extensive or numerous during wet seasons and the output of mosquitos is often on a gigantic scale. It is characteristic of some of these pools that *Aedes* larvae, including those of the predatory subgenus *Mucidus,* appear before those of *Culex.* The reason for this has been dealt with in the previous chapter; the drought-resistant eggs of *Aedes* have been lying ready to hatch, scattered among the dry leaves, whereas those of *Culex* are laid and then develop to maturity after the depression has become filled with water. There is thus a single sharp and short wave of *Aedes* activity with larvae all of about the same size and age, which may be succeeded by a more or less prolonged period of mixed *Culex* and *Aedes* breeding with larvae of various sizes and ages. It is sometimes possible to make a fairly accurate estimate of the age of a pool simply by an examination of its mosquito fauna. Once again we see how abstruse and innocent observations, this time on drought resistance in mosquito eggs, could have unexpected applications, even of a grisly and forensic kind.

Among the many species of mosquito larvae found in forest pools are the large and handsome larvae of *An. implexus,* which are rarely if ever found in any other situation. It has been suggested that it is the coolness of the water in these pools rather than the absence of light that is the important factor. *An. implexus,* and *An. obscurus* too, cannot be reared in the laboratory unless the temperature is kept down below a certain level. But one must be a little wary or even suspicious of so-called important factors. In nature there are usually so many interacting factors that all one can really say is that such and such is *an* important factor, but not necessarily *the* one. This is not to cast doubt on this particular suggestion but on attempts in general to find a simple answer, a common factor or single switching mechanism. We are all the time trying to cram the complexities of nature into the simple and usually undersized moulds that we cast for them.

Not far removed from the popular misconception that mosquito larvae are restricted to foul and stagnant water is the idea that their breeding places are usually dark and concealed. It may come as a surprise to learn that the larvae of the most important malaria mosquito in Africa, *An. gambiae,* live almost exclusively in open puddles fully exposed to the hot tropical sun. They are found in temporary puddles devoid of vegetation except for *Spirogyra* and other algae, and usually in water that is cloudy with colloidal clay. Puddles in the middle of an unmade-up road, brick-pits, borrow-pits, cattle- and other hoof-prints and footprints whether of man or elephant. Any bare collection of water that is exposed to the sun seems to be suitable. The breeding places are often man-made in so far as man has made the depression in the first place, or he has so denuded the ground of natural vegetation as to lead the way to the formation of animal footprints. There are obvious exceptions as, for example, footprints round an over-crowded water-hole, but by and large, with few exceptions, *An. gambiae* is associated either directly or indirectly with the activities of man. It is often hard to find the species round the shores of the great lakes until they have been disturbed or developed by man. This is partly because of the dense vegetation in the undisturbed stretches, making even hippopotamus footprints unsuitable for *An. gambiae*. Introduce man, and *An. gambiae* follows within a month. Yet it cannot really be described as a domestic breeder in the sense that some other species are; it is very rarely indeed found in discarded tins and vessels, and only a little less rarely in such places as waterlogged canoes or wheelbarrows. Essentially it prefers earth-walled containers.

Larvae may be found in a number of puddles within a surprisingly short interval after first being filled, and I have sometimes been tempted to think that either the eggs or more probably the larvae themselves are at least partially drought-resistant. The fact that third- and even fourth-stage larvae may be found in puddles which are only three or four days old leads one to wonder if these larvae have survived from a previous flooding, preserved in the partially dried mud. On the other hand these open and often very shallow breeding places are subject to extraordinarily high temperatures from their full exposure to the sun, well over 30°C commonly being recorded, and even 40°C at times. At these high temperatures larval development is extremely rapid and could well account for the appearance of mature larvae in so short a time. It is interesting to note that larvae of *Ae. vittatus* are characteristically found in rock pools, which are also often fully exposed to the hot tropical sun. Like those of *An. gambiae* these larvae tolerate extremely high temperatures. Andy McClelland has told me that he has recently shown that at 35°C a little over two days only is needed to complete all four larval instars, from first flooding of the eggs to the appearance of the first pupa.

The number of larvae in any one breeding place may be so large as almost to cover the surface. Thus even in the absence of permanent swamps or what is usually regarded as stagnant water malaria can become a serious problem and at times may reach epidemic proportions. A multiplicity of small open, bare pools and puddles is sufficient when the rainfall is just enough to keep them filled without flushing them out. *An. culicifacies,* an important transmitter of the disease in Ceylon, also breeds in exposed pools. During rainy seasons the rivers in spate automatically flush out any pockets of breeding, but when the rain eases or ceases altogether a series of residual pools may be left along great stretches of the rivers and these become the chosen sites of *An. culicifacies.* Dry season epidemics of malaria have been known to follow.

Just as there are very many species inhabiting forest pools, so there are many species of open pool breeders. We shall make no attempt to catalogue either of these but pass on to the next type—the open swamp and river breeders. Here again temperature and shade might be important factors, but living vascular plants seem also to be essential, whether growing through the water, floating on the surface or overhanging from the margins and banks. Whether they are essential in themselves or merely as providers of shade is not always possible to say with certainty. The requirements vary from one species to another. *An. coustani* in tropical Africa is found in almost any swamp one cares to search except for those above 2700 metres or so [11]. The species of plants associated with the swamp and the condition of the water within wide limits seem to be of little importance. Wherever you dip, whether in small swamps or large, shallow or deep, clear water or foul, the characteristic and conspicuous larvae of *An. coustani* will be found almost at once. They are so 'obviously anopheline' with their long large bodies lying parallel with the surface and are so readily captured that anyone with only the most superficial experience can pick them out and proudly say, 'Look, *Anopheles!*' As often as not no attempt is made to take the identification further, even though *An. coustani* (and its allies) is perhaps the most easily identified of all *Anopheles.* Yet, truth to tell, the adults of *An. coustani* rarely bite man, and play little or no part at all in the transmission of malaria. One wonders how often swamps containing this species have been declared malarious; one dare not inquire into how much money and effort have been spent in attempts at clearing and draining such places.

Very different from the ubiquitous *An. coustani,* both in its requirements and in its importance to man, is the elusive, small and hard to capture larva of *An. funestus,* a species that is second only to *An. gambiae* as a vector of human malaria in Africa. It requires coolness and shade, and is always associated with vegetation either growing through or on the water, or overhanging the water's edge. In other words, just the same as *An. coustani;* but there are clearly other factors that are difficult to put into words for it is surprising how many places appear on paper to be ideal for *An. funestus* yet you will not find it there. For one thing, the water itself has to be everything that one does not popularly associate with mosquito-breeding–clear, free from rotting vegetation and foulness of every kind. Favourite places for *An. funestus* are along the edges of great stretches of the White Nile and its tributaries, even in places where there is open water up to a kilometre in width and the current is very strong indeed. The larvae remain immune to the movement of the water by sheltering among the great mats of floating *Pistia stratiotes,* the so-called Nile or water lettuce, that occur along the edges. These plants may be in constant undulating motion with the general turbulence of the river, but as long as the larvae can float there feeding and breathing at the surface without being carried away, this up-and-down movement does not appear to bother them. If, as often happens along this part of the great river, sections of the mat of floating vegetation become detached and float away down the river, then the larvae go too, probably to perish.

It used to be thought that *An. funestus* were found only along the weed-choked edges of streams and rivers, but it is now known that they may be found also in still pools when the exacting conditions are fulfilled. In some of these still pools the water may, in fact, be quite cloudy with a grey white colloidal suspension, particularly in certain pools the surface of which is covered with *Pistia* or other floating plants. But the water itself in these places is never foul and is free from rotting vegetation. There are occasionally exceptions where larvae are found in atypical places. But then so too are there to the number of fingers possessed by man, yet most of us would accept that man is five-fingered.

Extensive areas of central Africa consist of swamps of papyrus, a handsome plant growing in shallow water and standing two to three metres tall. Actually many of these swamps are really slow-moving rivers. They may be a kilometre or so wide in places and extend for hundreds of kilometres in length, flowing eventually into the great lakes or into other rivers. The temperature of the water and the light value at the surface of the water are all that *An. funestus* could 'desire'. Yet, with very few exceptions indeed, these vast swamps are entirely free from this species. It may occur in pockets of emergent grass at the edge or in the middle of the swamps, but rarely within the papyrus itself. I, myself, know of only one papyrus swamp that harbours *An. funestus,* a lake lying at about 2000 metres altitude close to the Uganda–Rwanda border, but no doubt there are others. The water in which the papyrus is growing at the edges of this lake differs from that of all the many other papyrus swamps which it has been my lot to study. It is entirely free from the rusty-red flocculence so commonly associated with this plant. Now red ferruginous scums and rusty-red deposits are also sometimes present above the roots of emergent grasses, and again, when present, one may be sure that *An. funestus* will be absent. Other species will be there, including *An. coustani,* but if we are looking for *An. funestus* and find this rusty-red flocculence, it is better to move on, for we are wasting our time. It is tempting to say that here is the common factor; that the red scum or whatever it is that produces it, or perhaps the conditions favouring its presence, inhibit *An. funestus,* either as an egg-laying site or as a larval habitat. But really we know nothing about any causal relations; all we are justified in saying is that the two are associated–the presence of the red scum and the absence of *An. funestus.*

All this was discovered and published more than thirty years ago as a result of a survey of papyrus swamps made wherever they were encountered in the region known politically as East Africa, comprising Kenya, Tanzania and Uganda, an area as large as the combined areas of France, Spain, Italy, Belgium, Holland and Germany (East and West) or nearly three times the size of Texas [72]. Until recently the lack of threat that these swamps present has been largely ignored, despite its extraordinary consistency. But now it has been rediscovered as a result of a very detailed study of a papyrus swamp made near Makerere University College, Uganda.

In 1938 I was living in a place called Jinja, the town that has grown up round the spot where Speke had discovered the source of the White Nile sliding out of Lake Victoria over what was later named the Ripon Falls. There was a considerable amount of malaria in Jinja despite eradication or treatment of all known breeding places of *An. gambiae* and *An. funestus* along the miles of coastline. Although the main lake-shore had been cleared of vegetation, huge swamps of papyrus still choked the large bays at the edge of the town. No breeding of dangerous species could be found in the papyrus or, in fact, in the grassy area between the papyrus and land. It was difficult, however, to be certain that there were no pockets in this veritable forest of papyrus. Having failed to find any from the land side, or at least none with *An. funestus* in it, I decided to explore from the lake side. I set off into the open water by canoe with four African companions. The outer edge of the papyrus swamp is abrupt, it extends like a green wall two or three metres high. Here and there the continuous line is broken by hippopotamus tracks, and into one of these we pushed our canoe. There, sure enough, were isolated patches of grass growing through the water. The grass blades were in constant motion as they moved to and fro with the general movement of the lake. We decided to investigate. Pulling the canoe now by taking hold of the grass we moved into the centre of one of these patches and, after allowing things to settle down we proceeded to dip our soup ladles to see what we could find. We all of us found nothing. All of us that is except for one, Mr Francis Sserunkuma, who not only produced *Anopheles* larvae, although clearly not *An. funestus,* almost every time he sampled the water, but was the only one of us to do so. I was delighted that he was successful but I was completely baffled as to why he alone should be so. I was also not a little annoyed with myself that my own efforts were fruitless. I was so intrigued that I gave up trying to catch the larvae myself and took to studying Francis Sserunkuma instead. He always had his own way of going about things, did Francis, and I noticed that while the others carefully scooped up the water with a sudden sweeping movement as taught, Francis merely plunged his ladle into the water with a resounding splash. I tried the same and there, sure enough, were several *Anopheles* larvae the very first time. Not wishing to undo all that I had taught them for the successful capture of *An. funestus* I kept quiet about my little discovery, and not even Francis knew why he and I alone were successful, and he doubtless put it down to our innate skill.

I took the larvae back to the laboratory but I was unable to identify them, so I decided to rear them through to the adult stage. I placed each in a small glass tube partly filled with water and fed them on powdered dog biscuit and brewers' yeast. But when I came back the next day I found all the tubes empty. My heart sank. Then on looking further I found that each larva had worked its way up the side of the tube above the water, carrying a film of water with it. I was so concerned for their safety that my first reaction was to wash them down into the water again, but next time I looked there they were again half-way up the side of the tube. I wish now that I had thought of looking to see what they were doing there. Where did they feed? They certainly returned to the water for pupation, for the old larval skins were duly found in the water. But the pupae also climbed up the side of the tube. I decided to leave them undisturbed, and the adults that emerged proved to be those of *An. wellcomei,* a species whose larva had hitherto not been described. Now all became clear. It was evident that the larvae of this species must also climb above the water surface in nature, presumably up the grass blades. Francis Sserunkuma had stumbled on a method that dislodged them from the grass and caught them before they had time to dive. Thus not only do the larval habits and requirements vary, but the catching methods have got to be adapted to fit these habits. *An. wellcomei* is common enough in certain parts of Africa, but these peculiar habits have been responsible for the larvae remaining unknown for so long, and even now they are not often recorded.

Many species of mosquito larvae have taken to living in the smallest collections of water, not because nothing else is available but, it seems, by choice. This is not to deny that the choice may have been forced on them or their ancestors in days gone by, but nowadays these small container breeders are rarely if ever found in any other situation, even when water is available in plenty. Nor is mere size or volume of water the only deciding factor, for it is a curious fact that one rarely if ever finds these small container larvae in a ground pool, no matter how small it may be. Rock-holes and crab-holes are exceptions to this, but these for the most part have their own special mosquito fauna. It would seem as if the nature of the material forming the container is of more importance than the quantity of water contained, perhaps in the case of *Aedes* because it is chosen as the substrate on which the eggs are laid, above the water's edge. It is almost a golden rule that these species, including *Ae. aegypti,* are never found in earth-sided containers.

The best known of these special habitats is the tree-hole [83]. In England and in Europe generally the larva of *An. plumbeus* may be found in tree-holes at all times of the year, even in the depths of winter. In the tropics and elsewhere it is mainly species of *Aedes,* and certain species of *Culex, Haemagogus, Orthopodomyia, Toxorhynchites* and *Wyeomyia* that occupy this niche. Other small container breeders show a special preference for plant axils; *Ae. simpsoni,* the rural transmitter of yellow fever in Africa, breeds in the axils of certain varieties of banana, in dracaenas, *colocasias* and *Sansevieria* and even in the tops of pineapple plants. *Ae. woodi* is even more specialised; it is found only in the axils of a certain species of sedge. Each larval habitat holds only a few millilitres of water and rarely houses more than six to ten larvae. It is the sheer quantity of these water-bearing axils that allows huge but scattered populations to build up. Some species of the curious African genus *Eretmapodites* breed almost exclusively in plant axils, others show a special preference for discarded snail shells or even for the small quantity of water held up in the concave tops of certain toadstools. Others are found in fallen leaves, as are also some species of *Limatus* and *Culex*. The New World species *Wyeomyia smithii* breeds in the water contained in the pitcher plant. When the water freezes in the plant the larvae are caught in a solid block of ice where they are preserved for the winter. They resume growth in the following spring [4].

All but one of the half-dozen species of the anopheline subgenus *Kerteszia* breed exclusively in the axils of bromeliads, epiphytic plants that cover almost every branch of every tree in parts of Central and South America [78]. *An.* (*K.*) *bellator* is an important transmitter of human malaria, and is clearly not amenable to the more usual methods of larval control. Fortunately, although there are nearly 2000 species of bromeliad, only a few of these actually breed *An. bellator;* those that do, however, are present in fantastic numbers. The importation of bromeliads is understandably prohibited in many Old World tropical countries, but these plants nevertheless find their way into these, mainly as ornamental houseplants to begin with. If these should ever 'escape' and run wild, they could conceivably spread very rapidly indeed in some of these countries. Even if there are at present no species of *Anopheles* ready to take advantage of them, they would surely be taken as ready-made breeding places by some of the axil-haunting culicines; in South America these plants are also utilised by *Culex* and *Wyeomyia.*

In extensive areas of the tropics *Ae. aegypti* has taken to breeding in man-made containers of every description: tins, bottles, roof-gutters, water-pots, discarded tyres and disused cars. In other areas it is found in tree-holes, and in yet others in both natural and man-made containers. It is even found at times in rock-holes, usually the special domain of *Ae. vittatus.*

So far we have dealt mainly with breathing and feeding and with some of the many places colonised by mosquito larvae. We have only touched on locomotion in so far as it affects or is affected by these other aspects of larval life. In 1850 Sir T. Browne wrote, 'He that would behold a very anomalous motion may observe it in the Tortile and tiring stroaks of Gnat worms.' Most mosquito larvae, with the notable exception of those of *Coquillettidia* and *Mansonia,* are extremely agile in the water. Larvae of *Anopheles* are particularly so; they move with astonishing speed and agility if somewhat jerkily, whipping themselves along tail-first either across the underside of the surface to another position, or downwards to the bottom. Some *Culex* too are extremely fast movers; we have already had occasion to refer to *C. moucheti.* In some of the longer-bodied species such as *Ae. aegypti* the side-to-side lashing movements are not so rapid, being much more deliberate and more obviously serpentine. But they can nevertheless move quite fast. Most of the active swimming in this species is up to or down from the surface, since, except when suspended matter is plentiful, much of their feeding is carried out at the bottom or round the edges of the container. *Ae. aegypti* and other *Stegomyia* also at times glide through the water head-first without any bodily movements, the mouth brushes acting as a pair of tractor propellors. Usually members of this species are finely balanced in the water, and have both to drive themselves downwards or upwards by active swimming movements. Younger larvae, however, tend to be hydrostatically light, while their fully grown fourth-stage companions tend to be heavy. If larvae previously in hydrostatic balance stay long at the surface their buoyancy is increased, and if now disturbed they must swim to the bottom; such larvae can now return to the top by hydrostatic lift alone. On the other hand, if they are prevented from establishing contact with the atmospheric air they lose their hydrostatic equilibrium and sink to the bottom and can now only return to the top by active swimming movements. Indeed, if kept from air for long they can only reach the surface with difficulty. After a variable period, probably depending on the amount of dissolved oxygen in the water, the larvae are unable to reach the surface unless helped, and soon die. This period usually varies from one to three hours, but may be longer on occasion for the reason given, and perhaps on account of other unknown variables.

Larvae that have been forcibly deprived of air for twenty minutes or so usually remain at the surface for long periods when returned to normal conditions as if making up for lost time. Even the normally restless larva of *Ae. aegypti* that spends so much of its time browsing round the bottom and sides of its container will remain motionless at the surface after such an experience, and such an apparently surface-shy species as *C. moucheti* has been induced to stay quite still at the surface for forty minutes. This reaction following oxygen starvation, or carbon dioxide accumulation or both has been used as the basis of a method for studying living larvae before breeding them through to the adult stage [64]. For the purposes of identification it is usually necessary to have a prolonged look at the side view of the siphon and terminal segments. One can always do this, of course, by killing the larva and mounting it on a microscope slide in the right position, but it can be rather frustrating if the larva turns out to be something unusual, because one will never know what it would have produced had it been allowed to continue its development. But by placing each larva in a tube full of water and covering it with a glass slide in such a way as to exclude all air they can later be examined with ease; any interesting larva can then be returned to normal conditions and reared through to the adult stage.

Larvae of all mosquitos that come to the surface for air are extremely sensitive to any sudden change in light intensity. Any shadow that falls on a pool will be followed immediately by an exodus of all larvae from the surface. Even when the sun disappears behind a cloud, or in one's laboratory one turns out an electric light, all the larvae dive to the bottom. Similarly, if the sun suddenly appears from behind a cloud or an electric light is turned on, the larvae dive. It seems that it is the change in intensity and not merely shadow that produces this escape reaction. If there is no further change the larvae soon return to the surface, even if the illumination is now much lower (or higher) than previously. We shall see when we come to consider some of the enemies of mosquito larvae in Chapter 11, how this reaction protects them from attack. If the change in illumination is made repeatedly over a short space of time the larvae soon learn to ignore it, or to give it what Alan Watts would call an authoritative rumble—they rapidly habituate to the situation.

Some shade-loving species are stimulated to movement when exposed to continuous bright light; larvae of *An. maculipennis* will move about apparently at random until they find themselves in shade, where they will automatically accumulate. In northern Canada *Ae. communis* shows a preference for water at a temperature around 16°C. In sunlit pools the larvae form a dense clump at the point nearest to this temperature, the clump moving round the pool in a clockwise direction as the region of preferred temperature changes with the apparent movement of the sun. In shady pools larvae of *Ae. communis* do not clump together but remain dispersed [4].

We have mentioned that in addition to the characteristic side-to-side whipping of the larval abdomen and the resultant tail-first movement through the water, some larvae are able to propel themselves smoothly head-first by means of their mouth brushes. There is in fact a third mode of locomotion that is characteristic of mosquito larvae, but one that is not nearly so well known. Whereas a stranded fish jumps and kicks and wriggles in characteristic fish fashion, a stranded anopheline larva in a film of water on an otherwise dry surface does not attempt either of its aquatic modes of locomotion. Instead, it suddenly brings into play a reserve mechanism; without hesitation or need for practice it begins to crawl, caterpillar fashion, by successive waves of contraction moving forwards from the tail end to the head. The same is true of some other larvae, particularly those with very short siphons. They move like any ordinary land grub as if they had not had a fifty-million-year history of water-waggling behind them. Could this be the ancestral gait, the original dipterous larval locomotion on land? Generations may pass without any need to call on this reserve mode of locomotion, but when needed it is there ready for immediate use. Larvae of *Anopheles* in Europe have been observed crawling in this fashion from one hoof-print to another as each in turn dried up. And larvae of *Ae. kochi* have actually been caught in the act of crawling from one leaf axil to another.

Earlier in this book I pointed out that the fourth- or final-stage larva is just as much a specialised aquatic animal as was the first-stage larva. While this is true of its general shape and even of detailed larval anatomy, small changes have in fact been going on from the second stage onwards. Minute buds of embryonic tissue are, as it were, set aside for future development. These take no part in the general economy of larval life, but eventually develop as the typical organs of first the pupa, then the adult. They remain inconspicuous until near the end of the fourth stage, when a rapid increase in size occurs. The fourth stage lasts much longer than the earlier stages; in *Ae. aegypti* it occupies about three days compared with a day each for other stages. By the end of the third day it is hardly fair to call it a larva at all, and it is, in fact, often referred to as a prepupa; the thorax is greatly swollen with the development of the future flight-muscles [4], and the respiratory trumpets and other pupal organs can be clearly seen beneath the cuticle. The general shape and mode of locomotion remain typically those of the larva until near the end, and this tends to disguise the important developments that have been secretly going on, the small changes that set the stage for the dramatic events to follow.

4. PUPAE

No one who has handled or observed the aquatic stages of mosquitos can have failed first to be surprised and then fascinated by the pupae. In so many ways they are quite extraordinary. They have usually only a few days' existence and yet manage to achieve one of the most astonishing transformations in the whole insect world. Within this short period during which this quick change is enacted they are not only themselves extremely agile but move by a mechanism all their own. The mosquito pupa is certainly not the dull, torpid, dormant phase that familiarity with other insects might lead one to expect. It is entirely different from the larva, not only in shape and mode of locomotion but in its whole organisation; it even breathes now from the front end instead of through its tail. And, unlike so many insect pupae that retain some measure of mobility, it bears no obvious resemblance whatever to the adult insect to come. In every sense then the mosquito pupa is a distinct stage. It is extremely neat and tidy in appearance with its own specialisations, its own peculiar ways of going about things, one might almost add with its own personality.

Great changes have already taken place during the last day or so of larval life. It could not be otherwise since the pupa emerges ready made through a split that forms in the dorsal surface of the larval skin. The pupa resembles a three-dimensional mobile comma with a pair of long trumpet-shaped ears projecting from the top of the bulge. These are the respiratory trumpets through which contact is now made with the atmospheric air. It is not known for certain how water is excluded when the pupa dives, but it has been suggested that the sharp bend in the air passage at the base of each trumpet may serve to close the tube when the trumpet is deflected during the act of submergence [3, 4]. Unlike the larva, it at all times floats to the surface and when disturbed has actively to drive itself downwards. A large bubble of air within the thoracic region acts as a buoyancy tank and keeps it permanently light, and a pair of float-hairs on the dorsal surface of the first abdominal segment helps the pupa maintain its position hard up against the under surface of the water. In the pupa of *Aedeomyia* these float-hairs are reduced and only the tips of its trumpets make contact with the surface, where it remains suspended like a man hanging by his arms from a trapeze.

The pupa does not feed. It has no mouth. The bulbous part of the comma is made up of the head and thorax combined and is usually referred to as the cephalothorax. The ninth abdominal segment is modified to form a pair of large paddles. It is these that, acting in combination with violent kicking movements of the ab-

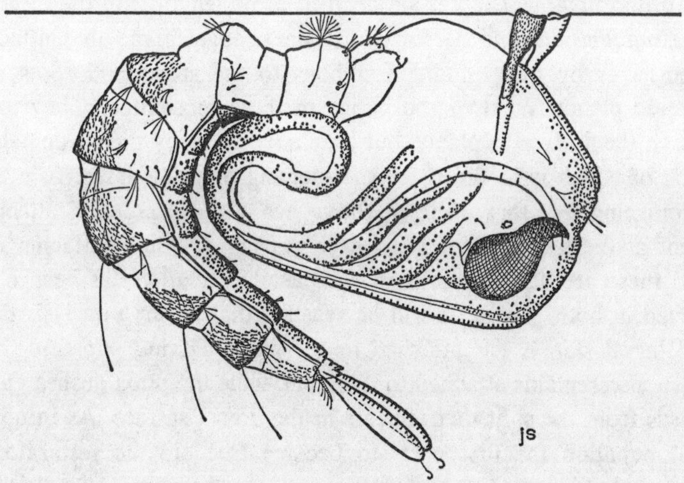

Figure 8. Pupa of *Anopheles,* showing the pair of respiratory trumpets in contact with the air above the water.

domen, propel the pupa extremely quickly through the water. Unlike the side-to-side whiplike movements of the larva, it is the sudden extension and flexion of the pupal abdomen that, causing the paddles to give a succession of backward kicks, drives the pupa forwards. Thus the abdomen is still the locomotory region, but it now functions entirely differently giving dorso-ventral movement in place of the lateral wave action of the larva.

Unless disturbed most pupae remain at the surface. But pupae, like larvae are extremely sensitive to changes in illumination and movement. A shadow falling across the surface of the water or a sudden movement is enough to send them down to the bottom, where they remain for a minute or two before floating rapidly up to the surface again. Actually the violent kicking motion transferred ultimately to the paddles merely serves to drive them in a forward direction. To dive they must tilt slightly downwards by the 'nose', like a helicopter in horizontal flight. Yet, once down on the bottom, pupae can remain there without any further effort. How this is achieved is not certain, but probably the various hairs and bristles that adorn the pupae make contact with objects and irregularities on the bottom and serve to anchor them down.

In the previous chapter we dealt at some length with the larvae of *Coquillettidia* and *Mansonia,* those strange larvae that attach themselves by their modified siphons to the stems and roots of aquatic plants. What of the pupae of these mosquitos? They too attach themselves to plants, but they now do so by their trumpets, each of which is modified for the piercing of plant tissue. It is an astonishing fact that while the larva has a single point of attachment at the tail end, the pupa now has two points of attachment and these are at the other end. The act of pupation has been observed in both genera. It will be recalled that at larval ecdysis the old larval skin is left attached to the plant. During pupation the larva also remains attached to the plant while the pupa pushes outwards from the split that appears in the dorsal surface. As in normal pupation the first parts to become free are the respiratory trumpets; the rest of the cephalothorax soon follows and finally the abdomen begins to be pulled clear. But just before it is entirely free the respiratory trumpets are thrust forward several times until contact is made with the plant. Then with the tail end still enclosed in the old attached larval skin the pupa begins to sink its respiratory trumpets into the plant tissue by movements resembling the action of a pair of pneumatic drills.

All this is achieved while the tail end remains safely inside the still attached skin. But once connection to the plant is secure the pupa whips out the siphon and with a few violent circular movements kicks off the larval skin altogether. Thus at pupation, unlike at other ecdyses, the old larval skin is detached from the plant, but not until the pupa is securely fastened by what may more appropriately be termed its two respiratory drills. It is a curious fact that, except for the siphon the fourth-stage larval skin is always turned inside out in the *Coquillettidia,* or at least in the African members of the genus. Strangest of all is the case of the beautiful blue and green larva of *C. metallica:* the larval antennae are enormously long and delicate and at pupation not only is the head capsule shed separately from the remainder of the old larval skin but the long slender antennae break away from the head itself. Each fourth-stage pelt is thus left in four separate parts [98].

We mentioned before the general slow pace of larvae of these two genera. The pupae are no exception; while in normal species pupal life takes only two or three days in tropical regions, in *Coquillettidia* and *Mansonia* it lasts from six to nine days. On the day before eclosion of the adult the pupa breaks itself free from the plant, and rises to the surface, leaving the drill-like tips of the trumpets still embedded in the tissues of the plant. The truncated trumpets now take over respiratory function in the normal manner, which clearly would not have been possible if the highly specialised tips of the trumpets were not left behind. The pupa remains at the surface for about a day before the adult emerges.

The sexes of mosquito pupae can be readily distinguished, the rudimentary sex organs of the male being different from those of the female. Moreover, the males tend to be smaller than the females and the duration of pupal life is shorter in the male by a few hours. Since, however, the larval life of males is also shorter the net result is that the adult males tend to emerge nearly a whole day before the females of the same batch. Naturally those males that are the first to pupate tend also to be the first to emerge as adults. What is extraordinary, however, is that those individual males that have the shortest larval life tend also to have the shortest pupal life—at least this has been found to be so in *Ae. aegypti*. It is one thing to make these bald statements, but quite another to do the work that enables one to make them. In this particular case 1400 *Ae. aegypti* were examined every hour, day and night from the sixth day after hatching from the egg until the emergence of the last adult five days later, a continuous piece of recording shared by three enthusiasts working two at a time, eight hours on and four off for a total of 117 hours [99].

The reader will remember that stranded larvae are able to crawl in grub-like fashion, provided there is a film of water surrounding them. If this film dries up the larvae are immobilised. They may survive in moist air for four or five days, but die at the time of pupation. Pupae, on the other hand, are not unduly bothered if they become stranded, and will usually give rise to normal adults provided the air surrounding them remains moist. Nor are they dependent on a free film of water around them; dry pupae not only survive, once the cuticle has properly matured, but they remain mobile, propelling themselves by sudden kicks of the abdomen which at times actually cause them to rise into the air. Indeed, unlike the larvae, movement becomes easier the drier the substrate. It is a surprising sight to see a group of a dozen or so pupae moving about with sudden jerks like a handful of jumping beans, even after two days without any free water. Thus pupae washed up by a wave or other disturbance come to no harm. They may by their own efforts find their way back to water. But even if they do not they will give rise to normal adults unless conditions become excessively dry. One may wonder why stranded pupae have these powers of locomotion if they in any case give rise to normal adults. It may well be, however, that stranded pupae are more vulnerable to the attacks of ants and other predators than those in their normal watery environment, and the jerking movements and small jumps increase the chances of reaching the safety of water once again.

We come now to what must be regarded as the basic 'function' of the pupa: the conversion of the legless water-living larva to the long and slender-legged flying insect. For its sheer strangeness and seeming improbability this transformation far exceeds the wildest flights of man's imagination; in a fantastically short period all traces of the former larva must be obliterated and the new adult insect sufficiently organised to allow it not only to escape from the pupal skin but also to fly away under its own power soon afterwards. Yet during this short period of hidden feverish activity the pupa has had its own life to lead.

Provided the temperature is sufficiently high the conversion is completed within two or three days. Actually this is not strictly true, for, as we have seen, changes have in fact been going on throughout much of the larval life, changes that are greatly accelerated in the fourth larval instar during the day before pupation, and minor changes continue also for a day or two after emergence of the adult [4]. But the nature of these early changes is strange; they are really little more than the cutting off of little pockets of more or less undifferentiated embryonic tissue ready for future development as legs, wings, mouth-parts and so on. Except for a certain amount of increase in size nothing much happens to them until the pupal stage. They are rather like savings put by out of hard-earned cash, savings towards some life insurance policy that will not be realised until retirement some time in the future. But when they do eventually mature, the results far exceed anything that might reasonably have been expected.

While there are changes going on both before and after the pupal stage, it is, of course, during pupal life that the major transformation occurs. Most of the old muscles are broken down or histolysed to be replaced by completely new systems [105]. Young pupae have an essentially larval musculature; old pupae an adult one. Larval muscles, however, are retained in the abdomen and continue to be used for locomotion until a few minutes before emergence of the adult. These are not actually broken down until the first day of adult life. The adult nervous system, on the other hand, develops mostly by accelerated growth of the old larval system [106]. There is very little actual histolysis, but greatly increased growth by cell multiplication. The fore-gut too is not broken down, the three diverticula of the future adult growing outward from the tenth hour of pupal life. The mid-gut, however, is entirely new. Like most of the musculature it is rebuilt by a process of degeneration followed by regeneration [104]. These changes actually begin during the last hours of larval life, but the whole process is greatly quickened up once pupation has taken place. Conversion from the larval to the adult form of hind-gut occurs from the sixth to the eighteenth hour of pupal life, the whole process thus being completed in the astonishingly short period of twelve hours [4].

When the time approaches for the final act the pupa darkens in colour very considerably. The change begins some hours before emergence, and the degree of darkening depends to some extent on the final colouration of the adult to come. Full mobility is, however, retained almost right up to the time of emergence itself. Then, only five or ten minutes before the beginning of this extraordinary

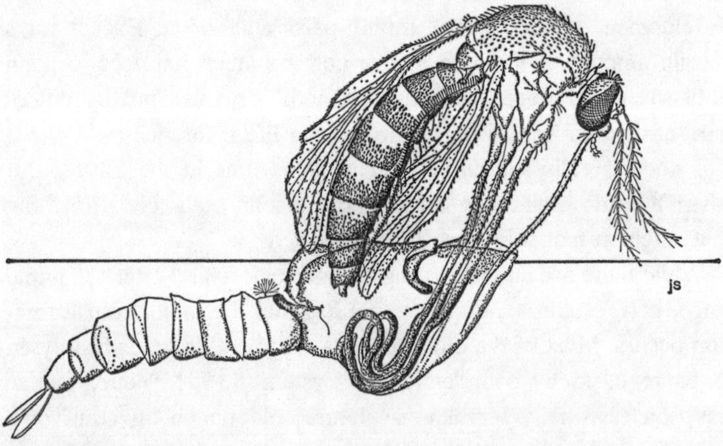

Figure 9. Adult mosquito emerging from the pupa-case.

process the pupa at last becomes reluctant to move, even when artificially stimulated. It lies as usual at the surface, but the abdomen now straightens out until the whole creature is lying almost parallel with the water surface. A layer of air begins to appear below the cuticle, thus separating it from the adult within and turning the pupa pale again. With a sudden upward movement the top of the cephalothorax breaks through the surface film. At this moment the dorsal surface of the cephalothorax splits open and the new adult begins to emerge [97].

The drama that now unfolds can only be described as one of the marvels of nature–cliché or not. It goes on free for all to see in a countless number of places. Yet how many of us, I wonder, have actually witnessed this astonishing event. First the thorax is freed, then the head, which by a backward movement steadily pulls out first the delicate antennae and then the long and slender mouthparts. Once these are clear the front and middle legs are drawn out. There is no rest period or interruption. There is no struggle or apparent effort. The whole process continues smoothly and steadily and as silently as a shooting star. The tarsi of the front and middle legs, once free, are laid gently on the water surface and then it is the turn of the hind legs to emerge. When these are free it only remains for the wings and abdomen to be brought clear of the old skin to complete the process. Once started eclosion goes on to the end in a matter of minutes. The whole process really beggars description; that it should happen at all is strange enough; that it should take so short a time is truly remarkable.

The newly emerged adult is a dull, soft and bloated affair, and aerodynamically not very promising. Were it not still soft it could never have expanded out of its pupal enclosure, for when all is said and done that is what it did. It neither crawled out nor drew itself out; once started the emerging adult swallowed air and pushed through the split in the pupal cuticle like an expanding inner tube through a rip in the outer cover of a tyre. Despite its unpromising appearance, the newly emerged adult is not only able to take a few steps on the surface of the water but within ten minutes it can, if necessary, make its first flight. At this stage this is always short, a matter of a few centimetres usually. It serves merely to take it to a safe resting place where it can hide up and complete the final touches in preparation for the new life that now opens before it.

5. ADULTS

I suppose there are some people living in the world today who have never been 'bitten' by a mosquito, but these must form quite a small proportion of the total human population. Indeed few people can have escaped throughout the whole period of man's history, for mosquitos were a going concern for fifty million years before anything that one may describe as man made his entrance. Perhaps the last glaciations of the Ice Age brought some respite, but even this seems doubtful judging by the arctic activity of *Aedes* mosquitos today. Most of the 3500 million of us that inhabit the world today have not only been bitten but receive the attention of these successful fellow-inhabitants every year of our lives. This cannot be said of any of the other blood-sucking insects. They are either restricted in their distribution to certain regions of the world (sand-flies and tsetses), and to specialised breeding sites (black-flies, midges and stable-flies) or they are restricted to certain levels of human social organisation (bugs, fleas and lice). But mosquitos, as we have seen in Chapter 1 are almost everywhere, from the arctic to the equator, in cities and in country places, in the forest and even in the desert.

I do not think I have ever met anyone anywhere in the world who does not know what a mosquito is, even if he knows nothing of its life-history or its dealings in death. I wonder if this can be said of many other animals: the dog with its fleas and other parasites perhaps, birds and, I suppose, the house-fly might qualify. It is possible that the people of the high Andes and any others that live in similar places do not know what a mosquito is, but I rather doubt it since mosquitos are known to occur and even to be a nuisance at altitudes of 2500 metres and more both in South America and in East Africa, while on the Himalayas mosquito larvae have been found at 3600 metres. Although the people who inhabit altitudes of 5000 metres or so may never come down to sea-level, many of them do visit levels low enough to make the acquaintance of these mountain mosquitos. In South America *Anopheles pseudopunctipennis* regularly transmits malaria at altitudes well over 2000 metres, epidemics occurring in Bolivia at over 2700 metres where the larvae breed in the warm springs [402]. But feeding on man, like shopping or its equivalent in human populations, is only one of the activities of mosquitos, and over the next few chapters we must see how this fits into the general pattern of their lives. Before we deal in detail with feeding and the other activities that make up the natural history of mosquitos it will be as well to spend a little time in summarising the general day-to-day happenings from the time of emergence till death, indicating as we go along some of the interesting variations and departures from the basic mosquito way of life.

Mosquitos that have fed well as larvae carry over sufficient reserves to last them through the pupal period with all its hurried metabolic activity and for the first few days of adulthood [4]. During this early period of adult life they must rest up in places that are sheltered from sun, rain and wind, while the final touches of adult development are completed. After a time they become ready and capable of taking their first meal which, it may surprise some readers to learn, is usually on nectar taken directly from flowers. They thus must be counted along with those respectable insects, the bees and the butterflies, whose visits to flowers are apt to distract our attention from some of their less pleasant activities. But the visits of mosquitos to flowers often occur at dusk or during the night and so are much less often observed. During the mid-1940s and early 1950s I spent much of my time 'on safari' in various parts of East Africa and I used to keep myself busy in the evening time when the day's work was done in collecting the Sphingid moths that visited certain nectiferous flowers. Each species had its own brief period of crepuscular activity and while I stood poised with net in hand I was time and again struck by the large numbers of mosquitos that settled on the flowers much as bees do in the daytime. Many of these were the brilliant shining golden *Coquillettidia,* or at least these handsome mosquitos were particularly noticeable.

Both male and female mosquitos feed on flowers, nectar as we shall see in the next chapter being a main source of energy. This side of mosquito nature is less well known than blood-sucking, because in nectar feeding it is we that must go to them if we are to witness it, and we must also be at the right place at the right time. Since, as we have seen, much of this activity takes place at twilight, or just after or just before sunset and sunrise respectively, it is not altogether surprising that we know comparatively little about nectar feeding in mosquitos, even though it is entirely characteristic and constitutes one of their main activities [126]. The mosquito flies off at the appointed time to seek out the right sort of flowers or other sources of sugar such as honey-dew. Little has been recorded about any selectivity on the part of the mosquito for certain flowers or about differences in attractiveness among the flowers themselves. It has been known for forty years that *Aedes nigripes* in northern Canada is an important pollinator of arctic orchids; the clubbed pollinia are frequently found adhering to the mosquito's eyes where they project forward like a second pair of antennae [364]. Yet mosquitos rarely seem to be mentioned as pollinators of flowers. Little, if anything, has been done to find out if, in our attempts to eradicate mosquitos, we run the danger of reducing still further the strength of this essential link in the economics of life.

There are some mosquitos that feed only on nectar and other sources of sugar and not even the females suck blood at any time. These belong to the genus *Toxorhynchites,* comprising some sixty species. The genus is widespread throughout the warmer regions of the Earth but extends as far north as South Carolina, USA and

Japan, and as far south as Cape Province, South Africa. It includes the real giants of the mosquito world, a large female resembling in size a small but stoutly built Tipulid or daddy-long-legs. Many of them incidentally, are extremely beautiful, comparing favourably to human eyes with any other group of insects for ornamentation and colour scheme. It seems likely that their conspicuous colouring is associated with their daytime visits to flowers. The reader may remember from Chapter 3 that the huge larvae of *Toxorhynchites* are predators feeding largely on other mosquito larvae and pupae, and thus any protein that they lack in their adult diet is compensated by the protein rich diet of their larval days. This common and widely dispersed genus is, then, a real ally of man both as a larva and as an adult.

When a mosquito feeds on nectar or on a solution of sugar and water it merely dips the tip of its elongated proboscis into the fluid and sucks it up. The fluid does not normally pass into the stomach, but is drawn off into the crop or ventral diverticulum, a large sac lying at the base of the abdomen and partly beneath the stomach [116, 133]. The crop is as large as the stomach itself. It acts as the fuel tank, small quantities of sugar solution being drawn off into the stomach for digestion as need dictates. There are also two other much smaller diverticula leading off from the base of the pharynx: these lie within the thorax. Some of the sugar meal also passes into them, but less is known about the part which they play and the fate of the sugar that enters them.

The mosquito, of course, is best known for its prowess as a blood sucker, and it is to this that we must now turn our attention. Before we do so, however, it will be necessary to deal in a little more detail with the structure of the mosquito's mouth-parts. I have referred to these as the proboscis as if there were one simple elongated snout used for sucking. The proboscis is in fact a most complex, delicate and beautiful affair. The part that we normally see, if we care to look, is really only the pulled out bottom lip or labium, a protruding partial sheath acting as a guideway for the complicated set of stylets and sensory structures that, collectively known as the fascicle, form the real business part of the apparatus. These structures comprising the fascicle consist of no less than six separate parts [168]. Two of these are tubular and are used as food and salivary canals respectively. The others resemble two pairs of elongated pointed and often serrated scalpels; they are the paired cutting and piercing organs. We need not bother with all the names given to these parts, but it will help if we remember some of them. Essen-

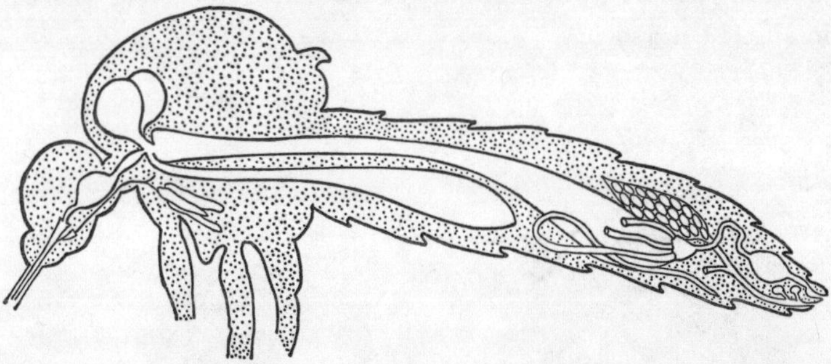

Figure 10. Main details of the alimentary canal of a mosquito (modified from Marshall). Note the paired dorsal diverticula in the upper part of the thorax and the three salivary glands lying below. Note also the long ventral diverticulum lying below the mid-gut.

tially there are two excessively fine tubes surrounded by two pairs of elongated cutting tools, all enclosed in the dorsal groove of the protecting sheath, in other words, a fascicle of six stylets lying within the gutter-shaped labium. The labial sheath itself is not as simple as I have implied, since it houses muscles for its own adjustment, nerves carrying information from the various sensilla and tracheal trunks carrying oxygen to all parts [3, 4]. All this applies to the female only, since the male does not suck blood and can manage with a much less complex affair.

Now let us see how this complicated piece of apparatus is put to work in order to suck blood. No longer can the mosquito merely immerse the tip of the fascicle into the fluid and suck. Tough tissue must be pierced and penetrated and the hidden blood supply tapped. Once a landing has been made on a likely host and a suitable site for penetration has been located, the tip of the proboscis, that is the whole apparatus, is applied to the skin of the host. With a downward thrust the tip of the fine fascicle now pierces the skin, and with continued pressure aided by alternating movements of the paired cutting stylets the fascicle begins to sink into the deeper layers of the skin. Only the fascicle enters, the tip of the labium remains pressed to the surface. Extrusion of the enclosed fascicle is made possible by an elbowing of the enfolding labium as shown in Fig. 11 *a* and *b*.

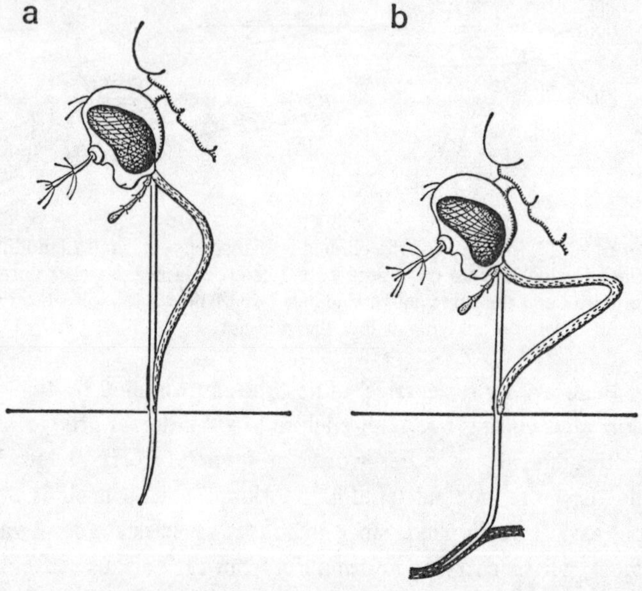

Figure 11. A mosquito's mouth-parts at the time of taking a blood meal (modified from Gordon and Lumsden): *a*. the proboscis at the beginning of the penetration stage; *b*. the proboscis elbowed back, allowing the very fine stylets to locate and penetrate a blood capillary in the skin.

Once inside the host's tissue the fascicle is partially withdrawn and then thrust in again, the tip making exploratory bends this way and that until a capillary is entered or pierced right through. If a capillary is entered, the tip of the fascicle, bending to about 45°, is actually pushed along for part of its length within the minute vessel itself [124]. The insect then taps the blood direct from the vessel. If, on the other hand, the vessel is merely pierced the blood leaking into the surrounding tissue is sucked up without actual entry of the vessel [148]. I said earlier that the stylets include two fine tubes; fluid from the mosquito's salivary glands passes down the smaller of the two and blood is drawn up the other. It is as if we took a drink using two straws one in front of the other, but managed to dribble down the one while sucking up the drink through the other, or rather the drink mixed with some of the saliva. The salivary secretion acts as an anticoagulant in some mosquitos but by no means in all. Its main function seems to be to increase the blood supply at the region of the 'bite' [4]. Whereas nectar was pumped mainly into the crop or ventral diverticulum, blood passes directly to the stomach. Sensory organs along the path determine which route the food is to take; if extra sugar is added to the blood some of it then passes to the crop.

The act of sucking is itself a complex business. There are two muscular pumps, one in the head at the base of the proboscis and one within the thorax surrounding the hind end of the pharynx. These two pumps work alternately and in such a manner as first to draw blood up into the head and then to push it on into the stomach. Valves placed between the two pumps and between the stomach and the thoracic pump determine the direction of flow and ensure that there is no regurgitation of newly arrived liquid food from the stomach. A curious thing happens to the blood once it has coagulated within the stomach of the female. A substance is secreted from the walls of the stomach and solidifies to form what is known as the peritrophic membrane. Thus the contents of the stomach come to be wrapped up by a loose-fitting cover which persists during the early stages of digestion. If another blood meal is taken during this period, it remains discretely parcelled off from the other [3, 4].

Female mosquitos take their blood meals from vertebrates of all kinds, but some species prefer one host species or group of species to others [163, 164]. Thus many species of *Culex* and *Coquillettidia* are predominantly bird feeders. Little is known about host-specificity among the bird-feeding mosquitos themselves; do they merely prefer birds to mammals or do they prefer certain species of birds to others? We do not know for certain in most cases, although we do know that some species of mosquitos feed largely on this species or group of birds or that [387]. But this could be an indication of host availability and not necessarily host-preference. For example, the swamp-breeding *Coq. perturbans* and *Culiseta melanura* feed largely on the wading birds that occupy the swamps. Many species of *Aedes* seem to show a preference for mammals, and some of these, of course, include man.

Many of these host preferences break down in the laboratory when various baits are offered at short range. Species that are known to feed mainly on rodents in the wild, as shown by precipitin tests on the fresh stomach contents, will feed on any host that is presented at close quarters in the laboratory. But not so *Ae. longipalpis,* a species occurring in the yellow fever belt of tropical Africa, that is suspected of playing a part in the transmission of this disease to certain species of monkeys. When I was working on egg-hatching in this species it was necessary first to catch some wild females and then to obtain eggs from them. To do so, of course, they had to feed on blood, but I found it extremely difficult to induce them to accept me, even when I tried all my tricks of persuasion. Nor was it merely me they objected to, since they were equally reluctant to feed on my two assistants who helped me so much, Gabriel Okech and Charles Ssaku. I then tried them on a rhesus monkey. Feeding was almost immediate and I never had any more difficulty in feeding this very handsome species.

Some species of mosquitos are particularly choosy about their source of blood, others are quite the reverse. *Ae. aegypti,* of course, is well known for the catholicity of its taste, and so are a number of species of *Culex,* but we shall have more to say about these in Chapters 10 and 13. The species that beats the lot in my experience is *Coq. fuscopennata,* the pest of every swampy district in tropical Africa. I do not know any vertebrate that it will not immediately attack. In addition to people, I have personally taken this species in the act of feeding on chickens, a frog, toads, monkeys, dogs, cattle, horses, donkeys and one elephant.

Most surprising in their feeding habits are mosquitos of the tropical genus *Malaya,* the females of which feed on ants [120]. Female *Malaya* hover in front of a tree or bamboo on which *Cremastogaster* ants are foraging. Actually 'hover' is hardly the right word to use since, although in flight, they move vertically up and down a short distance from the tree. When one of them perceives a suitable ant coming down the tree, presumably laden with honey-dew, the mosquito settles on the surface immediately in front of the ant, halts it and, inserting the tip of its curiously bulbous-ended proboscis in the mouth of the ant, proceeds to suck out the honey-dew which the ant has so laboriously obtained from the plant-bugs high up on the growing shoots of the tree. It is puzzling to know what makes these usually rather aggressive ants give up their hard-won prize so readily. Since these mosquitos are all extremely small and do not possess raptorial legs or other means of persuasion, one can only assume that the ants receive some benefit from this curious contact, or else do not recognise the intruder as a non-ant.

The name *Malaya* is rather unfortunate since there are known to be five species in Asia and the Pacific and six species in Africa. For many years these mosquitos were in fact placed by themselves in the euphonious genus *Harpagomyia.* They were well known by this name and perhaps there was a case for leaving them there even though someone had exhumed the older name, *Malaya.* Such changes are not uncommon, the most famous of all mosquitos, *Aedes aegypti,* having had its name changed no less than twenty-seven times over the past 200 years. It was for many years known as *Stegomyia fasciata* and first achieved its fame as the 'yellow fever mosquito' under that name, but someone unearthed an earlier record and *Ae. aegypti* it has been ever since. It is now so well known, not merely because of its sinister role as a transmitter of disease, but because, being an ideal laboratory animal, easy to keep and readily bred in captivity, it has figured in countless laboratory experiments. So well known is it now that it seems doubtful if its name would ever be changed again, even if a case were to be found for doing so.

While all mosquitos, both male and female, feed on sugar in one form or another, most females require a blood meal as well if they are to produce any eggs. We shall have more to say about this aspect of the blood meal in Chapter 8; all we need say now is that generally speaking each blood meal is followed by the development and laying of a batch of eggs. There are, however, some female mosquitos that can do without the blood meal, for the first batch of eggs at any rate. These autogenous females as they are called, make do with the food reserves that have been stored from larval life [329]. There are enough of these in most young adults that have been brought up on an adequate diet as larvae, but most mosquitos need the stimulus of the blood meal if these reserves are to be mobilised for egg production [244]. The best-known example of autogeny is found in the subspecies of the common house gnat *Culex pipiens molestus*. Females of *C. p. molestus* produce a single egg-raft a few days after emergence, but require the normal blood meal if they are to produce any more rafts.

While the blood meal usually provides the stimulus for egg production, it also supplies material that is used by the mosquito to nourish the developing eggs. But, like most eggs, at some stage they must be fertilised and in usual insect fashion this occurs just as the eggs are being laid. Mating, however, takes place some time beforehand and the sperm is stored by the female in special receptacles or sperm sacs known as spermathecae. In *Aedeomyia, Anopheles* and *Uranotaenia* there is a single spermatheca, in *Mansonia* there are two but in most mosquitos there are three arranged and looking very much like the three balls one used to see over pawnbrokers' shops. The sperm remains in good condition indefinitely and enough is transferred at a single mating to last for the rest of her life. Indeed, a single mating early in the female's life seems to be the rule, and even if she lays twenty separate batches of eggs, following twenty blood meals, each batch of eggs will be fully fertile. *Ae. aegypti* in captivity have been known to lay fertile batches of eggs following a blood meal taken 100 days after a single mating.

When the male emerges from the pupa his sex organs are the wrong way round and mating is impossible. This curious state of affairs lasts for about a day, the actual time depending on the species and on the mean ambient temperature [141, 153]. Fortunately for the male (and for the female) he does not find his position particularly frustrating since he is unreceptive to the female for much of the first day anyway. The problem is solved very neatly, if in rather a surprising manner; the terminal segments of the abdomen bearing the male organs rotate through 180°. The dorsal surface comes to lie ventrally and vice versa; back becomes front and front becomes back. It takes some hours to complete this about turn and males that become receptive before its completion attempt to copulate.

Receptivity to the female in most mosquitos depends on being able to hear her flight tone. While the antennae of the female bear rather sparse whorls of hairs at intervals along their length, those of the male possess thick whorls of long hairs and are bushy by comparison. Each of these hairs has its own nerve running from its base to join with those from the other hairs. Since there are hundreds of such hairs on each antenna, one can begin to have some idea how complex the antenna really must be. It is these hairs or fibrillae that act as resonators to the female 'voice', vibrating in sympathy when the correct pitch is given out. The two antennae, although based close together in front of the head, stand out at an angle from each other and are ideal for the stereophonic reception of sound. They not only receive the sound waves from the female and transmit them to the brain and other centres but being independently movable they can be used to locate her direction and whereabouts [170, 178]. The actual sound-receiving apparatus comprises much more than the innervated resonating fibrillae. There is a very complex organ situated at the base of each antenna known as Johnston's organ, which also plays a very important part in hearing; an account of the various views on its mechanism has been excellently summarised in Alan Clements' admirable book on mosquito physiology. Until the hairs have fully dried and stiffened following emergence they do not respond, and this may take much of the

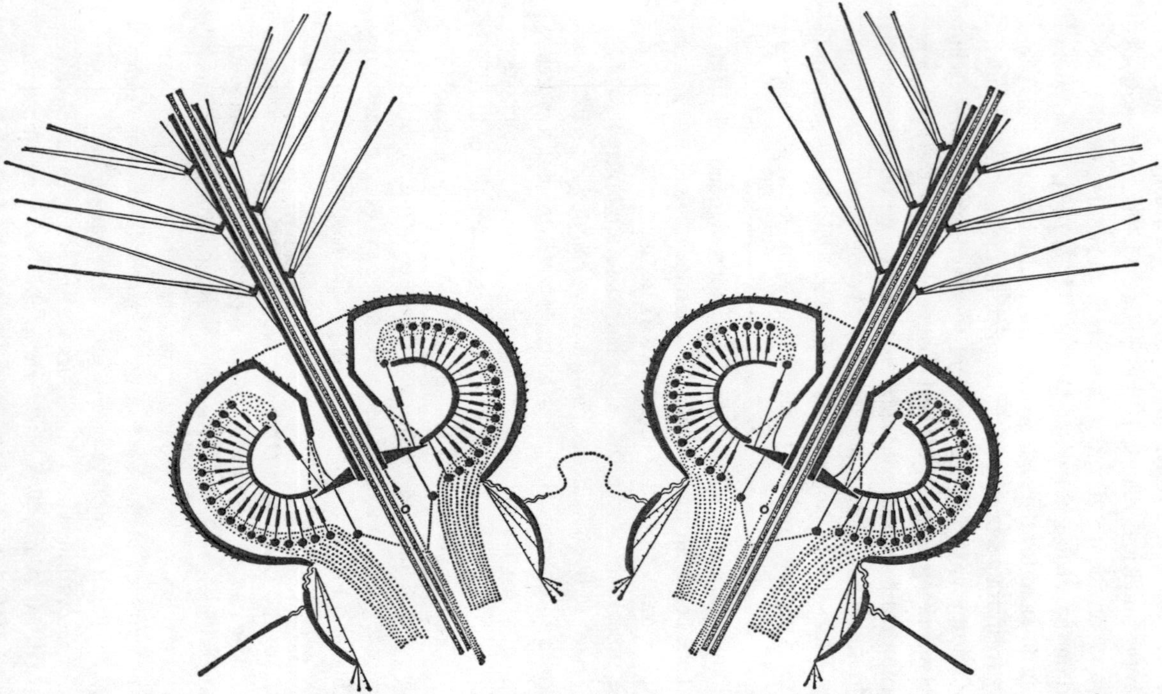

Figure 12. Simplified diagram of the two Johnston's organs located in the bulbous tori at the base of the antennae. The organs are stimulated by sound waves transmitted to them by nerves from the antennal hairs. These two hearing organs function stereophonically, allowing the male to locate the whereabouts of a flying female with great accuracy. (Modified from Risler.)

first day. In some species of *Anopheles, Aedes* and *Culex* the antennal hairs, even in mature males, lie almost flat against the shaft of the antenna for most of the time, and only become erect at certain times of the day; we shall have more to say about this and mating time in Chapter 9.

The response of the male to the female's whine sometimes has some unexpected consequences. More than forty years ago a new electric power station was set up, I think, in Canada. Shortly after it came into operation one after another of the new generators failed. When electricians were called in to investigate this costly failure they found that each generator was jammed or gummed up with hundreds of thousands of flies. No damage was done and the generators were duly cleaned and restarted. Again the same thing happened. This time, however, an entomologist was called in as a consultant and he pointed out that the piles of insects were all mosquitos and, moreover, they were all males. The hum of the generators had attracted the male mosquitos in vast numbers, the note or a component of it being close to that emitted during the flight of the female. The speed of the generators was changed slightly so as to raise or lower the pitch and the problem was solved.

Nowadays much more is known about flight tone and its effect on the male. The frequency of the wing beat is known in several species and tuning forks of the right pitch have even been used to separate males from cage colonies containing both sexes. When the fork is struck and held near to one side of the cage the males gather on that side; they can then be sucked into a tube leading to a suitable container. The wing-beat frequency of the male is higher than that of the female of the same species and, being out of the effective range, their own sound does not interfere with reception of the female note. This also means that males do not normally attract other males. In immature males, however, the pitch is somewhat lower, pending the hardening of the integument. During this period, which lasts about two days, the pitch is nearer that of the mature female and one sometimes witnesses males following and even trying to mate with other males. Such homosexual tendencies are short-lived and are replaced by normal behaviour on reaching maturity.

The female is not receptive during the first few days of adult life. No problem arises, however, since, as we have seen, she is also not attractive to the male during this period. Like the male, when she makes her first flights, usually towards some source of sugar, her flight pitch is lower than that of the mature female and is too low to attract the male [4]. It is only after two days or so that the pitch rises above the threshold of attractiveness. In *Ae. aegypti* kept at 22°C, quickest mating results are obtained with three-day males and two-day females. Since, as we have seen in the previous chapter, males emerge about a day earlier than females of the same age (measured from the time of hatching from the egg), the two sexes become reciprocally responsive at the same time.

Mating in most species of mosquitos usually but by no means always occurs before the female has taken her first blood meal. Males are attracted to the flying female who may find herself approached by several males at once. If many males are present they sometimes interfere with one another and even if a female is seized the pair usually breaks up without fulfilment. But provided she is not too heavily outnumbered the act may be completed entirely on the wing. One of the males manages to grab her with his legs and position himself below but facing her. Flight immediately assumes a characteristic series of circular orbits and within a second or two sexual contact is made. Very often, although started during flight, the pair settles on some nearby surface to complete the act, with the male still lying beneath and facing the female, the pair being supported by her legs alone. Even in the same species there is considerable variation, at least when the mosquitos are studied in a confined space. Many pairs that are made up during flight in fact alight before any attempt at sexual contact is made. On settling, the male raises his abdomen until the genitalia are brought into contact with those of the female. They remain coupled for a few seconds and then separate, the male flying off while the female remains behind. Sometimes he must make more than one attempt before successful contact is made, but if the abdomen has to be raised in this way more than two or three times it usually means that the two mosquitos are out of position, or that the female is unreceptive, and they separate without consummation.

In 1950 I carried out a long series of genetic experiments with *Ae. aegypti* during which it was necessary to mate individual males with individual females in order to be absolutely sure about the parentage of every batch of eggs laid [121]. The mosquitos were all given a number and the females were introduced one at a time to a cage containing a single male. The cage was small enough to allow me to keep the pair under continuous observation under a head magnifier and hand lens. Indeed it was necessary for me to do so because it was important for the work to know whether the pair had actually mated or not. I was thus able to make some fairly detailed observations on mating in this species. After mating the two mosquitos were returned to their individually numbered tubes and another male was introduced to the cage ready for the next female and so on. In this work involving more than a thousand mosquitos, rarely was any attempt made at contact until after the pair had settled. I noticed that when successful contact was made, as judged later by the production of fertilised eggs, the male and female genitalia were actually locked together, as shown by the difficulty the male had in extricating himself. The pairs remained interlocked in this way for periods varying from four to forty seconds and difficulty on separation could be used as the hallmark of success. I was able to observe also that when the male violently shook or vibrated his whole abdomen beforehand, proper locked contact was always made, and usually on the first time that the male raised his abdomen. This quivering overture on the part of the male was quite characteristic and, when careful observations were made, was seen to occur before nearly every successful mating. It is not clear whether the characteristic shaking of the abdomen served to excite his mate and bring her into a final state of readiness or whether it was a necessary prelude to successful mating on the part of the male. The latter seems the more likely as we shall see below and when we come to consider swarming towards the end of this chapter.

There are exceptions to the mate first, feed afterwards rule. In captivity fed female *Ae. aegypti* can usually be induced to mate at any time provided it is not too soon after the meal. Mature females can go on mating at intervals throughout most of their lives, although they usually become refractory for an hour or two after two or three matings in quick succession. But there is some variation; one exceptional female either lost or never possessed the appropiate keep-off signal and was observed to mate forty times with forty different males, one immediately after the other. It does not follow that the spermathecae of these more than once mated females come to store the sperm of many males and thus provide scope for multipaternality in a subsequent batch of eggs. There is some evidence suggesting that once filled the spermathecae are closed off to further instalments of sperm. Whether female *Ae. aegypti* mate more than once in nature we do not know. We do know, however, that the first mating usually takes place before the first blood meal for we can recognise a female that has not previously taken a blood meal and if we catch such females as they come to bite and examine the spermathecae they are almost invariably full of live sperm. There are exceptions: Christopher Teesdale has recorded a wild population of *Ae. aegypti* on the East Coast of Africa in which mating usually took place after the first blood meal [270].

Among different species there are considerable variations on the general theme that I have outlined above. In many of those that normally alight to complete the act the male releases his grip once sexual contact is made and hangs downwards like a trapeze artiste but suspended by his locked genitalia instead of by the backs of his knees. In some species of *Eretmapodites* the pairs remain coupled in this fashion for long periods often for more than half an hour and sometimes for over an hour, the males vibrating themselves violently at intervals. In view of what we have noted above it would seem fair to assume that these males ejaculate more than once during this prolonged contact. Sex behaviour in some *Eretmapodites* differs from the general picture that I have given in other important respects. Females of *E. chrysogaster* will not accept the male at all until after her first blood feed, even though she is clearly very attractive to the males at this stage. As she flies on her way to take her first blood meal she is followed by a group of males. These fly very closely just above and slightly behind the female like a small cloud of fighter aircraft tailing every movement of an escaping bomber. They maintain their position whether she is moving forward or hovering near her target. Usually they break off once she settles, but, as we shall see below, not entirely because the flight tone ceases. When she has completed her meal and begins her considerably less agile flight away from the host the males immediately resume the chase and one of them is usually successful. On settling this time mating takes place in the curious manner described above [122].

Sometimes one of the males that is following a host-seeking female actually settles beside her on the host and waits there while she is feeding. He is then quick off the mark when she takes off on completion of the meal. What is particularly interesting is that he apparently maintains interest in her during the whole of this period; it is not merely that he happens to be settled there on the host beside her and is restimulated when she starts to fly again. This may sometimes be the case, but at other times he remains close to her and can be seen to make approaches while she is still feeding. In such cases copulation usually takes place immediately she has finished the meal and while she is still standing on the host. Indeed I have occasionally witnessed the initiation of successful contact while the female is actually still sucking blood. In the mating behaviour of *E. chrysogaster* it is not at all certain whether the pre-fed female really rejects the advances of her pursuing males or whether, on coming close, the males do not receive the final go-ahead or all-clear. In other words, is it the female that refuses the male until she is blood-fed and less agile, or is it the male that, while stimulated to pursue by the flight of the female, requires the final stimulus provided by the presence of her full warm body? That the male beside her on the host takes no further action until her abdomen is almost full suggests indeed that it may well be the rise in temperature, or perhaps some other stimulus arising from the presence of the fresh warm blood in the female, that triggers off the final response.

Males of the Central American species *Deinocerites cancer* do not possess the characteristic bushiness of the antennae and thus cannot respond to the female's flight note. They walk over the surface of the breeding water searching for pupae by means of the minute olfactory sensilla on the surface of the antennae. Copulation takes place on the surface of the water as soon as the female has emerged. Recognition of her sex apparently depends on the smell of the pupa, since the males will sometimes attempt to copulate with empty pupal skins, but only if they are female. If they cannot find a female pupa, they test the receptivity of the already emerged mosquitos that are still standing on the surface of the water, by gently touching them with their front legs. Females that have already been inseminated raise their hind legs as if in protest [127].

More unusual still is the behaviour of *Opifex fuscus*. Like those of *D. cancer* the males of this New Zealand species also lack the characteristic bushiness of the antennae. They skim over the surface of the water in search of pupae. H. B. Kirk, who made a special study of *O. fuscus,* described in 1923 how he saw them 'skate' over the water with their heads turned downwards, sometimes actually dipping their heads beneath the surface. Unlike those of *D. cancer* they are not so particular about which sex they find. The male seizes any pupa as it breaks the surface, using its genital claspers to hold it captive while the pupa attempts to free itself. If the pupa is near maturity the cuticle splits open during this struggle and the male captor quickly inserts his claspers and, using them like a can-opener, extends the split backwards until he is able to insert the tip of his abdomen actually inside the pupal skin. If the now-emerging captive happens to be a male the two separate at once. But if not, copulation actually takes place while the abdomens of both partners are within the female pupal skin. Female *Opifex* are thus inseminated before they have even emerged. Since, unlike those of other species, the male must necessarily be the same way round as the female during copulation, rotation of the genital segments might be supposed to be an embarrassment and it is interesting to note that the males usually start their pupa-capturing behaviour when they themselves are only a quarter of an hour old [138, 140].

Certain other species, in which the males lack the characteristic bushiness of the antennae, are able to initiate mating in the absence of the female flight tone. Males of *Culiseta inornata,* a swamp-breeding species occurring in the United States, find their mates by touch. They respond to any mosquito which they touch with their legs, even to males and to mosquitos of other species. Touch is less discriminatory than flight tone, and they will even show the characteristic response when their legs are gently tickled. Copulation usually occurs with the pair at rest. The male first hovers over the resting female while jockeying into position and then, turning himself smartly round, comes to rest end to end behind her. Occasionally if a female flies very close to a flying male they will couple in flight [117, 156]. It seems probable that this response also results from stimulation of the legs of the male, the sensitive tactile sensilla perhaps responding to the slipstream of the flying female.

Males of the beautiful Central American mosquito *Sabethes chloropterus* also lack the antennal equipment to hear the female and, like *C. inornata,* use their legs. Pedro Galindo has made a special study of this species and has described how a hovering male taps one of the hind legs of a resting female with the tarsal segments of one of his own middle pair of legs. If she is receptive, which he finds out by testing first one side then the other, he does a half somersault and comes to rest upside-down beneath her. During actual copulation the mid-tarsi of the male continue to stimulate her further by gentle titillation of a pair of minute sensilla placed at the tip of each of her antennae [391].

While the females of *Ae. aegypti* usually become reluctant for a short while after two or three consecutive matings, not so the males. In captivity a single male will go on chasing and capturing one female after another, making proper connection each time. It is not the exhaustion of sperm that brings an end to these exploits but the exhaustion of his own fuel supply. Sperm is in fact no longer present in the fluid that is passed from male to female after three or four consecutive matings, and is not available again until some thirty hours later. The male, however, may go on mating and passing spermless fluid over and over again, even though all but the first few matings are sterile. One cannot of course argue from what happens in a colony of mosquitos, or for that matter of any other animals, to what may happen in nature. The individuals making up a colony are usually different from those in natural populations since the selective pressures are different right from the start. We shall have more to say about this important although often overlooked aspect of laboratory colonies when we reach Chapter 10. All we need add here is that observations such as we have been considering are almost impossible to make in the wild and one has no choice but to make them on captive populations.

Very few species of mosquito have in fact been successfully colonised, despite numerous attempts. One of the difficulties is the breakdown of the mating response in artificial conditions. Space is one of the critical factors. Space to fly freely in and for the formation of swarms that are so characteristic of mosquitos and their nematocerous allies. This problem has held back detailed genetic work on all but a few species for a long time. Fortunately for the geneticists, however, who often need to test the progeny of a known father, male mosquitos share with us an increased tendency to mate after losing their heads. While the intact male will not mate, even when the sex organs of each partner are brought together artificially, males deprived of their heads show none of these inhibitions, and mate immediately sexual contact is made. Such males, of course, cannot fly since decapitation removes all sources of directional information. But when held by the wings in delicate forceps or impaled on a fine needle and introduced to an anaesthetised but intact female, copulation follows immediately sexual contact is made. This technique of artificial insemination in mosquitos has been used widely in the maintenance of cultures of certain species that are wholly refractory to mating in the normal fashion under the conditions provided. The males are first anaesthetised with carbon dioxide and the heads removed; one after another can be carried out by a practised operator [112, 130].

One of the most fascinating and characteristic but nevertheless little understood activities of mosquitos is swarming. I suppose everyone of us has seen these gatherings of dancing mosquitos on warm evenings, all the individual insects moving rapidly and yet the whole swarm remaining stationary as if suspended in the air. Not all of these swarms are made up of mosquitos, since the habit of swarming is common in many of the Nematocera and indeed in Diptera generally. Most mosquito swarms consist of males only, but females also make up swarms and mixed swarms have been recorded. Usually, however, the sexes remain apart, although a female may fly into a male swarm whereupon she is seized by one of the dancing males and the pair drops out to continue in private.

Station is kept within the swarm by two methods, depending on the strength of the wind. In the absence of wind or when it is very light, each member of the swarm flies in a roughly circular orbit, but changes his direction of flight after every few completed orbits. The path becomes elliptic in a light wind with the long axis in the direction of the wind. On windy evenings, or indeed early mornings, for swarms usually occur at both twilight periods, the ellipse is pulled out in the direction of the wind until a straight-line course results, each mosquito flying upwind and drifting back downwind [144]. But there may be as many as a hundred mosquitos in the swarm and the movements of any one of them out of phase with the others. While each keeps station as far as the width of the swarm is concerned, there is no restriction on vertical movement, and individuals are constantly shifting their level within the column. Indeed, often the movement is characteristically up and down while they fly forwards and backwards within the confines of the column. The width of the column is not constant in any one species, the actual span or diameter depending on outside factors. Swarming usually occurs over some external marker such as a tree or chimney or over any dark or light object on the ground. It has been shown by experiments with various sizes of marker that cohesion within the swarm and the horizontal width depends on the marker and its size and not on any interaction between the individual members making up the swarm [144–7].

Ramachandra Rao and Paul Russell studied swarming in *An. annularis* over a tennis court in Madras. They showed that the swarms, mostly males, formed over their own heads and moved as they moved. They noted that mating occurred from time to time in these swarms, but it was not a regular feature. Males of *Ae. varipalpus* in the western United States also swarm over warm-blooded animals and move as the animals move. Females coming to feed on the animals are pounced on by the males and mating takes place immediately before the start of the blood meal [150]. On the other hand, lone males of *Ae. sexlineatus* hover over man and warm-blooded forest animals. These will approach any females that come to feed, even those of other species, but successful mating occurs only with *Ae. sexlineatus*. Sometimes more than one male may hover over the females' prospective host, but these small groups do not really constitute swarms in the sense that we have been considering them.

It is difficult for us to understand the function of swarming but this does not mean that there is not one. In 1937 the American malariologist, L. W. Hackett, suggested that swarming must be part of the general mating ritual. But F. J. C. Cambournac and Rolla B. Hill working on *Anopheles* in Portugal a few years later showed that mating in fact rarely occurred within the swarms, most of the females mating independently. They suggested that while the swarming ritual was preserved it had now lost its functional character. Swarming, however, is such a widespread and regular feature in the lives of so many insects, not only in the Nematocera, that such a simple explanation seems somehow slightly suspect. It is always easy to dismiss what one does not understand. Two interesting observations made by Ramachandra Rao and Russell on their tennis-court *Anopheles* may perhaps give a slight clue. They noted that all the males within the swarms appeared to be quivering or vibrating violently during flight. We saw above how male *Ae. aegypti* shake themselves violently immediately before coupling; could it be that the males need to wind themselves up to the right pitch before mating, and that dancing together in this fashion in some way helps them to do this?

Once the mated female has obtained her blood meal she rests up for a few days in some sheltered place. During this time, egg development, which started immediately after the blood meal, proceeds apace. In *Ae. aegypti* it may be as short as two days or as long as five. In some species of *Coquillettidia* the shortest time for egg development is not much less than a week, but we have already seen how the whole physiology of these peculiar mosquitos is extraordinarily slow. Egg development in mosquitos of temperate and northern climates is of course a much lengthier affair. We shall have much more to say about the whole question of egg development in Chapter 8. What concerns us now is that once the unfertilised eggs are mature and ready to lay the female has again to fly off, but this time in search of a suitable place for egg-laying. It is clear that female mosquitos can detect the whereabouts of even the smallest traces of water, although how they do it is somewhat surprising. It seems that the gravid female sets off at the appropriate hour for egg-laying (about which we shall also have more to say in Chapter 9) and, using her eyes, explores any region that is darker than the general surroundings. Actual detection of the presence of water comes, so to speak, at the very last moment; at close quarters response may be to reflection from the water surface or to the increased presence of water vapour above the water. Some gravid females have been seen preparing themselves for egg-laying when flying over a mirror [136].

Having found a likely place some mosquitos then test its suitability before laying by drinking some of the water and testing its salinity and so on [134]. That patches of darkness rather than the presence of water itself is the first thing to attract the gravid female makes it easier for us to understand how certain species of *Aedes* deposit their eggs in dry places liable to flooding. Such places are usually in hollows of one sort or another. Dark tree-holes will be explored and if the walls are damp the eggs will be laid. Depressions in the dry ground too will tend to be darker than their surroundings, and these too will be explored. It would be interesting to know if in England or North America, or in other countries of northern latitude more eggs are found on the southern slope of such depressions.

Three other activities need to be mentioned before we end this general survey of mosquito life. Or, rather, one more activity and two what might these days be called non-activities. These are dispersal, hibernation and aestivation. I have left dispersal till late in the story not because it is less important than mating, feeding and egg-laying, but because less is known about it. Also, although it is true that in some species a relatively long flight occurs soon after emergence from the pupa, even before the first mating, this is by no means always the case; dispersal flights are known to occur at various stages of adult life, according to species. Maurice Provost has for many years headed a team studying mosquitos in a rather wild part of Florida. He and his team, which for a number of years included the well-known Danish mosquito man Erik Nielsen, have made a special study of dispersal in *Ae. taeniorhynchus*. In a gigantic series of experiments they have shown that these mosquitos set off on a four-day flight when they are only about ten hours old. The males drop out on the way and rarely cover more than five kilometres from the breeding site. The females, on the other hand, travel distances of thirty kilometres or more. The flight is not continuous but is carried out each day at the normal period of activity for the species. Such dispersal flights are also well known in other species of *Aedes*, large numbers of *Ae. sollicitans* having been taken on a ship some 170 kilometres from the nearest shore [4]. *An. sacharovi* has been recorded as making flights shortly before hibernating for the cool season. These flights may also last for several days, the females even stopping for a blood meal *en route* [137]. For nearly fifty years it has been known that *An. pharoensis* may be found in the desert seventy kilometres or so from their nearest breeding places in the Nile. In Alec Haddow's six-year study of the biting times of mosquitos in the Semliki Forest on the Uganda–Congo border, he recorded how this species, usually taken only in ones and twos in the forest, would suddenly appear in huge numbers on a particular night. It seems that migrating clouds of *An. pharoensis* pass over or through the forest canopy from time to time on a flight of dispersal. Chris Teesdale, who worked for many years on the east coast of Africa, has described a most dramatic three-day invasion of Mombasa Island by hordes of *Ae. lineatopennis* at a time when this species was at its lowest ebb in the middle of an exceptionally dry, dry season.

The ubiquity of some species of mosquitos leaves one to suppose that dispersal flights are common even if they are difficult to detect and seldom observed. Given the right conditions for breeding, mosquitos will be there to take advantage of them. I have already pointed out that except for birds, mosquitos are as widespread as any other group of animals, and more so than most. The distribution of each species is limited only by the conditions imposed by physico-chemical factors such as temperature and the salinity of available water and by biological factors such as the degree of specialisation for breeding and feeding. *Ae. thibaulti,* a pest species in the southern United States, breeds exclusively in holes in sweet gum trees situated near streams and permanent pools. Only those holes which have openings so near the ground that they are subject to flooding from the stream or pool are suitable; ordinary rain-filled holes are no good. It is a remarkable fact that the adults of this species do not normally venture further than fifteen or so metres from the water which is essential for the filling of their specialised breeding places. They may move up and down the stream, but never far away from it [68]. Yet *Ae. thibaulti* occurs in isolated places over a wide area including Alabama, Arkansas, Georgia, Louisiana, Mississippi and Missouri. Clearly some means of dispersal must operate at times.

Where there are oceans or high mountains acting as physical barriers separating regions with otherwise similar conditions, then only those species that occupied both regions before the barriers appeared in geological time will be found in both, or those that can successfully make the passage between the two, if necessary assisted by man, as for example the domestic breeding strains of *Ae. aegypti*. But where there is continuity of conditions, then distance is no obstacle. When man creates favourable conditions for a given species in a hitherto inhospitable environment, it is not long before that species appears.

Where there is a harsh winter to negotiate, means must be found of allowing for it in the life-cycle. Thus only certain species have been successful in colonising the so-called temperate regions. Overwintering occurs in the egg, larval or adult stages. *Ae. vexans, Ae. excrucians* and several other species of *Aedes* hibernate as eggs; *Ae. rusticus, Coq. richiardii* and *An. plumbeus* as larvae, while *Ae. communis* may be found both as larvae and as adults from January to May. But in many northern species of *Anopheles* and *Culex* it is the adults that hibernate during the cold winter months. Some *Anopheles* rest for the winter in stables and other buildings occupied by warm-blooded animals and although they may take a blood meal from time to time, this does not lead to the production of eggs [171]. Others rest in empty outhouses and cellars and normally go without food for the whole winter. When the approach of winter brings a halt to the breeding of *An. freeborni* and *C. tarsalis* in the rice-fields of California, adults of both species take to houses for hibernation. But, whereas *An. freeborni* will sally forth on winter afternoons whenever the indoor temperature in its hiding place reaches 15°C, *C. tarsalis,* although a vicious biter during the summer, normally remains unresponsive until the winter is over. Many people living in Europe will have noticed numbers of small *Culex* entering houses, garages and other buildings during the autumn. These are *C. pipiens* that have emerged from pupae late in the season. They too rest up for the winter and do not attempt to bite even on unseasonably warm days.

These and many other species that hibernate as adults during the winter, resume normal feeding and egg production when the warm season returns. Detailed studies of hibernating mosquitos in the severe conditions of the Russian winter indicate that it is not so much the fall in temperature which determines that the females shall stop feeding on blood and seek shelter but the length of day during upbringing. Similarly it is the increase in photoperiod that occurs in the spring that breaks hibernation and sets in train the reactions that bring about the resumption of blood-feeding and the production of eggs. Actually things are not quite as straightforward as this, they rarely are where nature is concerned; while fall in temperature alone will not precipitate hibernation, reduction in photoperiod during upbringing loses much of its importance unless there is an accompanying drop in temperature [174].

Not unrelated to hibernation is the problem of tiding a species over long periods of drought. Although the one is associated with extreme cold, whereas aestivation as it is called is usually associated with extreme heat, in both water is at a premium; in order to survive the mosquitos must remain more or less immobile and be able to go for periods without food often extending over several months. Aestivating mosquitos shelter in minute cracks and crannies of all kinds and are usually extremely difficult to find. The onset of aestivation may be more or less gradual as the relative humidity falls on the cessation of the rains. The physiology of aestivating mosquitos, however, is quite unlike that of their hibernating northern cousins, since they usually rest with fully developed ovaries and when the favourable conditions return the mosquitos are immediately stirred into activity; eggs may be laid within a matter of hours and feeding resumed at the first opportunity [159].

I recently spent some time in Tanzania. I arrived in the country during the latter part of the main rainy season and *C. pipiens fatigans* was extremely numerous and very active. Every evening was torment because of the clouds of these mosquitos that managed to work their way even into screened houses. It was difficult to avoid them and everyone was complaining. Suddenly the rains ceased. It did not happen gradually; for days and weeks it had been raining for much of the time, then from one day onward there simply was no more rain. Over the course of the following week relative humidity dropped and so did the numbers of mosquitos. In fact so drastic was the fall in numbers that more than one local resident congratulated me on the success of my work. But I had done nothing. My work was not directly concerned with mosquito control. Where, then, had all the mosquitos gone? Had they died off leaving a 'vacuum' to be filled by recolonisation from places less drastically affected? The answer came later. The excessively dry season that followed was interrupted some six weeks later by a sharp but heavy rainstorm. Clouds of *C. fatigans* appeared that night. The walls of homes were still covered with them on the following morning and my new-found but unearned reputation was gone with the rain. Evidently the mosquitos aestivating in some sheltered spot were roused by the sudden rise in relative humidity and sallied forth for a feed at their usual feeding time that night.

It is commonly believed that mosquitos live only for a day or two or at most for a week. A moment's thought, however, will show that this cannot possibly be generally true, otherwise they would not take part in the transmission of disease, since multiplication of the pathogens to numbers that can be passed on by bite usually takes more than a week—and sometimes as much as three. Given the right conditions adult mosquitos seem capable of living for relatively long periods; they have been kept alive in artificial colonies for five months and more. In nature, however, things are very different [113, 115, 142]. Not many survive to reach old age followed by what we call death from natural causes. Most come sooner or later to a violent end. Bats, birds and geckos take their toll. Spiders, ants and predators of many kinds, even men play their part, killing off any that are not quick enough off the mark. It has been variously estimated that from ten to thirty per cent of the population is lost every twenty-four hours. If, then, we assume for the moment a middle figure of twenty per cent, a population will be about halved every three days and reduced to little over one-tenth every ten days. Thus a million mosquitos will be reduced to not much more than a thousand after one month and hardly any of those will exist after two months.

These are rough estimates only, but they serve as useful models, especially when we are considering the epidemiology of mosquito-transmitted disease [139]. In nature, of course, populations are constantly being reinforced by new emergence from pupae. There is also movement into the population by migration but this may be largely compensated by movement out of the population elsewhere. There are many other factors about which we know little or nothing. Not only do species vary from one another in their natural longevity but a single species varies under differing conditions. Adults of *An. atroparvus* in Spain live for at least six months during the winter, but summertime broods rarely live longer than six weeks [132]. It may be that some of the species that are well known for their vast numbers play no part in transmission of disease simply because insufficient of them live long enough to allow time for the development of the disease-producing organisms to take place. The swamp-breeding *M. africana* has been shown in the laboratory to be capable of transmitting yellow fever virus. Yet although present in fantastic numbers in much of tropical Africa, no yellow fever virus has ever been recovered from wild-caught specimens.

6. FLIGHT

Largely preoccupied with ourselves and other mammals, we do not perhaps always appreciate that, microbes apart, most land animals fly rather than walk. Something like a million different species of insects inhabit this world and almost all of them can fly when they reach the adult stage. Added to which there are some thousands of species of birds, most of which can fly and even among the 4000 species of mammals nearly a quarter of them are very efficient fliers indeed. Even when we include the many thousands of mites and nematodes among our non-flying land animals, crawling and walking animals are still heavily outnumbered by those that fly.

Mosquitos, then, are not unusual in moving through the air rather than across the ground. To be sure, like most flying animals they possess more or less efficient legs, but these are used as landing gear or for travel over short distances. They are used for assistance in mating and egg-laying. They are also used for the reception of sensory information; at rest the hind legs are carried characteristically in the air like a second pair of antennae. The front and middle pairs of legs probably also serve to inform the mosquito of its position in respect to gravity, for, unlike a fly, a mosquito on an upright surface always comes to rest right way up, with its head, thorax and abdomen in a vertical line. The special fascination and importance of flight in mosquitos is that it is through the air that they visit us, bringing sickness, misery and death, a distinction they share only with certain other Diptera and with the vampire bats, for apart from a few exceptional moths, the other blood-sucking arthropods either crawl to their prey or hop.

The principles of flight are the same for a mosquito as for an eagle or even a large modern aircraft, although there are special problems associated with size. The smaller the creature the higher the surface to weight ratio and the lower the power necessary for propulsion and aerodynamic lift. None the less the wings of a mosquito are extremely complex and appear highly sophisticated when compared with those of a modern airliner, for all their flaps, slots, airbrakes and the like. As in all nature's active flying-machines the wings of a mosquito serve the purposes both of propulsion and lift. The wing beat drives the air backwards giving forward motion to the insect, while the aerofoil surface, aided by wing stroke gives it lift. The wing acts as an oscillating propellor; on each down stroke the trailing edge of the wing is actively tilted upwards above the leading edge, whereas on the return stroke it is tilted below. This is brought about partly by the special muscles and associated structures at the base of each wing, which rotate the base of the wing first one way then the other with each double stroke, and by the structure of the wing itself. Each wing is supported by six main longitudinal struts or veins, some of which divide into two branches. Those at and near the leading edge of the wing are stouter than the others with the result that the trailing edge tends to bend upwards and downwards respectively with each down and up stroke, like the free edge of a fan. This greatly increases the efficiency with which the air is driven backwards.

The frequency of the wing beat remains remarkably steady throughout flight, irrespective of speed, varying from about 250 to 500 per second according to species. It varies in any one species with the age, size and sex of the mosquito and with the relative size of the wings [4]. It varies in any one individual with the load the mosquito is carrying as a result of feeding and egg development; the frequency and hence the flight tone rises in a fully fed or gravid female. Actually these remarks are not entirely true since recent work has indicated that mosquitos may at times make sudden increases in wing-beat frequency and that these are always accompanied by sudden bursts of speed [212]. By and large, however, the frequency remains pretty constant even when the mosquito is carrying out complex manoeuvres. It drops to some extent as fuel supplies get low and there is a steady drop with age [210].

Most changes in speed result from alterations to the pattern of the wing beats and from adjustments to their angle of attack. Indeed, since the wings provide vertical as well as horizontal thrust, in some respects the flight of the mosquito shares the properties of both aeroplane and helicopter. It can adjust its air speed from zero to bursts of four and even to seven kilometres an hour. It can hover in one position and move vertically up and down. It can pivot round in its own length and move smartly to one side or the other. It can also fly backwards, even in still air. It employs all these manoeuvres when making its final approach to its target, whether this is a flower for nectar, or a female for mating. One can see the full range of these aerobatics as a female approaches for a blood meal or makes its final selection of a site for egg-laying. It uses these and more besides when avoiding capture as anyone who has tried to catch a mosquito on the wnig, or even swat it, can testify. The tiny built-in computer making up the brain and associated ganglia allows the flying mosquito to make split second adjustments that outsmart us most of the time.

The main up-and-down movement of the wings results from the action of the indirect flight muscles of the thorax. As in all flying insects the thorax is essentially a stout-sided box containing two sets of powerful muscles, one working in an up-and-down direction the other fore and aft. The two sets contract alternately and deform the thorax in such a manner as to bring about the up-and-down movement of the wings. It is as if one held a small box between finger and thumb of each hand and alternately pressed in first the top and bottom, then the front and back. The transmission of the driving force then is really rather simple—even crude. But the control of frequency and amplitude of the wing beat and of the rotation of the wing base during each stroke together with the adjustments of flight attitude and angle of attack, converts this system of basic simplicity to one of exquisite refinement and precision.

The indirect flight mechanism in mosquitos is not in fact quite so straightforward as in many other insects. The wing-beat frequency is much higher than one would suppose judging by the rate at which the stream of nerve impulses reaches the driving muscles in the thorax. The stimulated muscles go into a smooth tetanic contraction which, combined with the elastic properties of the thoracic wall, set the wings vibrating up and down [3]. It is as if the mosquito flew in constant overdrive. It is interesting that this same phenomenon is also found in other two-winged flies, in the beetles (which are also essentially two-winged as far as flight is concerned) and in bees, wasps and other Hymenoptera–all insects with high wing-beat frequencies. The halteres, which replace the second pair of wings in mosquitos, and indeed in all Diptera, oscillate violently when the wings are in action. They have a gyroscopic function, correcting any tendency to yaw from side to side during flight [199]. Correction of roll is made by unilateral adjustments of wing-beat amplitude [212].

The normal 'cruising' speed of a mosquito is usually regarded as being maintained by sight. Ingenious laboratory experiments have indicated that when flying close to the ground, or only just above the tree-tops or other substrate, the mosquito adjusts its forward speed until the background, as seen by the mosquito, appears to be passing backwards at a speed of about half a kilometre an hour [194]. It follows that if the insect is flying high above the substrate errors of parallax will result, as when we view near and distant objects from the window of a moving train; flight speed will then be faster while giving the mosquito the same impression of speed. If there is a cross-wind, lateral drift will at once be detected by the eyes and the mosquito will turn into wind and increase its air speed until the background again appears to be moving backwards at the normal cruising rate. Flight thus tends to be made into wind and the mosquito's air speed can be adjusted to give a net forward land speed of half a kilometre an hour in head-winds of over seven kilometres an hour. If the wind exceeds this speed the mosquito gives up the struggle either by alighting or by turning about and flying with the wind instead of against it [4].

Flight into wind is characteristic of mosquitos and, as we shall see presently, it is important in the sequence of responses that lead them to the host. It is evident that visual information on drift helps the mosquito to orient itself into wind. Indeed, it is usually regarded as being essential. Many species in the wild, however, fly almost exclusively at night-time when visual information is minimal or even non-existent. I wonder, therefore, how much these generally accepted views are really generally applicable. Perhaps all we can be sure of is that when visual stimuli are present, then the mosquito tends to orient itself and adjust its speed in relation to the background. Perhaps other mechanisms become unlocked, such as direct response to air movement, when the eyes are unable to play their part. It has long been known that the large blood-sucking bug *Rhodnius prolixus* normally finds its prey by means of the warm-air receptors on its antennae. It makes no difference to *Rhodnius* whether its eyes are covered or not; they might just as well not be there. If, however, one cuts off both antennae there is a sudden awakening of a latent visual response [213]. Now the eyes take over the function of direction-finding that previously resided exclusively in the antennae. A similar release of function has been found in bees when the antennae are removed; if the hive is displaced normal bees return to and swarm at the original site, but following extirpation of the antennae they make straight for the hive in its new position. In most living systems that have been adequately studied there are found to be alternative routes to success, as with the auxiliary engines of sailing ships, to be used only when the winds fail.

Like all flying-machines, the range of a mosquito is largely determined by fuel capacity and consumption. Newly emerged adults carry enough reserve fuel to allow them to make their initial flights. The main fuel, however, in both sexes is sugar obtained from nectar; females are also able to utilise blood as an important supplement. The fuel for the flight is glycogen; it becomes available for the flight-muscles within minutes of taking a sugar meal; it is also 'drawn off' as required from the special storage organs known as the fat-body. Mosquitos artificially flown to exhaustion are able to take to the air again after a day's rest, by mobilisation of these reserves [187].

The maximum range on a single nectar feed is about thirty kilometres, but this does not mean that this distance is always travelled or that when it is, it is all in one direction or made non-stop. Flight activity in each species is usually restricted to a set time each day, often lasting only an hour or two. Wild *Ae. aegypti* are usually active in the late afternoon and early evening just before sundown. If they fly continuously for three hours at their normal cruising speed they will not travel more than a few kilometres on any one day, a distance well within their range. Considerable distances are, however, flown during dispersal flights, but these are usually made at some height above the ground and, as we have seen, the speed may be increased accordingly. At times of high winds such flights are made downwind and the distance covered non-stop on any one day or night may be very considerable, ten to twenty kilometres being wholly possible [188, 189, 192].

Having taken a brief look at some of the aspects of flight in mosquitos we must now turn our attention to what affects most of us: how do these tiny flying-machines set out, often in total darkness, and in rain, to find their way to ourselves or to some other unsuspecting host, for a drink of blood? Thirty years ago I was living in a tent on a small peninsula on the northern shores of Lake Victoria in Uganda. The peninsula was long and narrow, being about ten kilometres in length and varying from about a half to one and a half kilometres in width. Except for its free end the shores were all choked by enormous swamps of grass and papyrus. There was a large and varied mosquito population, but outnumbering all other species by far was the vicious-biting *Coq. fuscopennata.* That they appeared in clouds every night was no real problem since I would escape most of them by sitting close to a rather smoky fire during the evening and by sleeping under a mosquito net at night. Each early morning, however, I had to be up before dawn and things were very different.

Although most active in the early part of the night, there were still enough of them around just before dawn to make life very unpleasant indeed. They would pounce on me while I was shaving and had no spare hand with which to ward them off. I noticed, however, that a large number of them spent much of their time flying round the pot of hot shaving water, going in for the attack time and time again only to be repulsed at the last moment by the heat of the pot. The pot was necessarily within easy reach of me, and some of them would be attracted to me instead as they flew round it. Nevertheless, by the time I had finished shaving there were usually as many of these rather handsome mosquitos fussing round the shaving pot as there were around me.

It looked as if warmth was at least one of the factors attracting these mosquitos both to me and to the shaving pot, or perhaps the combination of warmth and moisture. My main concern at the time was avoidance, and so on future mornings I arranged for a large can of hot water to be placed on a table about two metres from me. The result was very satisfactory, most of the mosquitos hovering round the decoy while I shaved in comparative peace. The ritual of the large and small water-pots became a daily routine during the following three months and on many subsequent safaris over the next twenty years, when early morning mosquitos threatened to make things intolerable. Other species too are attracted in this way to warm, or warm and moist, objects. I have noticed *Culex fatigans* hovering round and even settling on warm dishes set out at meal-times. I have witnessed them time and again flying round a toast-rack and actually landing and probing the warm pieces of toast with the proboscis; the attraction of the toast quickly diminishes as it cools down. It has long been known in Italy that those who sleep with a pig in the room tend to remain free of malaria; perhaps the higher temperature of the pig attracts the visiting *Anopheles* away from sleeping man.

The importance of warm air emanating from the host as a signal for attack was demonstrated in a now classic series of experiments made on the South American kissing bug *Rhodnius prolixus* in 1934. Yet the part played by warmth in guiding the mosquito has rarely received the attention it deserves [196, 197]. With notable exceptions people working on this subject seem somehow to have favoured smell, carbon dioxide and light, the possible part of warm air being constantly played down. That *C. tarsalis* and other species of mosquitos in the United States can be readily caught in large numbers by means of light traps does not necessarily mean that light is the only stimulus concerned. It could just as well be the gradient of warm air emanating from the lamp. While it is evident that warm air or more likely warm moist air is an important factor in bringing mosquitos into the target zone, it is clear that it is by no means the only factor; *Coq. fuscopennata* and certain other species feed readily on toads, and *C. tarsalis* has been recorded as feeding on snakes. There has been a tendency to boost the part played by this factor and that; smell, sight and carbon dioxide have all had their spokesmen. In fact it turns out that all these factors are involved and these may vary in importance from species to species. If I have emphasised the importance of warmth it is because this factor has for too long been dismissed as relatively unimportant. It is likely that once again we see alternative routes to the achievement of the same end. Just as a cat normally uses its ears as well as its eyes in righting itself during fall, so mosquitos rely on various sources of information to allow them to home on the target. If one source is wanting, another assumes importance.

Enough is now known to permit us to piece together the main sequence of events of initial activation, exploratory patrolling, location and homing on the target zone and the selection of the final feeding site. Initial activation from a state of rest to flight is determined largely by the inborn circadian rhythm, the internal clock mechanism about which we shall have much more to say in Chapter 9. An increase in the concentration of carbon dioxide in the air also serves to awaken the insect and starts it off in flight [203, 216]. As we have already seen, a combination of visual information and wind will result in the mosquito's steering its course upwind, adjusting its air speed to give it a patrolling land speed of about half a kilometre an hour. If the patrolling mosquito now encounters a stream of warm moist air from man or from some other mammal or bird it continues its flight unaltered. Should its flight path now take it out of the host beam it immediately turns at about a right angle back into the beam again. It continues zig-zagging in this fashion, but turning only when its course takes it out of the beam. Naturally the stream of warm moist air becomes narrower the nearer the mosquito approaches the source and so the turns become more and more frequent and the zig-zag course tighter to fit the beam.

We have then an elegant mechanism bringing the marauding mosquito automatically into the immediate neighbourhood of the host. The details of the zig-zag homing on the host beam were worked out by R. H. Wright and his colleagues in Canada using *Ae. aegypti*. It is likely, however, that the various factors concerned differ in their relative importance from one species to another. The part played by the eyes is important in the day-feeders such as *Eretmapodites* and many species of *Aedes* [66]. But returning to what we may call our generalised mosquito, once the female is ready to land, what then? It does not follow that once near the target the flying mosquito has only to make a landing on a suitable part of the host and dig in. It seems that smell now takes over for the final approach, and unless this is right, the mosquito will, as it were, remain in circuit in the immediate vicinity, dipping and sampling the air but not actually going into the attack [200, 201]. With *Ae. aegypti* it seems that lactic acid formed by the host is important as a short-distance attractant, and if this is not present in sufficient concentration the mosquito will not feed even though she has made a landing. The amount of lactic acid increases of course after exercise, but then so does temperature.

The picture we have drawn is a gross oversimplification and is inadequate to account for all the events that lead the flying mosquito to its source of blood. Naturally when there is no wind the patrolling flight will be at random until it crosses the host beam, which is now likely to extend vertically above the host. We need to know much more before we can make all-embracing statements. It is difficult to erect a model that will allow for the sum total of information being received from moment to moment by the flying mosquito. There are undoubtedly variations on the main theme and secondary themes as well. There is ample evidence, for example, that movement of the host plays a part, either in stimulating a nearby resting mosquito into flight activity, or in guiding an already flying mosquito to the target area. This is particularly so with those day-feeding species that have a less well-defined feeding period each day. *Ae. simpsoni* will fly out to feed as soon as a suitable host arrives in a new place. They continue to appear during the next five or ten minutes and then disappear as suddenly as they came. Move on only ten metres or so and a further contingent is flushed out and comes flying round for a feed. This can be repeated over and over again and if the feeding mosquitos are captured one finds that the actual numbers coming to feed in each place are quite small–ten to twenty perhaps. One cannot be sure that movement is the stimulus for flight, it could just as well be warm moist air or smell, acting over a short distance only, or merely that the mosquitos have been mechanically disturbed from their resting places. Nevertheless, they do not come after dark.

An intriguing phenomenon has been repeatedly recorded in field studies on the flight to the host that shows how cautious one must be in applying the results of laboratory studies to what actually happens under natural conditions. Some night-feeding species appear suddenly each evening near the future feeding site well before the usual feeding time. *Anopheles funestus, An. gambiae* and *Coq. fuscopennata* settle on the walls of houses and in the surrounding vegetation some hours before taking off again on their actual feeding flight. Similar findings have been recorded with *An. aquasalis* and *An. leucosphyrus* in Asia. If one searches for these mosquitos one simply cannot find them until a few hours before the usual feeding time for the species. Suddenly they are everywhere. They have flown up in the darkness from their day-resting haunts and settled in thousands all around the outside and inside of the houses. This pre-feed gathering often occurs as much as four hours before they make their next move [206, 238].

This rest before the actual feeding flight is one of the mysteries of mosquito behaviour. It is almost as if they sense that having flown up from the swamp or other resting site under cover of early darkness they need a rest before making the hazardous next part of their journey, involving the final approach to a wary host, the taking of the meal without being detected and the difficult get-away carrying a load more than their own weight. Perhaps they are merely exhausted and the rest period is forced on them willy-nilly. Or is it that they know in their own mosquito way that the chances of a successful meal are greater once the host has gone to sleep? But if this were so, why not delay the first part of the journey? Perhaps it is easier to make this in the early evening before the ambient temperature drops and then make the final dash in the warmth of the home. We simply do not know. Perhaps having been drawn to the houses possibly by the light, they do not, so to speak, know what they have come for until the ambient temperature drops sufficiently to allow them to detect the warm air gradient coming from the occupants. It would be interesting to know if gathering before feeding time is also found in non-house-frequenting mosquitos. When I talk of mosquitos knowing this or that I do not wish to imply that they know these things in the sense that we do. It is a knowing at the species level–an instinct if you like, remembering what we had to say about instincts in Chapter 1; it is far removed from knowing at the individual level.

The actual approach to the host when, it seems, smell takes over for the final 'talk down', can be a curious affair. I have already had occasion to relate in Chapter 1 how the tiny *Hodgesia sanguinae* behave at this time, but this reluctance to land and seeming fussiness in selecting a site for feeding could be because man is not the normal or the preferred host. There is so much to find out. When I was doing field work in connection with the epidemiology of yellow fever on the Congo–Uganda border in 1941, I found that *Ae. simpsoni* fed almost exclusively on the head. Indeed these mosquitos, flying out of their place of cover in the maize and banana gardens to feed on me and my African companions in bright sunlight, would not only settle on parts of the head including the face but would often bury themselves in the hair on top of the head, almost disappearing when attacking someone with a particularly thick mop of hair. Alec Haddow followed this work up and showed that it was not the level above ground that determined this preference since *Ae. simpsoni* still chose the head region even when the subjects were lying horizontally on the ground and unclothed. He also showed that the forest-dwelling *Eretmapodites* flies in for its attack at lower leg-level and almost invariably feeds below the knee. This time, however, he showed that choice was indeed one of level and had nothing to do with the actual site since his horizontal subjects were now fed on all over. Thomas Aitken tells me that *Sabethes belisarioi,* the picture of which is depicted on the cover of this book (and in Plate 15), bites almost exclusively on the nose; this remarkable mosquito settles in a most unmosquito-like fashion with its wings outstretched sideways.

Flight in insects has sometimes been put forward as primarily a mechanism for dispersal. But almost everything that an adult mosquito does is associated with flight. From the first movement away from the breeding water, whether or not this is followed by dispersal over relatively long distances, flight dominates the scene. Swarming and mating, feeding and egg-laying, all are done on the wing, or at least partly so. A mosquito must first fly to a flower or to a man or other vertebrate before it can feed. Whether or not egg-laying is finally carried out from the air or on foot, again, the mosquito must first fly to the site. The mechanical basis of flight in mosquitos may seem relatively simple but the infinitely variable control over flight resulting in forward, hovering, backward, sideways, upward and downward flight is another matter. It could now profitably be studied by the cybernetician and perhaps only by him. In return we might learn much that would be useful in the way of natural control systems employing multiple feedback mechanisms, both those utilising nervous pathways and those at the biochemical level.

7. THE FEEDING CYCLE

Most of us are aware that mosquitos, even in the season, are not active throughout the twenty-four hours of each day. In the tropics and other malarious places it is only after dark that we need to surround ourselves with mosquito nets or retire indoors behind metal-screened windows and doors. This is because the species of *Anopheles* that transmit malaria along with many other species of less harmful mosquitos are active only after dark. Unfortunately not all mosquitos conform to this pattern, and day-feeders are numerous enough especially in wooded country, as every picnicker has found to his dismay. Nor is it a matter simply of day or night; certain times during the day and certain periods of the evening or night are chosen by this species or that. Indeed in each period of twenty-four hours some species may be active for less than one hour, while others may show a more diffuse pattern, biting for much of the daylight period or throughout most of the night.

Aedes aegypti is a day-feeding mosquito. In captivity or in domestic populations living and breeding wholly in houses the feeding cycle may not be very clearly defined. In outside populations, however, a distinct peak of feeding activity occurs during the last two or three hours of daylight; people sitting outside enjoying the evening coolness before the tropical night descends with its clouds of dangerous *Anopheles* often complain bitterly of being attacked by these mosquitos, sometimes attributing their sudden appearance at this time of day to the fact that they, the people, were either too busy during the rest of the day to notice these mosquitos or too active to give the mosquitos a chance to settle. But if one carries out critical catches making people available all day and indeed all night one finds that this simple explanation will not do, for while a few mosquitos may appear at intervals throughout the day and there may be a small peak just after dawn, far and away the majority come during the two last hours before sunset [236].

Ae. simpsoni in tropical Africa may be caught feeding on man during any of the daylight hours. On a still warm day, the first specimens appear within an hour of dawn and the last within an hour of sunset. When, however, one comes to count the numbers coming in each hour, by catching and removing from the population each mosquito that settles in preparation for a feed, one finds that they are not evenly distributed throughout the twelve hours or so of daylight. Apart from an initial build-up during the first two hours and a fall-off during the last, the number coming to feed in a given time remains fairly steady throughout most of the day. Then suddenly, in the middle of the afternoon the number coming to feed increases severalfold and during the two hours 1430 to 1630 almost as many mosquitos arrive as during the rest of the day. It is not merely that numbers build up throughout the day to reach this peak and then decline; the peak is superposed on a general level of activity. In some populations of this species there are actually two peaks, a lesser one at around 1100 hours and a main peak in the mid-afternoon [222].

That *Ae. simpsoni* and many other species come to feed throughout most of the daylight period does not mean to say that each individual insect is active for ten to twelve hours each day. The working day for any one mosquito probably lasts for only an hour or so. In most of the individuals making up a population of *Ae. simpsoni* this period of activity occurs around 1500 hours. There seems, however, to be considerable variation in the preferred feeding time or in the particular time of day that each mosquito becomes receptive, and this results in the rather diffuse feeding cycle that we have described. It also seems that some mosquitos are tempted out because they are already resting near the 'bait' and receive the signal to activity before their appointed time. During this period of desultory activity the mosquitos do not usually come singly in a steady stream; rather, they come at irregular intervals in twos and threes. It looks as though each time a female ventures forth, its flight sound stimulates nearby resting neighbours, who then join to produce a small wave of feeding activity. The others do not come at the same time as the first mosquito but follow very shortly afterwards. Could it be that the flight tone of the first mosquito changes as it prepares to settle on a promising source of blood, and that this change serves to inform the others?

Not all day-feeding mosquitos show these peaks of activity. Alec Haddow, whose continuous 24-hour catches by shifts of trained African personnel has contributed much to our knowledge of the feeding times of mosquitos in that continent, has shown how in the forested areas of Bwamba *Eretmapodites chrysogaster* feeds throughout the daylight period. Apart from a build-up in numbers soon after dawn and a fall-off just before sundown the numbers coming to feed remain fairly constant throughout the day. We shall, however, have more to say about *E. chrysogaster* when we come to consider strains and species in Chapter 10. *Ae. cumminsii,* another widespread species in Africa, is much less restricted in its times of activity than are most other species, females coming to feed at all hours of the day and night. Even so, nearly double the number come over the period of sundown as at any other period of the day or night. No corresponding peak is found at dawn. On the other hand, *Ae. domesticus* may be caught coming to feed in more or less equal numbers at any time of the day or night [226].

Night-feeding mosquitos show the same phenomenon: whereas some species are active for much of the night, others show distinct feeding peaks at certain times. *Anopheles funestus,* one of the two important malaria transmitters in Africa, starts to become active from around midnight onwards but shows a sudden escalation of feeding activity shortly before dawn. *Coquillettidia fuscopennata,* another nocturnal species, does just the opposite: most feeding activity occurs during the early hours of the night. In certain years when the numbers making up the populations of this species are excessive, owing to enormous expansion of the breeding grounds before their various enemies have caught up with them, activity may spill over into the daylight period. When one talks of the feeding cycle, or of any other facet of behaviour about which one is dependent on statistical data, one is chiefly concerned with the central tendency. The odd individuals that do not conform do not affect the general conclusions, even though their presence in small numbers may be the key to survival in a changeable world.

All these species that we have so far considered, and very many others besides, make their daily or nightly feeding excursions close to the ground. They are essentially ground-level feeders taking blood from terrestrial vertebrates. When, however, the hunt was on in 1944 for the forest vector of monkey yellow fever in Africa it became necessary to postulate the presence of mosquitos feeding at considerable heights above ground-level since some species of monkeys that were being infected lived almost exclusively in the canopy zone.

That mosquitos were to be found feeding in the tree-tops seems first to have been recorded by R. W. G. Hingston in his book describing the 1929 Oxford Expedition to the equatorial forests of Guyana. Hingston's companions included O. W. Richards and J. E. Duffield, who literally explored the forest in depth from the floor to the canopy. Although there are a few earlier records of actual climbing to the canopy of tropical rain-forest, including one by Sir Francis Drake in Panama in 1573, Duffield and Richards' work deserves a special place of honour, because theirs was the first really thorough exploration of this normally inaccessible region. Their methods of reaching the canopy and the difficulties which they encountered have been ably described and make fascinating reading in Hingston's book now, alas, out of print. These two entomologists made a study of the insect fauna of trees, even sitting up in the canopy all night. They recorded the presence of large numbers of mosquitos and Tabanid flies feeding in the canopy, and even recorded that one species of Tabanid showed a distinct biting cycle, with a sharp peak lasting only forty-five minutes around dawn [231].

This was truly pioneer work but unfortunately it was not generally known when the possibility of the existence of tree-top haunting mosquitos came to the fore some fifteen years later, first in South America and then in East Africa. Like the earlier explo-

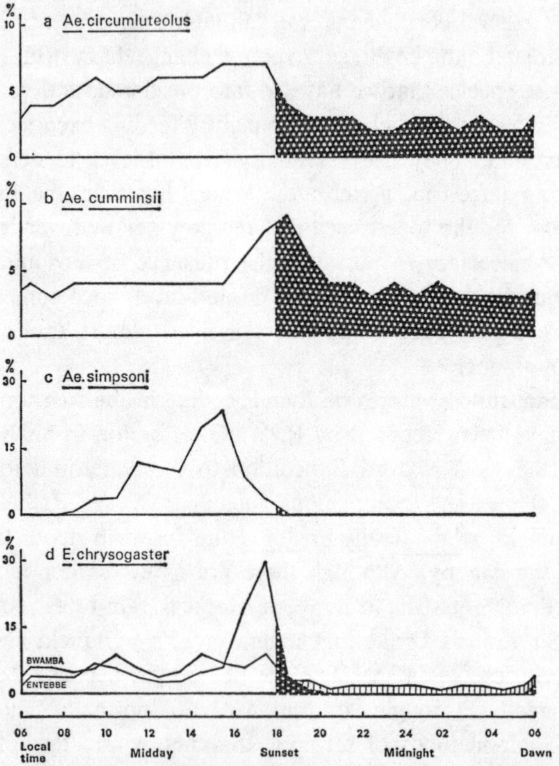

Figure 13. Feeding cycles of four day-feeding species of mosquito: data for *a*, *b* and *d* from Haddow; data for *c* from Gillett.

ration of the canopy by Duffield and Richards, the tree-top work in Africa also had its difficulties, and an account of some of these will, I hope, help convey to the reader some of the excitement of this work, and show to what lengths one may be drawn when studying the ways of the humble mosquito.

If yellow fever in monkeys was being transmitted by tree-top-feeding mosquitos as predicted, a way had to be found of studying the canopy and the various zones beneath. It would not be sufficient merely to reach these levels since each would require pro-

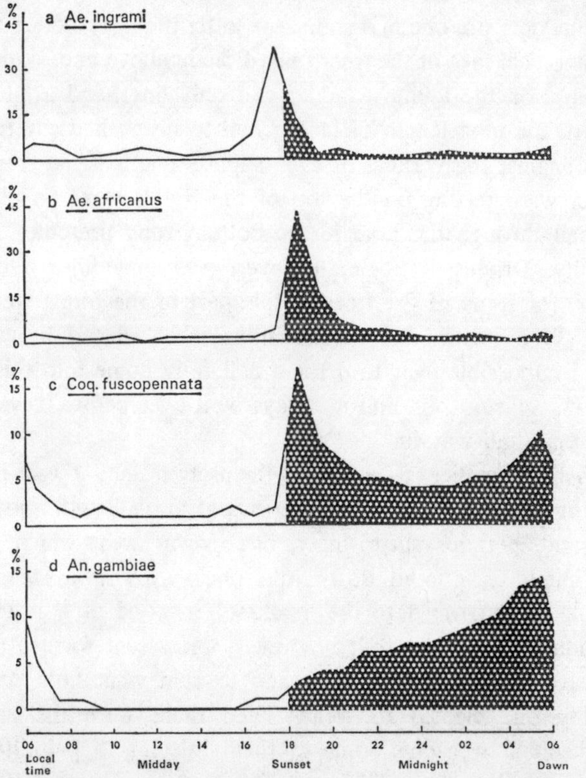

Figure 14. Feeding cycles of four evening- or night-feeding species of mosquito: data for *a* and *d* from Haddow; data for *b* from Haddow, Gillett and Highton, and data for *c* from Haddow and Gillett.

longed and continuous study. It was necessary, then, to set up platforms that would take teams of catchers working continuously day and night. These were at approximately five, fifteen, twenty and twenty-five metres, with an additional platform at ground-level to act as a control.

To reach the various levels in order first to erect the platforms and then to man them, a series of ladders had to be fixed one above another to the trunks of suitable trees. Each ladder was short and was easy enough to handle on the ground but high up on the side of a giant tree-trunk it seemed surprisingly unwieldy. Already five or six ladders up, one had somehow to fix the next ladder securely to the vertical face of the tree-trunk directly above one. With barely a toehold on the topmost ladder and only one hand available for support, the next length of ladder had to be pushed carefully upwards against the vertical or even slightly overhanging trunk until its feet were resting on the top of the last ladder. To place the first nail through the hole in the bottom rung presented no real difficulty. Driving it home, however, was sometimes a different matter, for some of the trees in this part of the forest were ironwood. Wood from this species sinks when placed in water. It proved impossible ever to drive a nail fully home into this wood, for however stout the nail it always bent over before it was much more than half-way in.

Other difficulties soon made themselves felt: these included wielding the hammer for long periods at arm's-length above one's head and, perhaps surprisingly, one's own sweat which, in that high humidity, poured down into one's eyes in a steady salty stream. Hammering into the much softer wood of a strangler fig was admittedly easier, but now every blow sent forth a spurt of corrosive latex to fill the eyes and to ruin what little protective clothing one elected to wear. Then there were the ants that swarmed all over one, some of them injecting a painful poison, others carrying off small lumps of flesh; these did not help either.

Having secured the bottom rung and reached up cautiously to get the middle nail in, or half-way in, before the weight of the new ladder brought it swivelling round to one side or the other, one had then to leave the now familiar security of the old ladder and venture on to the new one held only by its two nails, one through the bottom rung and one through the middle, in order to get the top nail in. Oddly enough it was the man whose turn it was to stay on the ground that came off worse. The denseness of the forest made it impossible to see the tree except from a place almost immediately below. In craning anxiously upwards to watch while his precariously perched companion attempted to hammer home that top nail far above, he developed, as the day wore on, an extremely painful ache in the back of the neck. So severe was this pain that the man at the top had eventually to be abandoned to his fate; he could expect no more encouragement from below.

Once the ladders were up the task of building the various platforms was begun. This work was supervised by Yovani Ssenkubuge, a tireless and experienced worker whose splendid example did so much to ensure the success of this and other experiments in the forest. During this work he fell from the fifteen-metre platform. Happily, he survived and later became an acknowledged expert in the building of platforms in trees.

The plan was to man each platform simultaneously from ground-level to the canopy with a team of four catchers, each man in the team working for four hours at a stretch. Any mosquito settling in preparation to feed was to be caught in a small glass tube and placed inside a large bag labelled with the hour and the level. At the close of each hour the bag was to be sealed and a new one brought into operation. By these means it would be possible to know the level and time of feeding of every mosquito caught. The 24-hour catching at all levels simultaneously was to go on in five-day stretches followed by two days off until a total of forty days and nights of catching had been achieved (twenty in one part of the forest and a further twenty a few kilometres away). During each five-day stretch the forty members of the catching staff, not to mention the supervisors, had all to be adequately fed, a none too easy exercise in logistics in that remote spot. They also had to be kept free from the dangers of boredom. On the official payroll were included some of the finest drummers in the land. They played their instruments, to the occasional accompaniment of a consort of end-blown flutes, to keep those on duty from falling asleep and to entertain the others. They also helped keep away wild animals and evil spirits, both of which were particularly numerous in that part of the forest.

In no sense was this work simply a matter of observation and record. A hypothesis had been put forward that monkeys become infected with the virus of yellow fever without their having to leave the forest canopy. The experiment was designed to test the prediction that followed from this; namely that at some time during the 24-hour cycle of day and night a species of mosquito capable of transmitting the virus feeds on primates in the canopy. This then was a true experiment in the usually accepted scientific sense, as rigorous in its discipline as any carried out at the bench —an ecological experiment on the grand scale [229].

The excitement of those early days did not cease when the erection of the vertical ladders and precariously perched platforms was complete. Indeed, to those of us who were passionately interested in the life of mosquitos and in the outcome of the experiment in particular the excitement was about to begin. Work started in the morning at all levels, but although large numbers of an assortment of species were taken on the ground platform and others were taken during the day in the middle storeys nothing whatever turned up in the canopy itself. After all the elaborate preparations over the past weeks it began to seem as if the prediction of a canopy feeding virus transmitter was perhaps erroneously based. Then suddenly a voice called out from far above, 'Iko mmoja'. Then another with, 'Iko mmoja ngini', followed immediately by 'Na ngini', which may be translated as 'Here's one' and 'Here's another', 'And another'. A great wave of activity followed in the canopy, all the mosquitos being *Ae. africanus,* a species that had already been shown in the laboratory to be fully capable of transmitting yellow fever virus. The wave was short-lived, activity dying down almost as suddenly as it had begun. All was over within an hour of sunset. A curious feature was that no similar peak occurred immediately before the following dawn. This pattern was repeated on every one of the forty nights of the experiment, a short sharp peak of feeding activity immediately following sunset. In later years Russell Lumsden repeated some of this work and by using much shorter intervals between bag changes, showed that the actual peak lasted for about twenty minutes only, the onset of activity depending on the actual time of sunset, which even on the equator fluctuates over a range of about thirty minutes in the year [412].

The original series of forty 24-hour catches at various levels in the forest has enabled us to build up a comprehensive picture of the distribution of mosquito feeding cycles both in time and level. There was no question of these results being applicable to one part of Africa only, for Peter Mattingly working some 3000 kilometres to the west, later carried out a similar series of catches and drew up an almost identical picture of separation in time and level [235]. These and the many other series of catches that have since been undertaken in various parts of tropical Africa have made important additions to our knowledge and helped to fill in some of the gaps. The full importance of these findings will become clearer when we come to deal with mosquitos as transmitters of disease in Chapter 13.

We know now that in many forested areas *Ae. africanus* is not the only species living in the tree-tops and having a crepuscular feeding cycle. Indeed, the full picture which emerges from all this work is quite astonishing. The various species occurring in any one area of forest tend to be separated either by the timing of their feeding cycles or by the level in the trees at which they show their maximum activity. Among the many species of mosquitos that may be found in the forested areas of central equatorial Africa there often occur four species of *Aedes,* the lives of which over the years have become particularly well known and serve to illustrate well the point I wish to make. Not only may these four species be found together in a single small patch of forest, but all four are sometimes found inhabiting an individual tree and even the larvae may all occur together in the same tree-hole. It seems also that all four feed as adults on monkeys. Two of these species belong to the subgenus *Stegomyia,* one of them, *Ae.* (*S.*) *apicoargenteus,* occupying the mid-storey zone, the other, *Ae.* (*S.*) *africanus,* the forest canopy. The remaining two species belong to the subgenus *Finlaya;* these extend vertically over both levels: *Ae.* (*F.*) *ingrami* is perhaps somewhat more plentiful at mid-storey and *Ae.* (*F.*) *longipalpis* is perhaps a little more so in the canopy (these tendencies are reversed in some forests). Summing up, then, we find one species of *Stegomyia* in the mid-storey zone and one in the canopy, while the two species of *Finlaya* are each found in both zones. We thus find three of the four species in the mid-storey (*Ae. apicoargenteus, Ae. ingrami* and *Ae. longipalpis*) and three of them at the top (*Ae. africanus, Ae. ingrami* and *Ae. longipalpis*).

Let us now look at both of these zones in a little more detail. The feeding cycles of two of the mid-storey species are of the diffuse type, extending over a comparatively long period of time. Nevertheless, the peak of activity of one of them usually occurs well before that of the other, *Ae. longipalpis* feeding in the morning, *Ae. apicoargenteus* in the afternoon. *Ae. ingrami,* the third species occurring at this level, shows the clear-cut type of feeding cycle with a short sharp peak around the time of sunset. In the canopy above, things are very different: while *Ae. longipalpis* continues to exhibit its diffuse type of feeding cycle in the late morning period, *Ae. ingrami* and *Ae. africanus* both show clearly defined feeding cycles around the time of sunset. Careful counts, however, show that *Ae. ingrami* feeds just before the sun goes down, while *Ae. africanus,* as we have seen, feeds just afterwards.

These two short sharp peaks of feeding activity at the same level, one following immediately upon the other, never fail to impress me. Sitting on the top platform high up in the canopy twenty to twenty-seven metres above the forest floor is itself an invigorating experience. One looks across (instead of up) at the tree-tops, watching moving troups of monkeys and forest touracos caught for a moment or two alongside in the horizontal rays of the setting sun. All is still and not a mosquito is to be heard. I have never sat there at this time of day without thinking that today there simply are no mosquitos to come. I have sometimes occupied the top platform with three African companions for three or four hours before sunset and not a single mosquito has come throughout the period. Then suddenly the first *Ae. ingrami* arrives, just as I am thinking that this time it is not going to work. Then follows another and then another, and within a few minutes a wave of *Ae. ingrami* arrives as darkness descends on the forest blotting out all views even from the top. Within fifteen minutes or so the wave is nearly over and just as I am beginning to have doubts about the arrival of the other species, the first *Ae. africanus* arrives. There follows about five minutes overlap when small numbers of both species appear together. But *Ae. ingrami* soon disappears from the scene and *Ae. africanus* now takes over and builds up rapidly to reach its spectacular peak shortly afterwards. Twenty minutes or so later and *Ae. africanus* too has dwindled to a few desultory specimens; save for a few more odd specimens at times neither species will be seen again until the evening of the next day.

Except when the wind is high these activity peaks occur like clockwork every day and every evening of the year at their set times in every patch of lowland rain-forest, however small or isolated from its neighbours. In drier regions *Ae. africanus* is replaced by *Ae. luteocephalus,* and near the tree-line on the slopes of the so-called Mountains of the Moon on the Congo border by *Ae. ruwenzori.* There are, as I have said, many other species occurring in the same forest as the four species of *Aedes* which we have been considering. Each occupies its own level in the forest and exhibits its characteristic feeding cycle. Unlike our four *Aedes* some of these others do in fact overlap both in level and in feeding time. *Coquillettidia aurites,* for example, is another crepuscular canopy feeder with its main peak of activity immediately following sundown. But *Coq. aurites* prefers birds to monkeys and so spatial and temporal separation from *Ae. africanus* is unimportant. Some ecologists, when considering insects that feed on the same food and live in the same or in a similar region in identical habitats, talk of 'homologous ecology'. I wonder, however, if such a concept is ever really justified when we come to study the time of feeding as well as the place.

While certain generalisations can be drawn about each species there are significant variations and departures from what we may conveniently regard as the basic plan for each species. For example even in *Ae. africanus,* a species showing such a clear-cut crepuscular feeding cycle in the tree-tops of most forested areas in which it has been studied, there are always a few individual specimens that appear not to know the 'rules' and feed at ground-level or at intermediate levels. It is interesting to note that these ground-level specimens usually come to feed just before instead of immediately after sundown and in some areas this species can be taken feeding at ground-level during the daytime. It is interesting to speculate on whether this means that there is a daily vertical migration of the population up the trees in the evening and down again as the sun rises on the following morning in this and in other arboreal species. It could be argued that while most of the population reaches the optimum level in the trees by the normal feeding time, a few individuals become attracted to bait before starting up on their journey and thus come to feed at ground-level or perhaps half-way up. Equally, however, these 'wrong'-level feeders could simply be regarded as non-conformers to the normal pattern. Deviants from the norm are usual in all populations. Indeed they are the very basis of survival as we have already had occasion to underline. I do not myself think that there is really sufficient evidence to allow us to state that vertical migration occurs or does not; such matters are extremely difficult to settle. These two views, daily vertical migration versus deviant individuals, are not however mutually exclusive. While carefully and properly collected and recorded data withstand the test of time, our interpretations of them are more often than not notoriously short-lived.

8. THE OVARIAN CYCLE

We have already seen that most female mosquitos need to take a blood meal if they are to develop a batch of eggs. The number of eggs produced after each meal varies from species to species, being as many as 500 in some species of *Anopheles* but usually under 100 in *Aedes aegypti*. The number also varies to some extent with the amount of blood taken; in *Ae. aegypti* the size of the blood meal may vary from a mere trace to nearly 5 mg, depending on whether or not the mosquito has been interrupted during feeding and on the size of the insect [111, 245, 268, 275]. Small blood meals of less than 0.5 mg do not result in egg production. Above this size and up to about 3 mg the number of eggs produced bears a direct relation to the amount of blood taken, but any further increase in the size of the blood meal does not result in any more eggs. There are, then, both lower and upper thresholds between which the number of eggs depends on the size of the blood meal.

Generally speaking, a single blood meal is all that is necessary for a mosquito to produce a batch of eggs. Not until these eggs have been laid will she again go out for a blood meal and thus play her part in the general feeding cycle of the population. The blood meal, then, can be considered as the first act in the ovarian cycle, the whole sequence of events leading from the entry of blood into the stomach to the development of the ovaries followed by the laying of a batch of eggs. It has long been known that the blood meal provides nutrient materials for the developing eggs but it is only comparatively recently that the factors controlling the timing and coordination of these events have become better understood.

In most female mosquitos the oöcytes that will give rise to the first batch of eggs to be laid actually begin their development during the early hours of adult life before any blood meal has been taken. This early development brings the first batch of oöcytes up to what is known as the resting stage. In all except autogenous mosquitos, in which as we mentioned the first batch of oöcytes develops to maturity without a blood meal, the oöcytes remain in the resting stage until the mosquito takes a blood meal. In some mosquitos, particularly in those coming from less well-nourished larvae, two blood meals are necessary for the production of the first batch of eggs, one bringing the oöcytes up to the resting stage, the other allowing development to go on to maturity. When the first batch of eggs is laid the next batch of oöcytes has already reached the resting stage and another blood meal is all that is necessary for the completion of its development.

At the time of the blood meal the oöcytes are normally at the resting stage. Indeed, they will remain indefinitely at this immature stage unless the mosquito does take a blood meal. There is no question whatever of the blood meal merely being timed to coincide with this stage of development of the oöcytes; except in those few species exhibiting autogeny, further development of the oöcytes is entirely dependent on the blood meal. The freshly drawn blood goes, as we have seen, directly into the mid-gut or stomach, which it distends to an enormous degree, the weight of blood taken being considerably in excess of the pre-feeding weight of the insect. Distension of the stomach provides the signal initiating ovarian development beyond the resting stage. The hundred or so oöcytes now begin to develop, each receiving its nutritive requirements from a group of seven special nurse cells, which accompanies each oöcyte within its follicle [243].

The nurse cells are only called upon to supply the developing oöcytes during the initial stages of development, thereafter either the oöcytes themselves or the follicles enclosing them are supplied direct with nutrients derived from the partially digested blood meal [4]. To enable this second step to be taken and to ensure that it occurs both at the right moment of development of the oöcytes and at the right moment in the breakdown of the blood meal, a hormone is released from neurosecretory cells that reside on the top of the mosquito's brain [250]. The stimulus for the release of this hormone from the brain seems to be the one that initiated the first step in ovarian development, namely the stretching that occurs when blood entered the stomach some hours before. But the part played by the brain hormone is an indirect one, for it in turn stimulates the release of a second hormone from a pair of endocrine organs situated in the thorax at the point where it is joined by the slender neck [244]. These endocrine glands are known as the corpora allata and it is the hormone released by them that now takes over responsibility for the development of the oöcytes, carrying them through to maturity. Thus, while ovarian development beyond the resting stage is initiated by entry of blood into the stomach, promotion of the activated oöcytes to maturity some hours later is under hormonal control [250].

The first suggestion that stretching of the stomach wall might serve in mosquitos to trigger off the hormonal mechanism that takes care of the later stages of ovarian development came from large-scale experiments which showed that release of the brain hormone was considerably delayed in mosquitos that had been permitted to imbibe fruit juice before taking the blood meal [251]. We have already seen how fruit juice does not pass into the stomach but is side-tracked mainly into the large ventral diverticulum that shares the abdominal cavity with the stomach. The partial occupation of the abdominal space by the greatly swollen diverticulum limits the size of the blood meal and so prevents full distension of the stomach. It would be interesting to study the time of hormone release in two groups of mosquitos, giving them large and small fruit feeds respectively before the blood meal.

Apart from a certain amount of technical difficulty in carrying out experiments of this sort because of the smallness and frailty of the material when compared with our own thumbs, there are certain inherent obstacles. For one thing blood in the stomach of the mosquito not only initiates development of the oöcytes and triggers off the release of hormones which will see them through the final stages of development but itself later provides the very substance on which this development depends. In an ingenious series of experiments Joe Larsen and Dietrich Bodenstein were able to study the effect of distending the stomach in the absence of a blood meal by giving some of their female mosquitos a milk enema. The milk clotted in the stomach and ovarian development followed in the mosquitos subjected to this somewhat undignified treatment. When, however, they used non-clotting solutions distension of the stomach was not maintained and ovarian development did not follow in any of the mosquitos treated. It seemed, then, that stretching of the stomach to be effective had to be maintained for a certain period [262].

The results of these experiments left room for some doubts, for who could really say whether it was the maintained distension of the stomach as postulated or the milk proteins present in the stomach that acted as the triggering mechanism? If a way could now be found of keeping the stomach distended but with non-proteinaceous fluid the problem would be neatly solved. Larsen and Bodenstein went about tackling this problem in a rather roundabout way. The reader will remember that nectar or fruit juice once inside the ventral diverticulum is drawn off at intervals and passes in small amounts to the stomach. No distension of the stomach results for the amounts are very small and are soon eliminated via the rectum. If, however, the anus were to be blocked then, they argued, these small amounts of fruit juice ought to accumulate in the stomach, eventually distending it. They duly sealed off the anus in a group of mosquitos with wax, fed them on fruit juice and waited. By the fourth day, sure enough the stomachs were blown out with the juice and on dissection a proportion of the mosquitos showed clear signs that ovarian development had started.

Evidence for the hormonal mechanisms that allow the already

activated oöcytes to proceed through the last stages of development to reach maturity has mostly been of rather an indirect kind and results from experiments of varying degrees of sophistication. Endocrine organs have been removed or tied off at intervals after the blood meal, either by cutting the head right off and so removing the effect of the neurosecretory cells of the brain, or by tying a tight ligature round the thorax immediately behind the corpora allata. Insects survive quite well without their heads: decapitated *Rhodnius* may be kept alive for many months, far longer in fact than their intact normal companions when they too are unable to have another blood meal. Mosquitos, admittedly, are much less robust than these large bugs and much less easily manipulated; for one thing their legs tend to drop off at the slightest provocation. But provided these operations are done skilfully mosquitos survive well and long enough to allow full development of the ovaries in the controls.

In other experiments the tiny corpora allata have been dissected out from fed mosquitos and transplanted into unfed recipients. Or, the ovaries themselves have been taken from one unfed mosquito and placed inside another, the transplanted ovaries developing when the new owner was later given a blood meal [261]. By such methods it has been possible to show that a critical period occurs a few hours after the blood meal has been taken; operation before this period prevents ripening of the ovaries, whereas the same treatment afterwards allows the development of a normal quota of ripe oöcytes. In mosquitos operated upon during the critical period itself there is a progressive increase in the proportion of mosquitos developing ripe oöcytes the later the operation is carried out. Transfusion of the insect's own blood or haemolymph by micro-injection methods from donors early in the critical period to mosquitos from which the head (containing the brain and neurosecretory cells) has been removed late in the critical period significantly raises this proportion [252]. The most impressive of these experiments have been those of Arden Lea in Florida, where I was privileged to see him at work. Lea first anaesthetised his living mosquitos and then actually opened up the head capsule and removed the neurosecretory cells from the brain. The living cells could be seen clearly standing out from the rest of the brain tissue, and by micro-methods could be individually dissected out while leaving the brain otherwise intact. His mosquitos recovered and lived to tell the same story as those given the somewhat less refined but albeit exacting treatment mentioned above [263, 264].

The speed at which all this proceeds depends on the ambient temperature. At high temperatures the interval between taking a blood meal and maturing of the oöcytes is about forty-eight hours but there is considerable variation between species. This interval in *Ae. aegypti* can be reduced to less than forty hours by raising the mean ambient temperature to over 30°C and increased to more than 120 hours at 22°C and below. As long as the endocrine system remains intact, as it normally does of course, ovarian development proceeds smoothly to maturity: initiation and promotion, the one dependent on the nurse cells the other on the hormonal mobilisation of material supplied by the blood meal, pass directly one into the other without a break, like the transition from one movement to the next in certain well-known pieces of music. The mature oöcytes have now to be released from the ovaries, this process of ovulation preceding the laying of the eggs or oviposition. The remarkable surgical prowess of Jack Corvard Jones and his associates in Washington has enabled us actually to see what happens inside the living mosquito. By inserting windows into the wall of the mosquito's abdomen T. J. Curtin and Jones noted that the oviducts leading from the ovaries undergo violent rhythmic contractions which carry the ripe oöcytes away from the ovary on the first part of their journey to the exterior and to independent life [246].

It is, as the reader will remember, during the passage of the ripe oöcyte between ovulation and oviposition that fertilisation takes place, sperm stored by the female since copulation now passing from the spermathecae to each egg on its way out. Now it is a curious fact that although actual fertilisation is, so to speak, left to the last moment, the oöcytes although ripe normally do not make this journey unless sperm is ready for them as a result of previous mating. Thus unmated females do not usually lay their eggs even though the ovaries have fully ripened following a blood meal. But it is not the presence of sperm in the spermathecae of the female which ensures that the eggs will now be laid, but the seminal fluid from the accessory glands of the male which accompanies the sperm. It is not certain whether the contribution from the male accessory glands has a direct effect on the female or whether it operates through yet another hormone produced by the female herself. We do know, however, that a virgin female with fully ripe ovaries following a blood meal will lay a full batch of eggs shortly after mating, even though the act is delayed until three weeks or so after the blood meal [304].

Once the eggs are laid the mosquito is ready for another blood meal which in triggering off the development of the next batch of oöcytes heralds the onset of a second ovarian cycle. In mated females one cycle may follow another without a break, as many as ten or even twenty ovarian cycles having been recorded before old age brings activity to a close. The brilliant work of a distinguished Russian pioneer in the reproductive physiology of mosquitos, V. P. Polovodova and the work of her equally distinguished colleagues T. S. Detinova and W. N. Beklemishev have made it possible to see on internal examination how many ovarian cycles any particular female mosquito has completed. At ovulation a relic of the follicle remains as an ovariolar swelling; the number of these swellings along each ovariole indicates accurately the number of completed ovarian cycles in the life of the mosquito. Indeed the number of completed cycles can be used as a criterion in assessing the mosquito's age; age in this instance is taken as the physiological age of the mosquito rather than the number of days or weeks, since the actual duration of each completed ovarian cycle will depend on the mean ambient temperature. It is clear that, say, whereas ten such swellings at low temperatures might indicate that the mosquito was at least fifty days old, the same number of swellings at higher temperatures could reduce this period to only twenty days or so [241, 248, 249, 267].

This work is of great importance since it enables one to find out what proportion of a population of mosquitos at any particular time of the year constitutes a threat to the health of man. A mosquito is not normally infectious when it bites unless it has already survived long enough to allow the multiplication and build up of the pathogens it has picked up at a previous feed. Since the rate of multiplication of the pathogens in the mosquito is also dependent on temperature, this means that the interval between the infecting feed and the mosquito itself becoming infectious can vary from one to three weeks. Physiological age, therefore, is a better measure of potential danger than age merely measured by the clock. Polovodova's method is now widely used in epidemiological studies and in all work where it is important to have information on the age composition of mosquito populations [259, 260]. Many workers in this field are also indebted to Mick Gillies for his timely translation of the original Russian papers into English; Gillies has himself made notable contributions to our knowledge of ovarian cycles in nature [253–7].

9. THE CIRCADIAN RHYTHM

We have seen how mosquitos tend to be active at set times of the day or night. In studying the factors that determine the timing of these periods of activity and rest, which play such an important part in deciding whether a species acts as a vector of disease or not, we come to one of the most fascinating and intriguing aspects of mosquito life. The initial exploration of this phenomenon in mosquitos was made independently of and in isolation from the main stream of work on other organisms that was going on contemporaneously elsewhere, yet the results of each have led to precisely the same somewhat surprising conclusions. It will be illuminating, therefore, if we set out in some detail the exciting events that led us to these conclusions in mosquitos and to the concept of what is now generally called the circadian rhythm.

Feeding activity is easy to measure in the field; given the manpower or the apparatus one has merely to draw off from the population all the individual mosquitos that settle on the host in preparation for a meal and to record the time of their arrival. One works on the assumption that one is dealing with a population of infinite size and that the removal of active mosquitos on one day will not influence the catch on the next, and so on. In practice this works very well; the numbers of mosquitos do not drop off as the days and nights follow one after another. But if one is to learn something of the underlying mechanisms it becomes essential to be able to vary the length of daylight or photoperiod or to reverse day and night. This inevitably carries the investigation from the field to the bench. Under laboratory conditions feeding activity is not easy to measure because in a population of limited size it usually involves the steady removal of fed mosquitos and rapidly leads to a state of diminishing returns. If, however, we take egg-laying as a measure of activity this difficulty disappears, for we can remove the eggs at regular intervals without affecting the size of the adult mosquito population that laid them. Indeed, in some ways, egg-laying is superior to feeding as a measure of flight activity since each female, first having flown from its place of rest, lays many eggs at a single oviposition; it needs only fifteen egg-laying females to produce more than a thousand eggs. Irregularities brought about by the comparatively small numbers that can be used in the laboratory tend thus to be smoothed out, a mere 500 females producing some 35,000 eggs following each blood meal.

The first step in this investigation then was to find out if there is in fact an egg-laying cycle comparable to the feeding cycle, about which we had so much information from field observations. A culture of *Aedes aegypti* was established with several hundred males and females in a single large cage. Continuous opportunity was provided for fruit- and blood-feeding; mating and oviposition could occur at any time. Now it would have been reasonable to suppose that the females would lay their eggs as soon as these matured following the blood meal, and, since feeding would occur at set times according to the feeding cycle of the species, a regular egg-laying cycle would follow—the interval between the feeding cycle and the egg-laying cycle being determined by temperature. Indeed, one would have been aided and abetted in this supposition by the standard textbooks of the time. The reader will remember that the peak feeding time for *Ae. aegypti* occurs in the late afternoon, and sure enough the first thing that this experiment showed was that the eggs were all laid over a short period three days after the first blood meal had been taken. The egg-laying peak occurred at the same time of day as the feeding peak, and moreover a similar peak occurred in the late afternoon of each succeeding day. This could, of course, still be interpreted as a false oviposition cycle imposed on the population by the females participating in the feeding cycle three days before and that the mean temperature of the room in which the experiment was carried out happened to be just right to ensure the maturing of the oöcytes in exactly seventy-two hours.

Thus far then the textbooks could still be right. But surprise number one: egg-laying occurred at the same time each afternoon irrespective of temperature [287]. Indeed mosquitos fed on ourselves at a known time and kept at a temperature that would allow the oöcytes to reach maturity just after sundown two days later did not lay their eggs until the appropriate laying period late on the following afternoon, some twenty hours after reaching maturity. On the other hand, control mosquitos fed and kept at a temperature that ensured maturity of the oöcytes early in the afternoon, duly laid their eggs before sundown on the same day. In other words, egg-laying activity in a population of mosquitos occurs at the same time each day irrespective of temperature, but while the timing of this cycle coincides with that of the feeding cycle one is clearly independent of the other.

Since egg-laying activity was restricted to the late afternoon one might suppose that the mosquitos were either responding directly to the falling light intensity or falling temperature or both or to changes in humidity [225]. Or that having been kept from activity by the previous conditions they were now released from this inhibition. Up till that time these were in fact the current theories concerning the feeding cycle and it surely seemed reasonable to extend these to include what was clearly a comparable egg-laying cycle. But, surprise number two was about to be sprung —a surprise that could never have occurred had the study of cyclic behaviour continued to be made under field conditions only. The experiment was repeated under conditions of constant temperature and by substituting artificial lighting for the natural 12-hour photoperiod of the equatorial day. The lights were merely turned off and on at 12-hour intervals, providing a 12-hour light (L) period alternating with a 12-hour dark (D) period, which for convenience can be expressed as LD 12:12. There was now no

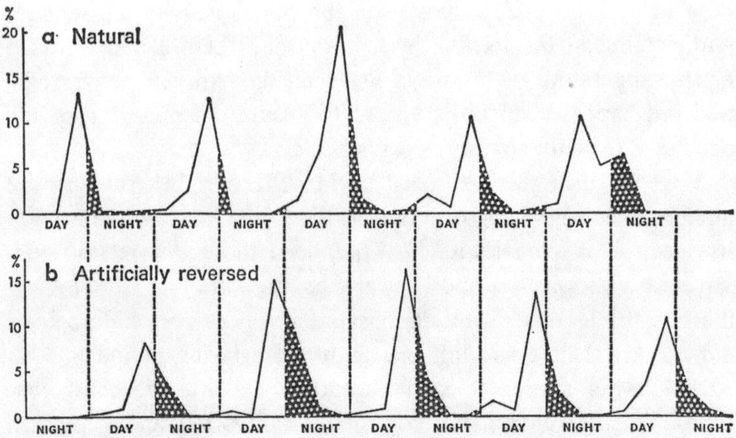

Figure 15. Circadian rhythm of egg-laying activity in *Aedes aegypti: a.* under natural equatorial conditions of twelve hours day and twelve hours night; *b.* under a reversed regime of twelve hours artificial dark and twelve hours artificial light. Note that in both cases the peaks of activity occur immediately before light-off (data from Haddow and Gillett).

twilight period—the lights were either on or off—and no change in temperature or humidity, yet the eggs were almost all laid just *before* the lights went off. Moreover, when the photoperiod was reversed the egg-laying cycle was also reversed; the peak of activity still occurred just before the lights went out, but this period was, of course, now shortly before the natural dawn [287]. In fact the peak could be positioned at any time of day depending on the timing of the artificial photoperiod. Nor did the duration of the photoperiod affect the result: within wide limits it made no difference if we increased or decreased the photoperiod per twenty-four hours, the eggs were always laid just before the close of the light period [281, 284].

This, then, was surprise number three: whether the duration of light per day was lengthened or shortened the eggs were always laid just *before* the return to darkness. It made no difference if the mosquitos were reared through their aquatic stages beforehand in alternating light and darkness of varying LD ratios or even if they were reared in constant light or in constant darkness. If, however,

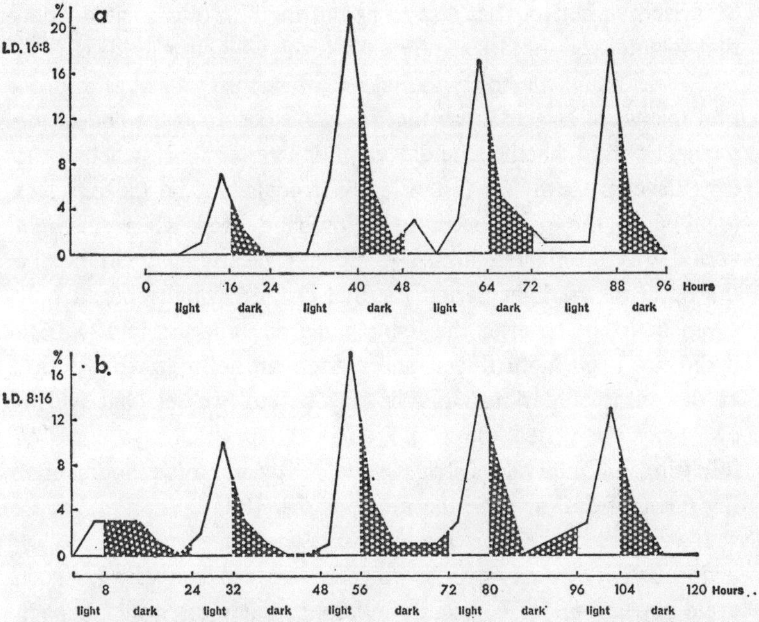

Figure 16. Circadian rhythm of egg-laying activity in *Aedes aegypti* under *a.* artificially lengthened 'day', and *b.* artificially lengthened 'night'. Note that the peaks of activity occur immediately before light-off in both cases (data from Gillett, Haddow and Corbet).

the mosquitos were kept in constant light or in constant darkness during adult life then the picture was entirely different: cyclic behaviour was not shown at all, even when they had been brought up through larva- and pupa-hood in a normal regimen of LD 12:12. Some of the later experiments were carried out at a higher temperature so as to cut down the time spent on these lengthy and exacting experiments. The higher temperature had the effect of telescoping the preparatory period of each experiment, first when the cage populations were being reared in the rigid regimens of light and darkness and later when the newly fed adult mosquitos were digesting their blood meals and developing eggs. But the higher temperature had no measurable effect whatever on the rhythm of egg-laying, which remained as before on a 24-hour basis.

Surprise number four? It was to us out there in our isolated situation remote from libraries and out of touch with current ideas.

The next step was to try to find out what it is that serves to position this wave of activity so that it always occurs at the end of the daylight period whether the daylength is long or short, whether the eggs develop slowly or fast. Clearly it could not be the onset of night since the peak occurred beforehand even when the light period was terminated abruptly. Equally clearly it could not be the onset of the light period for in LD 8:16 activity began only seven hours or so after the lights came on, whereas in LD 16:8 it did not begin until fifteen hours afterwards. But when we look at the onset of darkness on the day before, we see that whichever way we varied the photoperiod the peak always occurred following an interval of twenty-one to twenty-three hours after the onset of darkness on the previous day [284].

If onset of darkness is the cue for a wave of concentrated egg-laying activity twenty-two hours later, we can see why no such cycle appears under constant conditions. But what would happen if mosquitos reared and kept in total darkness later were subjected to a single exposure to light? If the onset of darkness is indeed the cue then the return to darkness should be followed by a peak of activity twenty-two hours later. To test this prediction a cage colony of about a thousand *Ae. aegypti* reared in total darkness was subjected to a once-in-a-lifetime experience of twelve hours of light. The colony had been going long enough to show that until the light came on egg-laying was completely acyclic. Shortly after the light came on egg-laying ceased abruptly and almost no eggs were laid for the following thirty-four hours. Then, suddenly, about twenty-two hours after the colony had been returned to darkness, sure enough, there was a great burst of laying activity, just as predicted. But, and here was surprise number five, a similar peak appeared at about the same time on the next day and another on the day after and so on for each of the first five days following the return to darkness. No such cyclic behaviour followed the reciprocal experiment of rearing them in continuous light and subjecting them to a single experience of twelve hours of darkness [289].

This experiment was repeated many times exposing each culture of mosquitos to a single light period of different duration; no matter how long or how short the single exposure to light acyclic activity was converted to cyclic, with the first peak always occurring about twenty-two hours after light-off. A single exposure to light of only five seconds in an otherwise all dark existence was enough to bring about this astonishing transformation [282].

It would seem from these experiments that there is an inborn or endogenous 24-hour rhythm of activity and rest, but that this rhythm only becomes manifest following exposure to light. Since, however, it seems to be the return to darkness at the end of the exposure that acts as the time-cue for the rhythm and not light itself or its duration, it should follow that a single change from continuous light to continuous darkness would have the same effect. The critical experiment was planned: bring up a cage-culture of mosquitos in continuous light from the time of hatching of the eggs onwards and then, after egg-laying by the adults has been established, plunge them into total darkness.

About 500 *Ae. aegypti* of each sex were duly reared through in continuous artificial light. On the face of it this would seem simple and straightforward enough to arrange, simpler in fact than rearing them as before in total darkness. But appearances can be deceptive; the experiment was in fact extremely difficult to carry through and the unusual nature of the obstacle encountered justifies a slight digression. From the time of hatching of the eggs to the time when the lights would eventually be turned off, four or five days after acyclic egg-laying had been well established by the newly emerged adult mosquitos, would require a total run of some twenty days of an uninterrupted supply of electricity. A single break in that supply whether long or short and whether by day or by night would immediately invalidate the whole experiment. But this experiment, like all the others described above, was carried out at Entebbe, Uganda, a place subject to frequent thunderstorms of particular violence. During these storms it is not unusual for the overhead power-lines to be struck by lightning and the supply of electricity halted. Indeed, the first thing that had to be done in planning this work was to consult the meteorological records and to choose a time of the year when the probability was highest of having a run of three weeks without a storm.

After one false start (a thunderstorm in what is usually the calmest time of the year, but fortunately early on in the experiment) the work got under way. From just before the eggs were placed in water containing substances that would act as a suitable hatching stimulus, through the four larval stages and pupahood, the mosquitos experienced nothing but continuous light and constant temperature and humidity. When the adults emerged they were free to mate and feed at all times. From the fifth day after the bulk of them had emerged eggs began to appear at odd intervals in the specially designed laying-pots. Egg production built up over the next day or two and then remained at an irregularly fluctuating level, just as one had come to expect. Five days after the first eggs appeared the lights were switched off. All egg-laying ceased. Examination of the laying-pots continued as before. About twenty-two hours after the single change from light to dark egg-laying was resumed. It built up to a sharp peak during the twenty-third hour and died away again. But the peak was repeated cyclically at 24-hour intervals over the next five days of darkness [282, 289].

If proof were needed here it was. There could now be no doubt whatever that the daily egg-laying cycle in a population of mosquitos had at least a strong inborn inherited element, and that onset of darkness operated as an external time-cue for the timing of the cycle. How else could a single change from continuous light to continuous darkness convert irregular acyclic laying to a beautifully regular 24-hour cycle? But the rate of development of the oöcytes following a blood meal is, the reader will remember, dependent on temperature and may take anything from forty to one hundred hours. How is it then that eggs are laid at the same time every day?

To answer this question groups of eggs of *Aedes aegypti* were

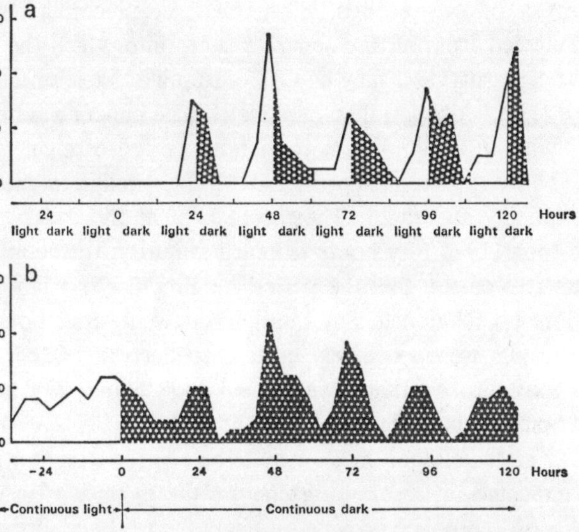

Figure 17. Circadian rhythm of egg-laying activity in *Aedes aegypti: a.* the normal cycle in alternating light and dark; *b.* the conversion from no cycle at all to a regular 24-hour rhythm when constant light is replaced by constant darkness (data from Gillett).

placed in water at short intervals over a period of a few days, the resulting larvae being kept in a strict regimen of LD 12:12. Male and female pupae were harvested from these cultures at intervals of five hours or so and a few from each crop set up in small specially designed containers, one male and one female in each. The population thus became divided up into single pairs, each having continuous opportunity for mating, fruit- and blood-feeding. Since by these precautionary measures it was possible to ensure a scatter of age over a period of a few days, no spurious cycle was imposed on them as a result of simultaneous feeding; harvesting the pupae in small groups every few hours ensured that the adults became ready to feed at different times.

Each small cage was designed in such a way as to allow the removal of eggs with minimum disturbance of the single pair inhabiting it; a wedge-shaped piece of hardened filter paper provided for egg-laying could be slipped out and replaced by another in a matter of five seconds. When all these pieces of paper had been collected from all the separate cages into which the population had been divided, they could be examined at leisure to see if there were any eggs and if so how many. The papers were removed at four-hour intervals over a continuous period of eight days and nights. The cages were kept in LD 12:12, similar to the normal equatorial day and night. They were kept at 29°C at which the ovaries took about fifty hours to reach maturity; it was important, of course, to ensure that the period required for ovarian development bore no relation to any multiple of twenty-four hours.

The results were extremely clear cut: since in each individual female mosquito ovarian development took about fifty hours for completion no single female could possibly lay its eggs until the third day after its blood meal. Any female harbouring mature eggs at the expected laying time just before the lights went out would lay them on the paper provided. Others, as we have already seen, had to wait till the next laying time came round on the following day. What was particularly interesting was that some mosquitos were caught out half-way through and the laying period came to a close before they had got rid of the complete batch. These waited till the same time on the next day to complete the batch. It is rather like people waiting for a bus that runs only once a day; anyone missing the bus on one day has to wait till the proper time on the next. Synchrony of laying in a population then is achieved by delay on the part of each mosquito, the regular daily cycle being a statistical effect of large numbers. The people may gather at the bus stop in ones and twos, but the poor service ensures that each bus is full [280].

What was still far from clear was whether each mosquito making up the cage population had in fact a 24-hour rhythm of activity of its own all along and the external time-cue merely served to bring all the independent rhythms into phase, like the synchronising of 500 clocks that hitherto, although wound up and working accurately, had all been out of time with one another. Some way had now to be found of testing the alternative, which was that none of the mosquitos in the population showed any cyclic tendencies at all until the single external event in their lives —the change from light to dark—triggered off a 24-hour rhythm in each. This alternative hypothesis then was not merely synchrony but the simultaneous winding up of many independent clocks. The net result of either is synchrony, but the one depends merely on bringing many pre-existing 24-hour rhythms into phase, the other the simultaneous starting up of many 24-hour rhythms where none existed before. The next experiment then was designed to answer the question: does the cycle of activity merely depend on bringing many already established independent 24-hour cycles into phase?

A final experiment was set up along the lines of the previous one in which the population was split up into separate breeding pairs, only this time they were reared and kept in continuous light until the fifth day after laying was established, after which they were plunged into darkness for the rest of their lives. Again there was the ever present threat of unexpected thunderstorms upsetting the whole experiment. As in the other experiment precautions were taken against simultaneous feeding by ensuring that maturity of the adults was reached over a period of a few days. Egg counts were made at four-hour intervals for ten continuous days and nights—five in the light, five in the dark.

The results gave a clear answer to the question asked: in the first part of the experiment when the mosquito pairs in their separate containers were kept under continuous illumination each separate female laid its eggs as soon as the oöcytes were mature, which at the temperature provided was about fifty hours after the blood meal. Since in the first place feeding had been spaced over three days as each female became mature following emergence from the pupa, egg-laying from the whole population was also scattered in time. By the time the last mosquito to feed had laid, the first to lay had already taken a second blood meal and later added a second contribution of eggs and so on. When, however, the lights were turned off and continuous light was replaced by continuous darkness laying ceased altogether, even by those females just about ready to lay. Then, twenty-two hours after light-off those that had been caught out almost ready to lay and all those in which the oöcytes had matured in the meantime combined to lay all at once to form a small but clear-cut peak. This was followed by an enormous peak on the next day at the same time, and by medium-sized peaks on the three days following. It is clear, then, that there was no question of the single change from light to dark merely bringing a collection of independent 24-hour rhythms into phase. No mosquito showed any tendency whatever to rhythmic activity until the lights were turned off. Without any external sign each mosquito laid its eggs just as soon as its ovaries became mature some fifty hours after its blood meal. Each mosquito in its individual cage laid independently of its neighbours in theirs. But the one event provided by light-off changed arrhythmic behaviour into rhythmic and since all received this external time-cue simultaneously the newly established rhythms were naturally in phase with one another. From now on the mosquitos would have to wait till the proper laying periods came round at 23- to 24-hour intervals [280].

Let us see if we can now summarise in simple terms what we have so far found out. Each mosquito has its own inborn or endogenous rhythm of activity and rest, one replacing the other in a cycle of about twenty-four hours. This is now usually called the circadian rhythm because the urge to activity overtakes the mosquito at intervals of about, but not exactly, twenty-four hours. It is as if there is a not very accurate, near 24-hour clock within each mosquito which needs to be wound up and put right by some external signal or time-cue. We have seen that it is the change from light to dark that serves this function and that a single change in the life of the mosquito is enough to set the clock going for about five days. In nature of course all the mosquitos in a population receive the signal simultaneously and so individual circadian rhythms once started are always in phase with one another. Furthermore, the signal is received at the end of every day, thus resetting the clock and correcting any tendency to drift. Resetting each day also compensates for changes in sunset time associated with the seasons: since there is a time-lag of about twenty-two hours following sunset on one day and egg-laying on the next, the peak will always occur just before sunset, whatever the time of year [280].

We have then considerable information on the mechanisms underlying the egg-laying cycle in mosquitos that fits well into the general framework of ideas that has been built up over the past twenty years or so on organic rhythms in general. Circadian activity cycles seem to be general throughout the living world [276]. We ourselves are no exception and even cells in isolated tissue cultures kept for years in bottles show the same phenomenon. The circadian rhythm is an adaptation of living things to the special conditions found on this planet rotating on its axis every 23 hours 56 minutes and 4 seconds and moving through space at more than 100,000 kilometres per hour in its orbit round the sun.

Other activities of mosquitos are also influenced by the circadian rhythm but these have not been so amenable to study quantitatively as the egg-laying cycle. The sugar-feeding cycle has been studied by allowing mosquitos to feed on sugar solution from a graduated tube attached to a special head [279]. This arrangement allowed many mosquitos to imbibe the solution simultaneously; feeding activity could then be measured by showing the drop in level of the solution at regular intervals throughout the day and night. In a regimen of LD 12:12 *Ae. aegypti* shows two peaks–one, as might be expected just before light-off, and another just after light-on. Now, in our previous studies on the egg-laying cycle it was not uncommon to find a small supernumerary peak following light-on, but it was considered to result from the direct disturbance of the mosquitos brought about by the sudden change from dark to light [281]. In the sugar-feeding cycle the 'early morning' peak was a regular and very conspicuous affair, comparable in size to the 'late afternoon' peak. This was at first puzzling since it was different from everything that we had experienced in our long series of experiments on various species of mosquitos. The interesting thing is, however, that when we came to look at the records of sugar-feeding in nature, particularly those of J. S. Haeger who had made a detailed study of nectar feeding in *Ae. taeniorhynchus,* it was a common feature for a species to show both morning and evening visits to flowers. This was important and encouraging for it indicated that the conditions of our laboratory experiments were not too far removed from those found in the wild. The early morning sugar-feeding peak thus seemed less likely to be a laboratory artifact than the small bumps up in activity following light-on which we had previously found in the egg-laying cycle. In a long and exhausting experiment in which the mosquitos were reared and then kept for the first part of their adult lives in continuous light before being plunged into continuous darkness, arrhythmic sugar-feeding was converted into rhythmic, the free-running cycle in darkness persisting, as before, for several days. Now, however, only the 'evening' peak appeared [285].

The circadian rhythm is not the same in all species. To be sure, by definition, it always shows a cycle of about twenty-four hours, but the peaks of activity occur at different times of the day or night. The pattern in species active by night differs from that found in species, like *Ae. aegypti,* which are normally active by day. It is not known for certain whether each species has a single period of potential activity in each 24-hour period which is controlled by a single circadian rhythm, and that the actual activity shown, for example mating, feeding or egg-laying or indeed none at all depends on the physiological state of the insect at the time, or whether each of these activities has its own rhythm. In the notorious night-biting *Coquillettidia fuscopennata* feeding reaches a peak soon after sundown but is continued at a fairly high level throughout the night. Egg-laying also occurs throughout the night but reaches a peak just before dawn. It is as if the general circadian activity rhythm persists throughout the night, but different parts of this period are exploited for different ends–egg-laying shortly before dawn and feeding shortly after sundown on the night following [288]. In some nocturnally active species the interval between the light-off signal and onset of activity appears to be much shorter.

Nowadays circadian activity in mosquitos is being studied by somewhat more sophisticated means. Mike Jones has developed apparatus that uses the sound emitted when a mosquito is flying to trigger a remote recording device which can run continuously for several weeks on end. The apparatus runs automatically and the results can be read off afterwards. The method has obvious advantages. For one thing it relieves the experimenters of extremely tedious work. More important, it removes the necessity of having man present at all in the experiment; his continual presence for recording purposes cannot but have a disturbing influence on insects that are dependent on him for their very livelihood. Jones and his colleagues have been able to show that there is actually a free-running cycle under conditions of continuous light as well as in continuous dark, although it is not nearly so conspicuous. But whereas in the dark the free-running cycle has a period of about twenty-two hours, in continuous light it is delayed and appears to be about twenty-six hours. What this means in terms of the normal cycle found when light and darkness follow each other in a regular 24-hour pattern is not yet known [292–4, 298].

In nature the conditions are very much more complex than those provided in the laboratory. That the inborn circadian rhythm is an important element in bringing about the typical cyclic behaviour that we have learnt to expect of mosquitos seems certain; one found no change in behaviour whether the mosquitos experienced natural conditions of day and night with an intervening twilight period, or whether artificial light was abruptly replaced by darkness. It is certain, however, that the circadian rhythm is not the only factor concerned. Overt behaviour in nature is likely to be the result of a complex interplay of endogenous and exogenous influences: light, temperature, humidity and wind will all have their effect, just as local weather conditions at an airport will affect the actual time of departure whatever the printed timetable may say. Nevertheless, it is an interesting thought that if any mosquitos come to worry us tonight, it may well have been yesterday, twenty-two hours ago or so that they were wound up and the alarm set for tonight's performance. Local conditions at the time may make flight impossible; it may be too windy or too cold. Even if flight is started a suitable host or egg-laying site may not be found before the period of activity comes to an end and forced inactivity clamps down again. The circadian rhythm not only serves to brief the mosquito on its take-off time, but determines the duration of the flight period, keeping it within the normal fuel range. Those that do not find a suitable target within the allotted period return safely with fuel to spare which, as we have seen in Chapter 6, is now topped up from reserves in the storage organs, ready for tomorrow's sortie.

The reader may well wonder what bearing all this detailed knowledge about egg-laying can possibly have on our fight against mosquitos as bringers of human misery and disease. In fact it is of very great significance. We have seen how the ovaries of mosquitos ripen following a blood meal either quickly or slowly according to temperature. At 30°C, for example, ripening of the ovaries may take just under forty-eight hours whereas at 28°C it may take a little over this period—a difference of two or three hours only. But at the lower temperature the eggs will not be ready for laying when the circadian egg-laying period or drive comes round and they will be retained by the mosquito for a further twenty-four hours before being laid. Thus an average drop of only two degrees can delay egg-laying by twenty-four hours, even though ovarian development itself has been delayed by only two to three hours.

Since the next blood meal will not be taken until the eggs have been laid, it follows that blood-feeding too will be similarly delayed. In other words, a slight drop in mean temperature can result in a disproportionate lengthening of the gonotrophic cycle and vice versa—a factor surely of the utmost importance in the epidemiology of malaria and other mosquito-borne diseases. At larger temperature differences the intervals between feeds will be doubled in length, while the frequency of transmission of disease will be at least halved. Much has been written about gonotrophic cycles and epidemiology, but in nearly all this work the importance of the circadian rhythm has been entirely overlooked.

1. A cluster of *Mansonia* eggs on the underside of a floating leaf of *Salvinia*.

2. Stereoscan electron micrograph of a small portion of a *Culex* egg. The picture shows clearly the micropyle through which sperm enter the mature egg as it passes down the oviduct on its way to the exterior.

3. Stereoscan electron micrograph of a small portion of the antenna of *Toxorhynchites* sp., magnified 170 times.

4. Stereoscan electron micrograph of a portion of the compound eye of *Toxorhynchites* sp., magnified 880 times.

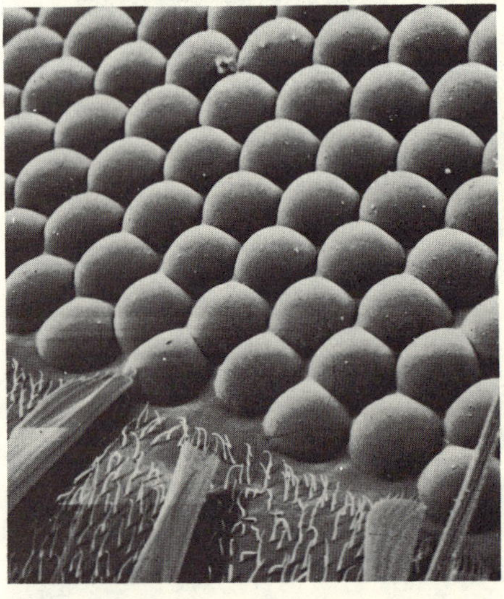

5. Stereoscan electron micrograph of scales and other surface details on a small portion of the thorax of *Toxorhynchites* sp., magnified 1800 times.

6. Tire dumps left after military occupation of the Pacific island of Guam; tires retain water for long periods and are common breeding places for *Aedes aegypti* and, in some places, for *Culex pipiens fatigans*.

7. Extensive breeding ground of arctic *Aedes* in Labrador.

8. *Mansonia africana* in the act of feeding; note the typical culicine stance with the abdomen more or less parallel with surface. The sugar-filled ventral diverticulum can be clearly seen lying below the blood-filled mid-gut.

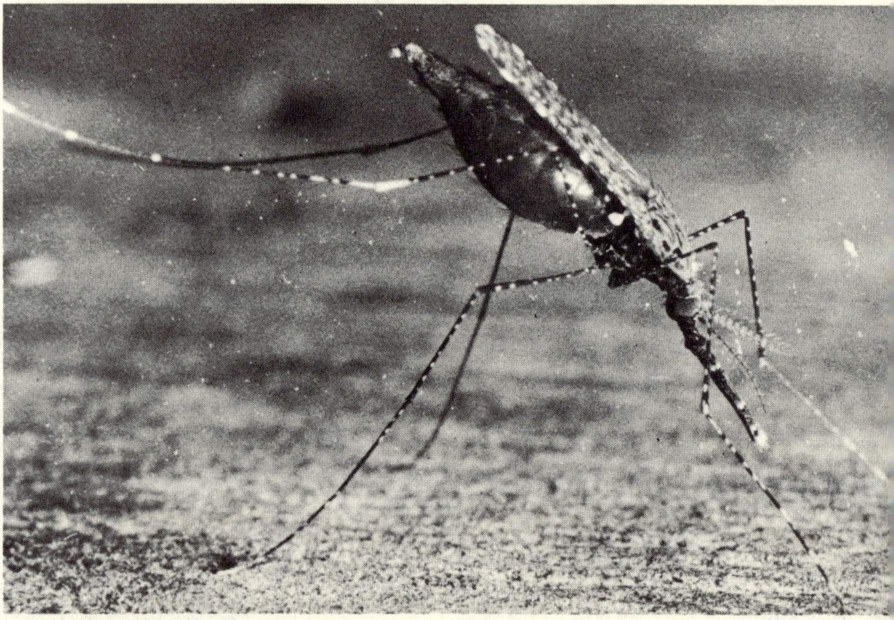

9. *Anopheles balabacensis* resting after a blood meal; note the typical anopheline stance—"almost standing on its head."

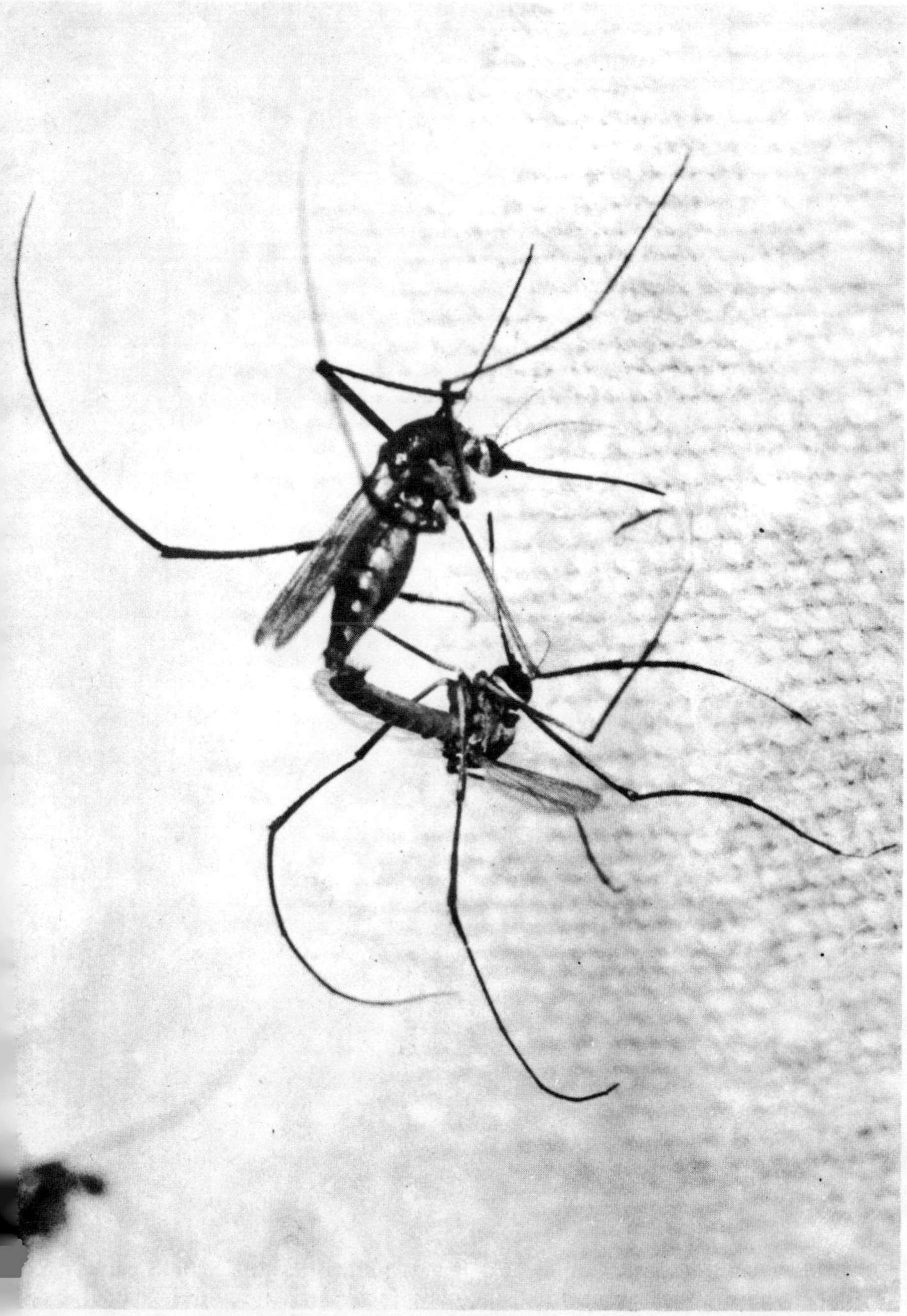

10. *Sabethes chloropterus* mating. The male is upside down underneath the female.

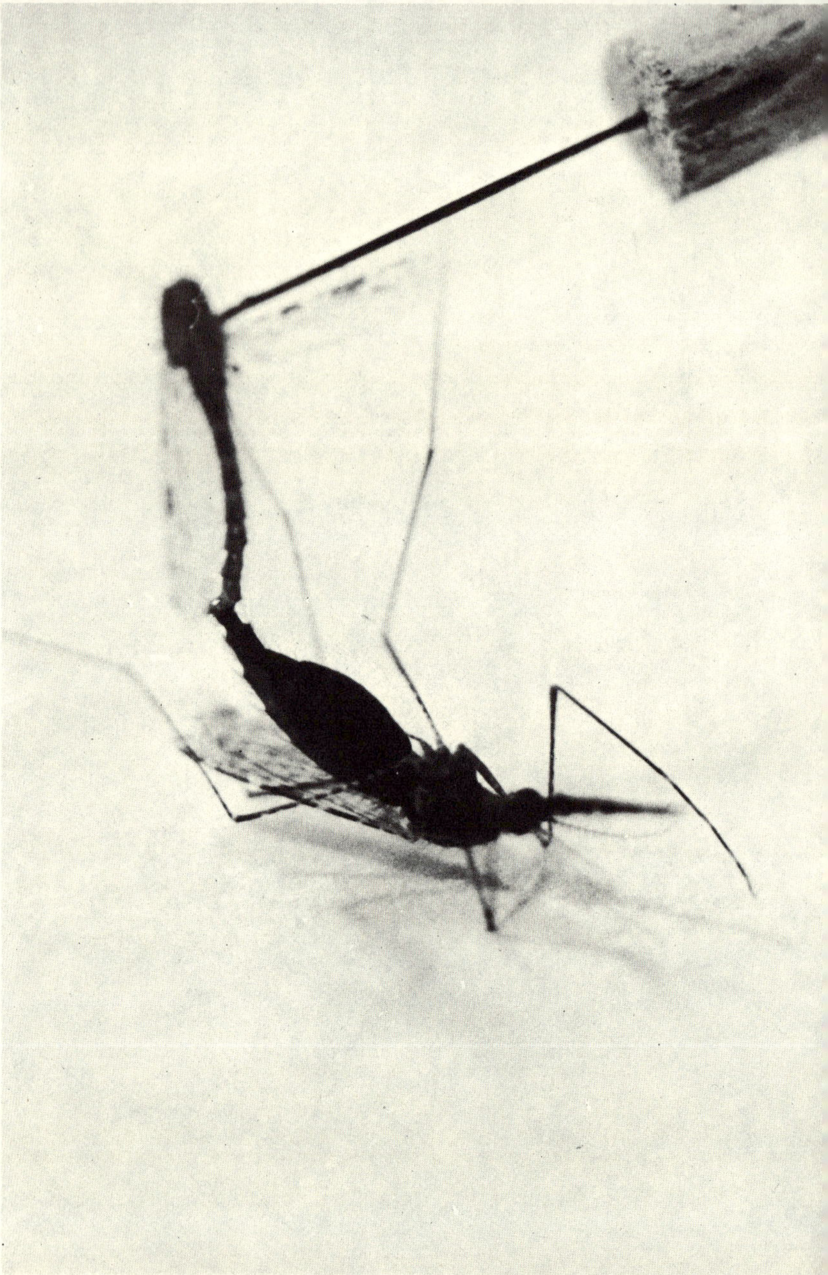

11. Artificial insemination of a female by headless male. Both partners have been anaesthetized; the male (above) has been decapitated and, for convenience of handling, impaled on a fine needle. Copulation is then automatic as soon as the male and female genitalia are brought into contact (the male, still under anaesthesia, is killed and discarded immediately after insemination is complete).

12. Some species of mosquitos feed almost entirely in the treetops and are responsible for the transmission of yellow fever virus from monkey to monkey. The picture shows a catching platform in the canopy twenty-five meters above the forest floor, Bwamba, western Uganda.

13. The author's arrangement for studying circadian activity in mosquitos. A small population of mosquitos was separated into pairs (I♂ and I♀) and provided with continuous opportunity for mating, sugar-feeding, blood-feeding and egg-laying (each mouse "working" for a single 4-hour stretch only). Observations, manipulation and recording had to be made every four hours over periods of up to ten consecutive days and nights.

14. The more modern setup devised by Dr. M. D. R. Jones at Brunel University achieves similar results automatically; the apparatus allows continuous recording of individual mosquito activity through thirty-six independent channels which may be left more or less unattended for ten consecutive days and nights.

10. STRAINS AND SPECIES

When we speak of *Aedes aegypti, Anopheles stephensi, Culex pipiens* or of any other species of mosquito, or indeed of any animal or plant, we are usually fairly clear in our minds about what we mean. Yet when we come to look closely at any one of these species we find differences between individuals, colour and pattern being the most easily noticed. Moreover, on even closer examination, we find that the proportion of one form or another in any population of the species tends to be different in certain regions or in certain situations. In this chapter we are going to have a look at some of these differences, for some of them are not only of academic interest but have far-reaching consequences in the affairs of man.

The momentous discoveries which showed that mosquitos were responsible for the spread of filarial worms to man and for the twin scourges of malaria and yellow fever made us realise how little was known about mosquitos in general. The common species were known and had been described and named, but our knowledge of mosquitos on a world-wide basis was pitifully inadequate; not enough was known about the group in general to provide sufficient material for an all-embracing classification. It fell to F. V. Theobald to take on this huge task and in the first ten years of this century he produced his now classic *Monograph of the Culicidae*. In 1901 he described an extremely pale form of *Ae. aegypti* from Queensland, Australia. It was so distinct that he named this variety *Ae. aegypti* var. *queenslandensis.* Later work showed that this pale form was not confined to Queensland but occurred widely in the warmer regions of the world.

Typical *Ae. aegypti* are brownish with characteristic brilliant silver and white markings, but in Africa an extremely dark form occurs, which was first described by F. Walker way back in 1848. Walker made a special study of the mosquitos in the collections of the British Museum, and noted a particularly dark form that came from Sierra Leone. He named it *Culex formosus* on account of the striking beauty of its contrasting black and silver markings. A hundred years later Peter Mattingly took a further look at the collections in the British Museum, which by now were much more comprehensive, and decided that there are indeed three main forms of this widespread species: the brownish *Ae. aegypti* type-form, the black *Ae. aegypti* var. *formosus* and the pale *Ae. aegypti* var. *queenslandensis.*

Everything would be easy if all dark or pale *Ae. aegypti* were equally dark or pale as the case may be. But this is far from being true. While the two dark forms (brown type-form and black *formosus*) tend to come from different regions of the world, there are places where every degree of intergrade exists between them. Andy McClelland, who has studied the genetics of this species has recognised more than twenty grades and it is sometimes clearly impossible to determine whether a particular mosquito belongs to one or the other. Nevertheless, the recognition of these forms is important even if they only become meaningful when we are dealing with the extremes of a continuous spectrum of variation.

Similar difficulties arise when we look closely at the pale form of *Ae. aegypti,* the so-called *queenslandensis:* while some of the pale forms are very distinct indeed, every conceivable intermediate is known. Indeed there comes a point at which the experts say something like this: I shall define *queenslandensis* as having such and such a number of pale scales with such and such a distribution. How meaningful it is to do this is another question. If the 'true' *queenslandensis* is merely the extreme of a whole spectrum of variation, then the point at which one proclaims that a single individual mosquito is this or that is purely arbitrary. Even if we really are dealing with two geographical subspecies which hybridise wherever their ranges overlap one still has to make a subjective judgment on which is the 'true' subspecies *queenslandensis* and which the hybrid; where does the one leave off and the other begin?

All this variation would be of little more than academic interest if there were not also differences in behaviour associated with it. As long ago as 1921 G. F. Hill noted that in Australia the pale forms were usually found inside buildings, whereas the darker forms came from larvae found in tree-holes and other outside containers. Hill actually found the pale forms inside the hospital at Townsville, Queensland, and the dark forms outside in the hospital compound [309]. The interesting thing is that similar findings have since been recorded from several other parts of the world, and wherever both dark and pale forms coexist it is the pale forms, or populations containing a high proportion of these forms, that are the trouble-makers. Paleness in *Ae. aegypti,* it seems, is linked to domesticity and to an accompanying propensity for feeding on man. In 1941 I spent some months in Bwamba in western Uganda during an epidemic of rural yellow fever there. Surprisingly, although *Ae. aegypti,* the classical transmitter of the disease, was present in the area the adults were not feeding on man, or so rarely as to be safely discounted as playing any part in the spread of the disease, which, as it turned out, was being undertaken by another species of mosquito altogether. Indeed, the presence of *Ae. aegypti* could only be judged by finding the larvae. These were in fact common enough, but were hardly ever to be found in the usual domestic containers, whether inside the houses or outside; they were almost entirely confined to tree-holes. Even when tree-hole-breeding of *Ae. aegypti* was going on in the village itself, the adults rarely attacked man. We know now that the Bwamba *Ae. aegypti* were the very dark *formosus* and similar non-man-biting populations of this variety have now been recorded from other forest and rural areas of Africa.

It is important to decide how far these differences in colour and scale distribution are a result of the inherited make-up of the population, that is the local gene-pool, and how far they are the result of upbringing. Hill found that his pale forms from inside the hospital bred true, whereas when he took the outside dark forms and reared them inside these produced both dark and light forms and a whole range of intermediates. It is a pity that he did not do the reciprocal experiment of rearing his hospital pale forms under outside conditions. One cannot really tell from his experiment whether outside and inside conditions selected out and removed from the population pale and dark forms respectively, or whether darkness and paleness are merely phenotypic effects. Both effects could be operating simultaneously. Happily, other work has helped to clarify this important question. Dark *Ae. aegypti* from Lagos, West Africa, reared indoors remained 100 per cent dark even after ten years. At the same time another indoor colony of *Ae. aegypti,* coming from Newala, East Africa, with a high proportion of *queenslandensis* of varying degree of paleness, maintained their distinctive paleness for at least three years, after which the colony was discontinued. The pale Newala colony was reared alongside the dark Lagos one. Under these indoor conditions there was no drift towards one form or the other; the differences were evidently genetically determined, both forms breeding true.

Colour and pattern are not necessarily of particular importance in themselves. They are, however, easily observed and serve as convenient markers from one generation to another in any genetic study. But we have already seen that differences in behaviour appear to be associated with these differences in colour. When we were dealing with the hatching of *Aedes* eggs in Chapter 2 we mentioned that females from East Africa tended to lay their eggs in a series of sub-batches over a period of days, whereas those from West Africa tended to lay them all at once in a single batch. We saw also how this difference in laying behaviour was correlated with a difference in the hatching response; eggs of the East African strain tended to hatch as soon as they reached maturity, whereas those of the West African strain hatched in small instalments over a period of many weeks. The difference was not absolute. Whereas eighty-five per cent of the East African eggs hatched during the week following completion of embryogenesis, only some four per cent of the West African eggs hatched during the same period. In the East African eggs, which were laid in instalments on successive days, hatching tended to be spatially scattered. In the West African eggs, which were laid together all in one place, hatching was scattered temporally.

The East African mosquitos came from Newala and as we have seen contained a high proportion of pale forms of varying degrees of paleness. The West African mosquitos came from Lagos and were all dark. It would have been interesting to know if the quickness of hatching of the eggs in the Newala strain was associated with the degree of paleness. Breeding experiments showed that the differences in the hatching response of the Newala and Lagos mosquitos was maintained from one generation to the next; they bred true. When, however, quick-hatching Newala males were crossed with slow-hatching Lagos females a range of hatching resulted from one hundred to four per cent; that is as high as the Newala parents and as low as the Lagos parents and everything in between. A similar picture emerged with the reciprocal cross. It is clear that these differences in behaviour between two strains of the same species are genetically determined [31].

In Chapter 8 we noted that although egg development follows the taking of a blood meal and is independent of mating, the mature eggs nevertheless remain unlaid unless the female has mated. At least this is true for the Newala strain of *Ae. aegypti*. Fertilisation in insects, we remember, does not occur until the moment of laying when sperm stored by the female pass to the eggs as, one by one, they pass the opening of the spermathecal duct. It is clearly advantageous for unmated females to retain their eggs on the chance that mating may occur later. Nevertheless, the Lagos strain of *Ae. aegypti* does just the opposite: unmated females will lay their mature eggs at the first opportunity regardless of the fact that they will never hatch. Once again, however, and this is the important point, this difference in behaviour between the two strains of the same species is not absolute. Even the unmated females of the Newala strain may lay up to about eight per cent of the eggs which they have developed, while those from Lagos usually retain up to about fourteen per cent of theirs [304]. By reciprocal crosses and by back-crossing the hybrids both to pure Lagos and to pure Newala individuals it has been shown that retention of the mature eggs by virgins is an inherited property. It seems that whereas in the Newala females production or release of an ovulation hormone is triggered off when the female receives the secretion of the male accessory glands at the time of mating, in the Lagos females this hormone is produced or released independently of this stimulus [306].

It is evident that there are deep-seated differences between these strains of *Ae. aegypti,* differences that affect their physiology and behaviour as well as their outside colour and pattern. We have seen in Chapter 2 how egg-laying behaviour in American *Culiseta morsitans* differs from that of their European representatives. In Bwamba, wild *Ae. aegypti* do not feed on man, and *Ae. simpsoni* has taken over the role of domestic mosquito. It breeds in the small quantity of water held up in leaf axils of pineapples, cocoyams or *Colocasias* and of certain varieties of banana, all three of which crops are grown alongside houses. *Ae. simpsoni* also breeds in the axils of *Dracaena* and *Sansevieria;* these are also planted close to houses, one as shelter and for demarcation of boundaries, the other for making string. But the important thing is that *Ae. simpsoni* not only breeds in enormous numbers near to human dwellings but the adult mosquitos feed on man. It was in fact *Ae. simpsoni* that was responsible for the 1941 epidemic of yellow fever which we mentioned earlier.

The plants are common enough crops around human habitations in many parts of Africa, including much of the rest of Uganda, and *Ae. simpsoni* breeds in them just as it does in Bwamba. But in many of these other places the adult mosquitos simply do not feed on man. Thus, as with *Ae. aegypti,* we find that there are man-feeding and non-man-feeding populations of *Ae. simpsoni* [394]. So far, however, no morphological differences have been found between them. This is not to say that there is no way of telling them apart, only that we have so far failed to find one. Long and careful biometrical studies could conceivably reveal some minute but constant difference or that such and such a variant tended to be commoner or even to predominate in this population or that. Whether they do or they do not the fact remains we have a difference in behaviour that is of first importance to man. *Ae. simpsoni* has recently been responsible for another epidemic of yellow fever [427]. This time it was in Ethiopia and the deaths alone have been estimated at 30,000. Yet in the museums man-biting and non-man-biting *Ae. simpsoni* must continue to be labelled as a single entity. One begins to wonder sometimes at our own priorities and prejudices.

It has been suggested that *Ae. simpsoni* feeds on man only in those places where some preferred host is absent. This indeed may sometimes be the case. But the distribution of *Ae. simpsoni* is discontinuous; it occurs in patches of cultivation here and there, some large and some very small. Some of these islands of cultivation are separated from the next by many kilometres of surrounding bush country. It may well be that host preferences have arisen in the first place owing to differences in availability in these isolated populations, but such preferences will in time be built into the population by the process of natural selection; individual insects with an innate preference for the most abundant host will naturally tend to produce more offspring than their less fortunate companions and in the fullness of time come to make up the bulk of the population. It could also be that in some instances these differences have arisen by the chance colonisation of distant small areas of cultivation by a single mosquito carrying the genes for a preference for this host or that [395].

Ae. simpsoni has until recently resisted every effort at laboratory colonisation, and so genetic experiments have not been easy to carry out. Indeed, when we come to think of it, only a handful of species has been successfully colonised, the usual obstacle being that the mosquitos will not mate under the artificial conditions provided. This difficulty, as we have seen in an earlier chapter, can be circumvented by resorting to the technique of artificial insemination. But the method is usually too cumbersome and laborious to be used for routine breeding. Nevertheless Louis Mukwaya has successfully employed this technique to carry out a series of genetic experiments with *Ae. simpsoni* and these indicate that the tendency to feed on man or not is indeed an inherited character. When offered a choice of alternative hosts, females reared from material originally collected in Bwamba usually choose man, while those originating from other parts of Uganda choose rats.

These differences between populations of a species, whether physiological, behavioural or simply morphological, are rarely absolute. There are usually exceptional members of a population that do not conform. This is turning out to be a recurring theme throughout this book and rightly so. It is evident that we must consider populations and not merely the individual insects. For whether we are dealing with colour, with man-biting or non-man-biting or with egg-laying or egg-hatching or with any other property it is the statistical balance within the population that counts. When we speak of man-biting we really mean *mainly* man-biting. It is wholly unimportant in this context whether a few individual insects feed on man in an otherwise non-man-biting population, or a few virgins lay their eggs. The non-conformers may make up the residual pool providing the material basis for survival when conditions change, but as things stand they are of no importance.

About the time that Hill published his important observations on dark and pale *Ae. aegypti* in Australia some surprising things were happening elsewhere. *Anopheles maculipennis* had for long been known as the main malaria mosquito in Europe, the disease being common enough in the summer months in Italy, Greece, Albania, France, Holland, Poland and even in places as far north as Denmark, anywhere, it seemed, that provided suitable breeding places for *An. maculipennis* and shelter in which the adults could hibernate for the winter. It was noted, however, that in certain regions there was no malaria even though the species was present in sufficient numbers to maintain and spread the disease. As so often happens this discovery was made independently and more or less simultaneously by more than one observer. Grassi in Italy and Roubaud in France decided that there must be man-biting and non-man-biting strains of *An. maculipennis,* while Wesenberg-Lund, whose monograph on the mosquitos of Denmark has become a much sought-after classic among the early mosquito literature, thought that the species had become adapted to feeding on horses and cattle. It turns out, as we would now expect, that the former interpretation was nearer the mark. Slight morphological differences have since been found associated with these behavioural differences (mostly confined to the eggs, larvae and males) and subsequent breeding and genetic work has shown that the man-biting and non-man-biting forms must really be considered as different species. In fact the mosquito that used to be called *An. maculipennis* turns out to be a complex of six distinct species with important differences in behaviour and genetic barriers of increasing effectiveness separating one from another.

The six species that are now recognised as making up the *maculipennis* complex are *An. labranchiae, An. maculipennis, An. sacharovi, An. atroparvus, An. messeae* and *An. melanoon.* These are kept separate in nature by sterility barriers and thus remain distinct even where their ranges overlap. For example, if male *An. atroparvus* are crossed with female *An. messeae* most of the eggs that result fail to hatch and those that do, produce larvae that die soon afterwards. If, on the other hand, male *An. atroparvus* are mated with female *An. sacharovi* the females of the F_1 generation all die during development and one is left with a population consisting entirely of males. Since these males are in any case sterile they cannot even reproduce if they subsequently mate with the parent generation. If male *An. atroparvus* are crossed with *An. maculipennis* the F_1 adults of both sexes are sterile, but if these males are crossed with *An. melanoon* or with *An. labranchiae* a proportion of the F_1 females are normal and can be back-crossed to *An. atroparvus* males. Marston Bates who made a special study of *An. maculipennis* and its allies in Europe observed that the members of this species complex differed profoundly from one another in every respect except appearance [1, 331].

In England *An. atroparvus* breeds in coastal waters often of high salinity. It is this 'form' that sometimes takes a blood meal in the middle of winter, an unduly warm day rousing it from its state of partial hibernation in a house or stable. It is also the member of the *maculipennis* complex that is responsible for occasional outbreaks of malaria in coastal regions. *An. messeae* on the other hand is found near inland streams and pools, and hardly if ever feeds on man. On the approach of winter blood meals are no longer followed by egg production but are switched to developing the storage organs known as the fat-body which serve to sustain it for the whole winter without any need for a further meal [4].

A similar story emerges almost every time a species is adequately studied. In the *maculipennis* complex only *An. atroparvus* males will mate under artificial conditions, and hence all the hybrids listed above had this species as the father. In the Pacific region the *An. punctulatus* complex includes *An. farauti* and *An. koliensis*. Again it is difficult to separate one from the other, although there are slight morphological differences, especially in the larvae and pupae. Yet there are clear-cut differences in behaviour: so far it has not been possible to persuade *An. farauti* females to feed in captivity [4]. Similar cryptic forms are to be found in the *An. funestus* complex, which now includes no less than eight species [11, 72], and even among such well-known names as *An. stephensi* and *An. gambiae*. Typical *An. gambiae* and the dark form known as *melas* will hybridise either way, but the F_1 males tend to be sterile. The *gambiae* complex is now recognised as including three distinct species: *An. gambiae, An. melas* and *An. merus;* two others have yet to be named [301].

In the forests of Bwamba *Eretmapodites chrysogaster* feeds at a fairly steady rate throughout the hours of daylight, but near Entebbe it behaves quite differently: while there is a certain amount of feeding activity during the day this is crowned by an enormous burst of activity just before sundown. It is not known if this rather astonishing difference in feeding time has a genetic origin or is merely the result of differences in the forest environment [308]. It should be mentioned that *E. chrysogaster* has also been broken down into a complex of closely related species (*E.*

chrysogaster sensu stricto and eleven others). The members of this complex are hard to tell apart except by the male genitalia. Do the Entebbe evening-feeders constitute yet another member of the complex, even though the males are indistinguishable?—a strain perhaps or a cryptic species? Perhaps we are wrong in giving the members of this complex specific rank merely because of slight differences in the male sex organs; could they not simply be variants in a single polymorphic species?

Artificial, self-perpetuating colonies, whether of cells in tissue culture or of mosquitos are likely to differ from wild populations. The new living conditions are hardly likely to resemble even remotely the complex conditions found in nature, however superficially they may appear to simulate them. The requirements for survival will be different right from the start and by the time the colony has been established the battle for survival has been won; those individuals that are fitted to withstand or are favoured by the new living conditions and by the new regimen imposed on them will eventually take over. The selection process will be greatly accelerated and after a short time, even a generation or two, the genetic make-up of the new population is likely to have undergone a shift in one direction or another. It follows also that, unless the conditions in two colonies of the same species are truly identical, one is likely to differ from the other. Thus, in addition to wild strains associated with different regions, we have various laboratory strains to take into account. Until recent years the importance of the strain used in experimental work has not always been realised. *Ae. aegypti* and other colonised species all too often figure as experimental animals and the assumption is made that the results of the experiments are directly applicable to the species in the wild. It may well be, however, that the results cannot even be repeated in the laboratory when other strains are used.

While species tend to become split up into strains, varieties and subspecies whenever we come to study them closely, or even into separate species to make up a species complex of the kind we have been considering, this is not always the case. Indeed, in the *Culex pipiens* complex, the best known of them all, the events leading to its recognition have happened the other way round: what were once regarded as separate species have now been brought together as subspecific divisions of a single almost world-wide species. *C. pipiens pipiens* is the well-known house gnat of Britain and most of Europe. It is also found in the southern parts of Africa and South America. The larvae of this mainly bird-feeding member of the complex are found in the summer-time in rainwater butts, drinking troughs and ornamental pools; the adults enter houses in large numbers for shelter shortly before the onset of winter. In Australia *C. p. pipiens* is replaced by *C. p. australicus,* while in Japan *C. p. pallens* is the local representative [314].

In the warmer parts of the world, including both the New and Old World Tropics, *C. p. pipiens* is replaced by *C. p. fatigans,* which for a long time was considered to be a distinct species. It differs from *C. pipiens sensu stricto* in several respects, its man-feeding habits making it one of the most important mosquitos in the world; in the Americas it transmits the virus of St Louis encephalitis, while in tropical regions it is one of the main transmitters of a filarial worm to man, multiple infections of which may lead to the grossly disfiguring and incurable condition known as elephantiasis. The larvae of *C. p. fatigans* may be found in water of the foulest sort. In urban areas they occur in gulley traps taking house waste of all kinds; in rural areas they thrive in pit latrines.

C. p. pipiens and its tropical representative *C. p. fatigans* are so widespread and so common and indeed familiar that each has been named over and over again by well-meaning enthusiasts. During the last 200 years *C. p. pipiens* has received more than twenty different names, and *C. p. fatigans* in somewhat less time has received no less than forty; in the United States *C. p. fatigans* is known as *C. p. quinquefasciatus,* a name given to this mosquito in 1823.

In Europe and throughout much of the range of *C. p. pipiens* there occurs a member of the complex now known as *C. p. molestus.* While this subspecies is almost if not identical in appearance with *C. p. pipiens,* it differs in other important and rather strange ways [300]. Whereas *C. p. pipiens* mates on the wing—or at least the pair first connect during flight, males of *C. p. molestus* seek out and mate with resting females. The type form rarely if ever feeds on man, but *C. p. molestus,* as its name implies, not only feeds on man but at times becomes a real pest, as those who tried to sleep in London's underground tube stations during the Second World War will remember. But stranger than all these differences is the ability of *C. p. molestus* to produce its first batch of eggs without having to take a blood meal. This autogenous production of eggs is, as it were, against all the rules. In the same year that Theobald described his pale forms of *Ae. aegypti* in Australia he also noted that newly emerged *C. pipiens* (as he called them) will sometimes lay eggs before they have had their first blood meal. Thirty years then passed before Roubaud in France showed that autogenous *C. pipiens* bred true and that the species could be divided clearly into autogenous and anautogenous populations.

The mechanisms involved in autogeny are particularly interesting and have been studied in detail by Alan Clements at Cambridge and by Don Twohy and Lloyd Rozeboom at Johns Hopkins. Well-fed mosquito larvae, as we have earlier seen, carry over sufficient reserves to provide material for the production of some eggs, but mobilisation of these reserves depends on the release of a gonadotrophic hormone. Autogenous *C. pipiens* or, as we must now call them, *C. p. molestus,* release this hormone without the need of the stimulus that is provided by the blood meal. We have already seen how the Lagos strain of *Ae. aegypti* come to lay most of their eggs without the stimulus provided by mating; the two situations are analogous in every respect.

The kind of differences with which we have been dealing—and we could go on listing many more—make work distinctly awkward for any taxonomist who likes his species cut and dried. It is less difficult when differences in behaviour can be tied to differences in morphology, but when they occur alone without any clear-cut morphological differences associated with them, they are awkward indeed, and their importance has tended to be played down [312,

313]. Like our fellow-primates we are essentially creatures of vision, obsessed by what we can see. There is, however, no reason whatever for supposing that an extra white spot here or the absence of pale scales there, or whether the larvae are pale-coloured or dark, has any greater genetic significance than differences in the hormonal regulation of ovulation or the tendency to feed on man. All are determined ultimately at the biochemical level. The predilection of the geneticist for visual markers is mainly one of convenience and not entirely uninfluenced by tradition. Coloured larvae are so much easier to sort and to trace from one generation to the next; they can merely be counted or removed, whereas differences in behaviour or physiology must be tested at every step in the procedure, a long, tedious and often unrewarding business.

In this intraspecific variation, whether merely of colour or of behaviour or both, we are witnessing evolution in action. One variant is favoured here, another there. Whether these differences will be maintained depends on how exacting are the conditions favouring the one or the other and on how long one population remains isolated from the other. It also depends on how far each has deviated from the common plan. These variants then, and indeed species themselves if you like, can be considered as stages in the evolutionary process. Some of them will be passing stages, others more or less permanent–permanent, that is, until drastic alteration in the living conditions either produces further change or blots them out altogether. They may thus be the last stages or what we might call climax species. In accepting the concept of evolution by natural selection we see that this intricate jig-saw of strains and species making up groups and complexes in space as well as in time is exactly what we would expect to find. Evolution is not a static business, a thing of the past. It is a dynamic process and is happening all around us; a process that can be comparatively fast or incredibly slow.

If differences in the environment acting through natural selection can produce shifts in one direction or another leading to the establishment of strains, subspecies and even to new species, it does not follow that constant conditions will necessarily lead to genetic uniformity. On the contrary, a population made up of individuals with a mixed genetic background is sometimes better suited to local conditions than one that is more homogeneous. This phenomenon, known to geneticists as heterosis, usually manifests itself by a general increase in fitness to the environment: for example, resistance to disease or success at mating under local conditions or merely longevity. Sometimes, however, increased fitness is brought about by the maintenance of more than one form in a local population, a phenomenon known as polymorphism.

We have seen how survival in tropical *Aedes* depends on maximum variation in the hatching response of the eggs; the depth and duration of the depressed hatching response varies from egg to egg even within a single batch. Hatching depression is an inherited property and we can see how this maximum variability comes about. Inbreeding by batch-mate brothers and sisters is prevented or minimised since three months or more may separate the first and the last to hatch. When two or three eggs hatch from a single batch during a particular flooding of a tree-hole, plant-axil or other small container a similar number will also hatch from hundreds of other batches scattered around. Some of these will be early hatchers, others late and others anywhere between the two. While this may result in a very large output of adults at any one time, their parentage will be mixed in the extreme; rarely will males meet females from the same batch.

It is evident that, whereas in arctic *Aedes* we have a high measure of genetic uniformity leading to a prolonged period of diapause or depression and fitting them for the rigours of the northern winter, in the tropical representatives of this huge genus we have just the reverse: maximum heterogeneity of response fitting the eggs for survival in small and precarious breeding places. It is an extreme but elegant example of a balanced polymorphism producing maximum variation in a single population.

11. PARASITES AND PREDATORS

Mosquitos, like the rest of us, have their fair share of enemies. Many of these play an important part in the natural control of mosquito abundance, and with fair regularity advocates have appeared on the scene to press the case of this one or that as an ally to be encouraged. Usually one particular enemy is selected in total disregard or even ignorance of the vast array of other enemies that confront the mosquito at all times. Anyone working on living mosquitos soon discovers for himself some of the more conspicuous of these enemies. Predators usually come to his attention when they are inadvertently introduced to a colony of mosquitos; within a day or two a few healthy looking beetle or dragonfly larvae come to replace what was a thriving culture of mosquito larvae, or a single bloated spider now sits in the corner of a cage that was previously the home of adult mosquitos. Before long the unwary worker to whom all this is still fascinatingly new finds himself side-tracked into a study of the feeding capabilities of the predator with an eye to harnessing it as an agent to be pressed into service in man's fight against the mosquito.

The argument usually goes something like this: let us first find out how many mosquitos a single predator can consume in a day. This is usually worked out by presenting the dragonfly larva or the spider, or whatever, with a plentiful supply of mosquitos as food. If the predators happen to be other mosquito larvae, such as one of the many species of *Toxorhynchites* that are commonly met with throughout the tropics and subtropics or a member of the subgenus *Lutzia* such as *Culex tigripes* in Africa or *C. halifaxii* in the Solomon Islands, so much the better. Now we must make the assumption that because these are eaten readily in the laboratory (usually no other food is provided) they will also be the chosen prey in the wild. A little simple arithmetic is all that is needed to complete the exercise and, eureka, an estimate can be made of the number of predators that need to be released in order to eradicate the local population of mosquitos or so reduce it in size as effectively to break its contact with man. Over a hundred articles were published more or less in this vein during the first third of this century, and they have continued to appear in a steady drip ever since.

Nothing, of course, could be further off the mark. Conditions in the laboratory usually bear extraordinarily little resemblance to those in nature. For one thing we must now reckon with many different kinds of predators, not just a single species. In any case the size of the predator population depends largely on the number of prey available. While it is true that a rise in the number of predators will be followed by a fall in the number of prey, a fall in the number of prey is automatically followed by a fall in the number of predators; there is not enough readily available food to support the larger population. This would not matter if a new low level of both predator and prey were to be maintained. Unfortunately, as soon as the number of predators sinks below a certain level, recovery of the prey population sets in and the stage is again set for another natural rise in the number of predators. The two waves go on chasing each other endlessly in this fashion.

Various factors interfere with the smoothness of these waves. Weather conditions favourable to the predator or to the prey or to enemies of the predator or to other enemies of the prey, will make for sudden and irregular bumps and secondary troughs in the waves. From time to time a combination of all these variables and others that we know little or nothing about will result in unprecedented numbers of mosquitos, a veritable plague of this species or that. They will also at times combine to spell near disaster for a particular species, at least over a limited area until reinvasion restores the situation. Such events may be separated by a hundred or a thousand years or a hundred thousand years; the life-span of a human being is as insignificant in these matters as a single second in a day.

Some of the factors affecting prey-predator relations are sufficiently well known to allow the erection of mathematical models, which can lead to fairly accurate predictions. While it is possible to demonstrate these elegantly enough in the laboratory, difficulties arise when one attempts to apply them under field conditions where the prey has to contend with a multiplicity of predators and each predator has other sources of food at its disposal and its own enemies as well. Nevertheless, there have been some successes and the detailed work of Marshall Laird deserves special mention: with the sponsorship of the Vector Biology and Control Section of the World Health Organisation, he made a thorough study of the natural enemies of mosquitos on a world-wide basis.

Biological control agents have, of course, been used with notable success in combating various pests of crops. Usually, however, it is necessary to ship the predator, parasite or pathogen from some faraway country, otherwise counter-agents will be found already to have established themselves long before. But it is one thing using imported enemies on pests of our man-made monocultures, quite another on mosquitos in their natural strongholds. If I have tended to overstate the case against man's possible use of parasites and predators it is because I am anxious to put the subject into realistic perspective in order that we may take a more balanced look at this fascinating but difficult subject. We are now going to consider particular examples of enemies of mosquitos and try to gain some idea of the overall problem from the mosquito's point of view.

Found in most parts of the world, but curiously absent from South America, is an aquatic fungus known as *Coelomomyces*. Some twenty species of this fungus are known and nearly all are parasitic on mosquitos. They are commonly found naturally infecting larvae of *Aedes, Aedeomyia, Anopheles, Armigeres, Cu-*

lex, Culiseta, Psorophora and *Uranotaenia.* The mycelium of the fungus branches out in the thorax of the host and spreads to the abdomen where it produces fruiting bodies or sporangia. These invade and come to fill pretty well the whole body cavity of the posterior segments of the larva. Heavy infections lead to destruction of various organs including the imaginal buds, those pockets of undifferentiated embryonic tissue destined to give rise to the adult organs [341]. The individual sporangia are about fifty microns in diameter and are usually yellow in colour. A larva may be so packed with them that it takes on their colour and can be recognised as infected even with the unaided eye. The sporangia escape from dead larvae but, like the eggs of *Aedes* and *Psorophora,* usually fail to develop further unless they have first been dried. Indeed, the similarity is even closer, for only a few may germinate each time the sporangia are submerged.

Certain bacteria are sometimes found in the haemolymph of mosquito larvae, where they multiply by division and may come to distend the whole larva almost to the point of bursting; larvae of *Armigeres suballatus* seem to be particularly susceptible to heavy infections and on being touched with a needle burst open like an over-ripe boil to exude a characteristic milky fluid.

Protozoal infections of larvae are not so common, but they include ciliates and microsporidia. A larva may pick up the resting stage of one of the ciliates while feeding. Once inside, the parasite is stimulated into activity; it burrows through the wall of the gut and multiplies in the host's body cavity. Ciliates have been found in larvae of *Aedes, Armigeres, Culex, Culiseta* and *Wyeomyia.* Infection with these parasites and with microsporidia usually lead to the death of the larva. Heavy external growths of other protozoa and of algae too are found from time to time. Large areas of a larva may be completely forested with the fascinating stalked cups of *Vorticella;* these, however, are really epibionts, receiving support and transport from the larva but otherwise doing it little or no harm. They are left behind when the larva breaks out of its old skin at the time of pupation [1].

If the parasites of larvae are relatively few, the predators more than make up for them. Indeed, there are so many that one sometimes wonders how any larva ever manages to complete its aquatic life. *Hydra,* planarians, rotifers, leeches and mayfly larvae all exact a steady but conspicuous toll. But the real villains among the invertebrate predators are various species belonging to seven separate families of predacious bugs: the water-boatmen, water scorpions, pond-skaters and other aquatic Hemiptera. These capture the larvae and suck them dry. Then there are the dragonfly larvae with their intricate grabbing mouth parts working like tiny captive catapults; these too make larval life a hazardous affair. Add to these no less than four families of water beetles and larvae belonging to five families of Diptera, and we begin to see why so many eggs are necessary. The dipterous predators include some chironomids and even some tipulids, both of which are close relatives of mosquitos; larvae of the Brazilian crane-fly *Sigmatomera shannoniana* live in tree-holes and feed avidly on all mosquito larvae inhabiting such places [1]. Then, of course, there are the larvivorous mosquito larvae that we mentioned in Chapter 3: *Toxorhynchites,* the subgenus *Mucidus* of *Aedes* and the subgenus *Lutzia* of *Culex.* Some of these, unfortunately, are pests themselves in their adult stage, as for example *C.* (*L.*) *halifaxii,* but most of them do no harm to us at all.

Despite the disappointing picture drawn earlier in this chapter certain predators have been used to keep mosquito-breeding in check and with eminent success [341]. Many kinds of tropical freshwater fish that one sees in ornamental aquaria will feed readily on mosquito larvae if they are given half a chance. A hundred larvae introduced to the average-sized tank disappear in a matter of seconds as they are sucked into numerous gaping mouths like so much fluff into a battery of vacuum cleaners. In nature, too, freshwater fish are important predators of mosquito larvae. The genera *Gambusia* and *Lebistes* and many species of *Cynolebias, Nothobranchius* and *Aphyosemion* seem to be especially fond of mosquito larvae and when introduced to comparatively weed-free reservoirs, open storage tanks and ornamental pools quickly settle the fate of any mosquito larvae present. So efficient are they at mopping up all the larvae that some additional food must usually be provided if the number of fish is to be kept up and the colony is to have any permanence. Indeed, it has been known for the owner of an ornamental pool actually to add mosquito larvae in order to keep his fish colony alive. This is not so bizarre as it may seem, since it is possible to add larvae of a species whose adults rarely if ever attack man.

The speed with which fish will eat up all the larvae present depends, of course, partly on the larva/fish ratio and partly on the amount of cover provided for the larvae by aquatic plants. It is sometimes possible to strike just the right balance of numbers of fish and amount of plant cover to keep the fish alive without supplementary feeding. The important thing about fish is that they are comparatively long-lived and are easier to manipulate and to control than, say, dragonfly or predacious mosquito larvae. Their successful use, however, is mostly restricted to man-made collections of water.

Deliberate supplementary feeding is not always necessary even when plant cover is wanting. In parts of Thailand the houses are raised on stilts and stand over accumulations of water that have no opportunity of draining away. These pools take on a heavy load of faecal pollution and become the favoured breeding sites of *C. pipiens fatigans,* one of the important transmitters of human filarial infections. The larvae, however, can be successfully controlled by adding to the water some of the little larvivorous fish popularly known as guppies (*Lebistes*); one female guppy can cope with more than 100 large larvae in a single day. Once the larvae have been reduced in number these fish take to feeding on the human faeces, a commodity that is rarely in short supply. Manabu Sasa, who studied the feeding capabilities of these South American fish in their Thailand habitat, showed how a single guppy eats about 0.25 gm of faeces per hour. The interesting thing is that, even when the larvae have been eliminated, the fish live on in their rather special environment ready to cope immediately with any new invasion of larvae.

Apart from fish, vertebrates are not very important as predators of mosquito larvae. Newts on occasion take a fairly heavy toll and there are odd records here and there of freshwater turtles playing a part. Ducks, too, and other water-birds and waders feed on larvae when opportunity offers [1].

No respite can be expected for the mosquitos that manage to complete their aquatic life unmolested and at last take to the air as flying adults. While some of them carry over infections acquired during larval days, for the most part the old enemies are replaced by entirely new ones. Waiting for them from the moment they emerge and at every turn throughout adult life is a horde of new parasites and predators. In many respects the threat is now greater than ever; the mosquito that reaches us safely for a meal is fortunate indeed, but even she has still to make her getaway without detection. That, however, is a subject for the next chapter.

Our old friend *Coelomomyces* is sometimes found in adult mosquitos, for from time to time an infected larva manages to pupate and produces an infected adult. The parasite now spreads its activity to new sites: sporangia are produced round the mosquito's brain or round the ovaries or both; infection of the ovaries prevents the development of eggs. But the number of adult mosquitos carrying these grim reminders of their past are few; of more than 9000 *An. funestus* and *An. gambiae* examined only fifty-six were found to harbour *Coelomomyces*.

Most of the parasites of adult mosquitos are picked up when they feed. They occur in the blood of vertebrates, including ourselves, some of them being well known as the causative agents of tropical and other diseases. These include the viruses, some of the protozoa and a few nematodes and will be dealt with in some detail in Chapter 13. Many of these organisms do no harm at all to ourselves or to our domestic stock and it is these that concern us now. Some of them also do no apparent harm to their mosquito hosts, but others are highly pathogenic and destroy them.

The viruses that have so far been recovered from mosquitos are usually only detectable when mosquito tissue containing them is transferred to susceptible vertebrates or to cell cultures made from these vertebrates. They do not harm the mosquito in any known way even after multiplying many thousandfold within the host, an infected insect living as long in captivity as an uninfected one. There may well be other viruses that neither harm the mosquito nor the inoculated vertebrate; there is at present no way of knowing. Indeed the hundred or so different viral agents recovered from mosquitos to date may well be only a small proportion of the total.

Lying somewhere between the viruses and bacteria is a group of microbes known as the Rickettsiae. These are best known as the causative agents of louse- and tick-borne typhus. While mosquitos are innocent of the spread of typhus or of any other Rickettsial disease, organisms resembling Rickettsiae are nevertheless found from time to time in the ovaries of *C. pipiens pipiens* and of its autogenous subspecies of *C. p. molestus*. The significance of these findings is not known [347].

True bacterial infections seem to be rare in mosquitos but, as Marston Bates has pointed out, no one has really studied the subject thoroughly. Spirochaetes have been found in the salivary glands of *An. l. atroparvus* in Europe, which suggests that these organisms may be transmitted when the mosquito feeds, for we have seen earlier that a mosquito pours saliva into the wound before sucking up the blood. We know nothing, however, of any vertebrate host of this parasite; *An. l. atroparvus* is fairly catholic in its taste for blood, and although it feeds readily enough on man it really prefers pigs. This spirochaete, incidentally, is not closely related to the one causing syphilis in man [346].

Bacterial infections sometimes follow when in the course of work on virus isolation whole mosquitos are triturated, suspended in serum and inoculated into the brains of mice. We do not know the exact source of these bacteria; they could arise from internal infections of the mosquito or from external contamination. It is interesting to note, however, that although recovered in this way from *Aedes, Anopheles, Culex, Haemagogus, Mansonia, Psorophora* and *Sabethes* mosquitos, they are particularly common from *Mansonia* both in Africa and in South America [1].

Fungi, including various yeasts, are not uncommon in adult mosquitos but, except for the rare infections with *Coelomomyces* mentioned earlier, the importance of most of these to the mosquito is little understood. *Entomophthora culicis* causes fatal symptoms and sometimes appears in epidemic form.

Apart from harbouring and allowing the multiplication of many species of malaria parasite derived mainly from man, other mammals and birds, adult mosquitos are commonly found to have numerous other sporozoal protozoa. Little is known of the life-histories of these or any part which they may play as parasites; many of them may in fact be beneficial or at least neutral to the mosquito's economy, like the millions of microbes we ourselves harbour about our bodies. Sir Ronald Ross, of malaria fame, seems to have been the first to find haemoflagellates associated with mosquitos, parasites related to those causing sleeping sickness in man; in 1898 he described these parasites from the mosquito gut. In fact three genera of trypanosomes (*Crithidia, Herpetomonas* and *Leptomonas*) have been found in mosquitos, infections with some species of these leading to sterility by attacking the ovaries. We do not know how these infections occur. It has been suggested that the parasites pass *via* the ovary to the next generation, but F. G. Wallace, who made a special study of these parasites, found no evidence for this. Adult *Ae. aegypti* and *C. pipiens* can pick up infections of *Crithidia* if supplied with food contaminated with this parasite. What is difficult to understand is that mosquitos harbouring *Crithidia* may pass the infection to other mosquitos without the intervention of an intermediate host. Wallace collected mid-winter, hibernating *C. pipiens* from a cave on the banks of the Mississippi near Saint Paul, Minnesota; eight out of twelve of these were infected with these parasites.

Turning now to the larger parasites, nematode worms are commonly found in mosquitos. Filariae and mermithids may both do considerable harm to the host but, except for those species of filaria that are also parasitic in man, again we know little of their life-histories. Mosquitos also harbour trematode flukes. Characteristically these parasites have a most complicated way of going about their lives, choosing different hosts at various stages in their life cycle. The adult fluke lives in the lungs of certain frogs, while

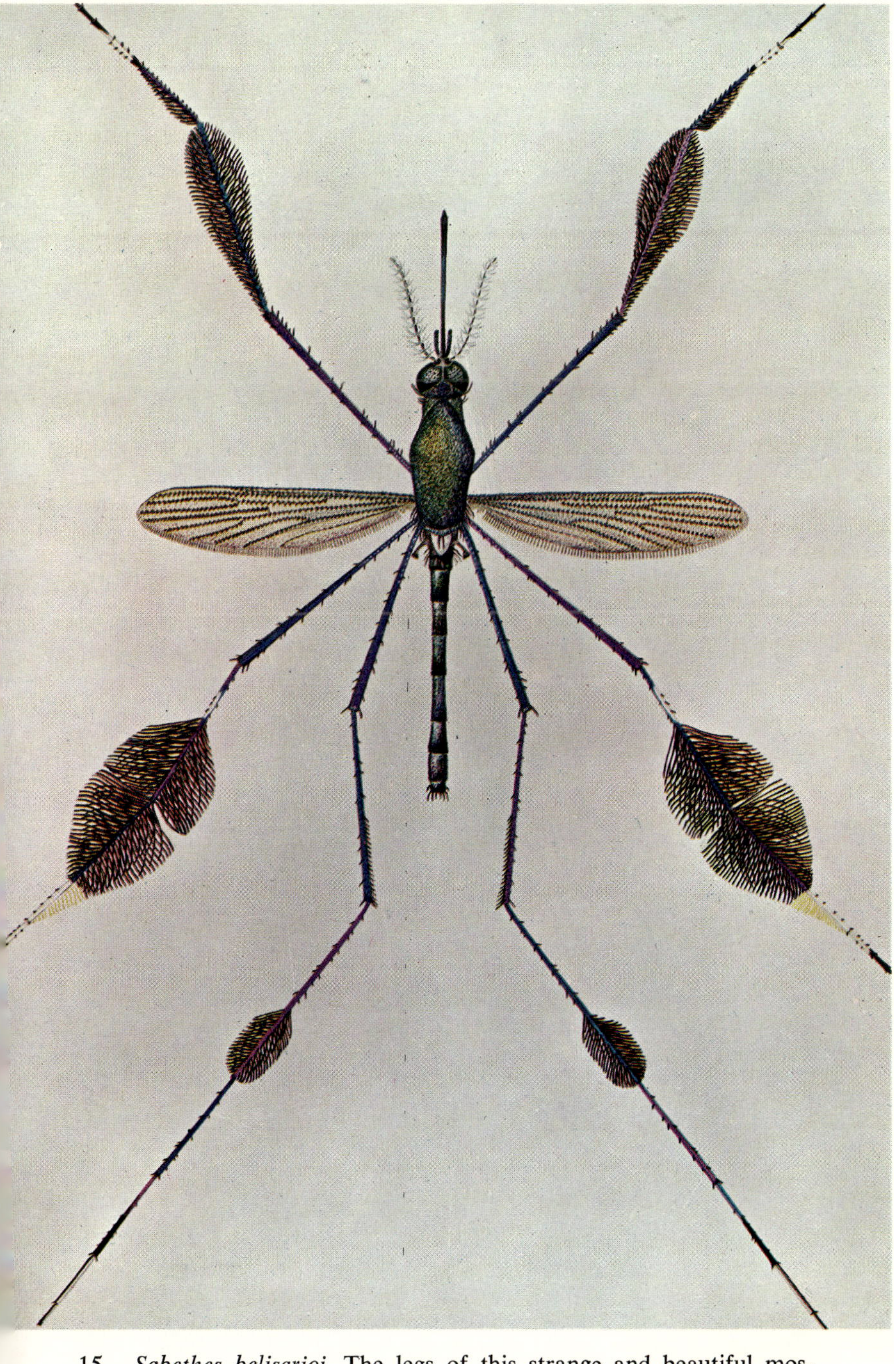

15. *Sabethes belisarioi.* The legs of this strange and beautiful mosquito bear a strong if superficial resemblance to those of some of the water-skaters; could it be that *S. belisarioi,* and other species of *Sabethes* with these curious expansions, also paddle themselves rapidly over water surfaces? Or, do these hairy patches serve as sex-signal receivers?

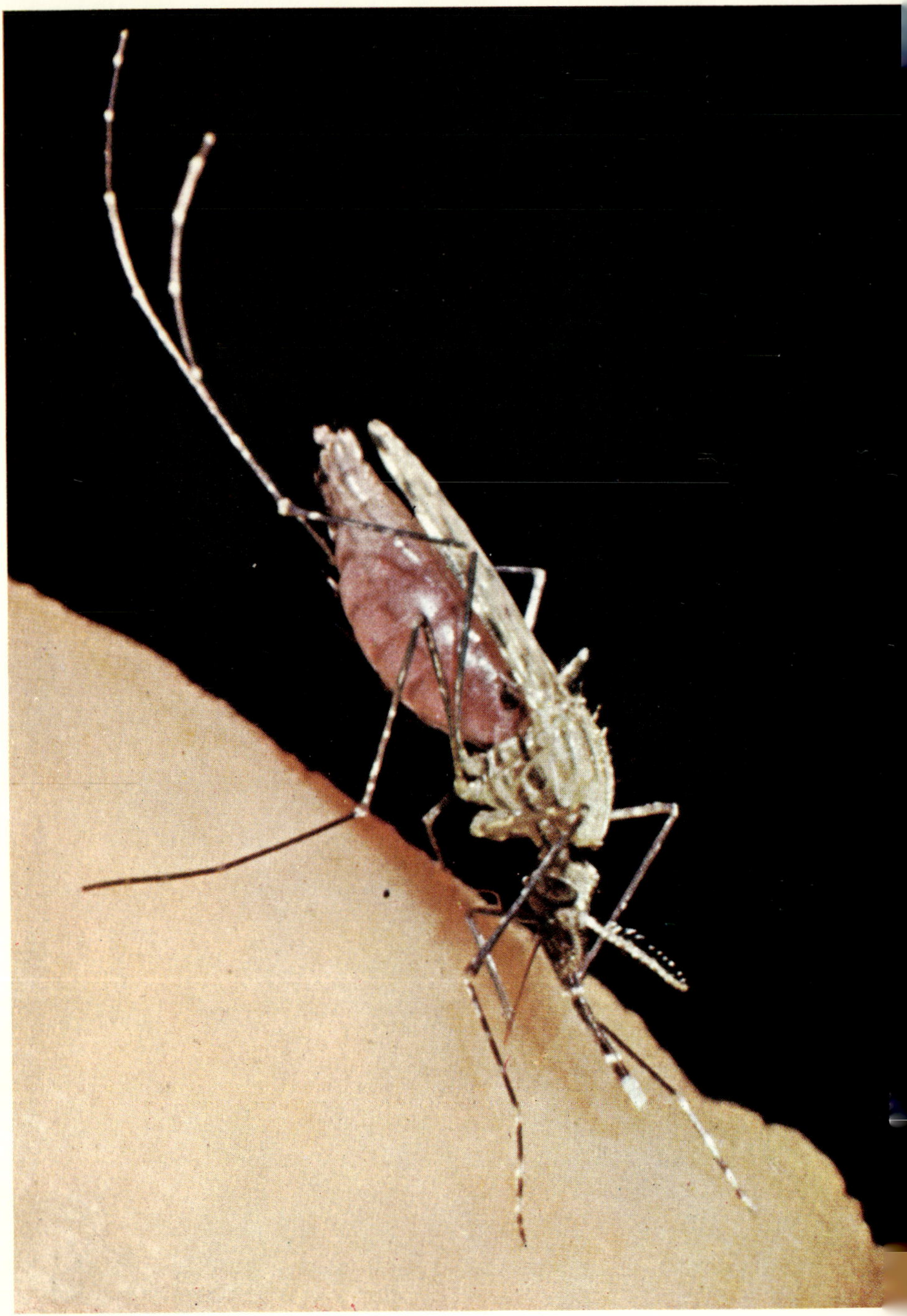

16. *Anopheles gambiae* in the act of feeding. Note the red color of the mouth-stylets as the blood passes up the proboscis to fill the stomach.

17. Breeding places of *Aedes simpsoni,* the rural transmitter of yellow fever in Africa. Larvae and pupae of this species live in the small quantity of water held up at the base of the leaves of these peri-domestic plants. The picture shows three kinds of such plants: in the background—bananas, in the middleground—dracaenas, and in the foreground—colocasias.

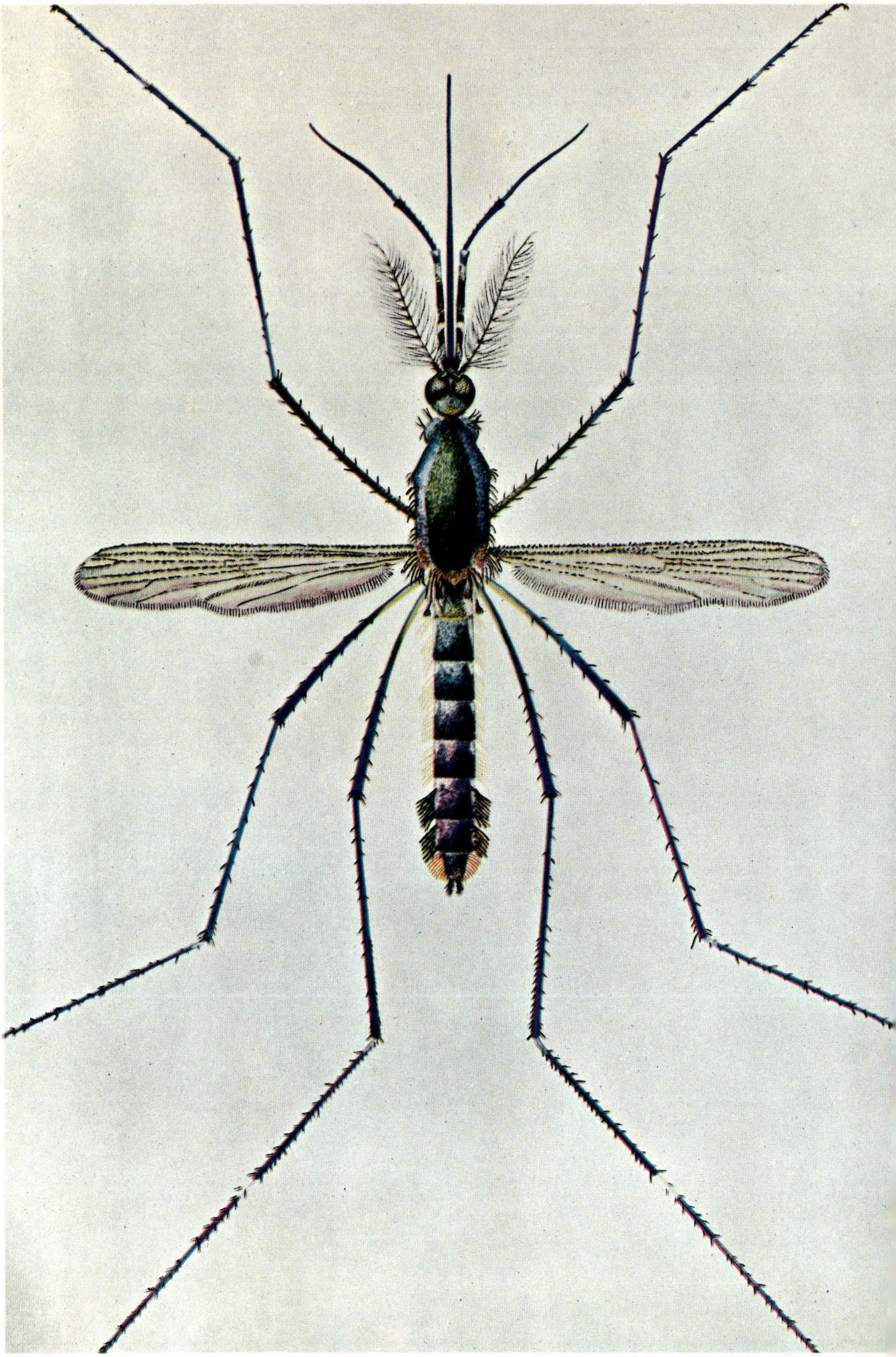

18. *Toxorhynchites brevipalpis.* The outer half of the proboscis of these very large mosquitos bends sharply downwards (not shown in this picture). The larvae are predators and feed extensively on the larvae and pupae of other mosquitos.

the younger stages occur in freshwater snails. Larvae of *An. maculipennis* apparently pick up infections from water containing infected snails and the parasites are carried over through the pupa to the adult. How the fluke gets back to the frog is not known; presumably the mosquito is eaten by the frog. Infections in adult mosquitos are common enough, recent work in Bulgaria with a large sample of wild-caught *An. maculipennis* showing an infection rate of sixteen per cent.

Most conspicuous of all the parasites of adult mosquitos are the aquatic mites that are found clinging to the thorax and abdomen; as many as forty of these creatures may sometimes be found on a single mosquito. Some species of these parasites are bright red in colour and heavy infestations can be noticed even when the mosquito is in flight. One may wonder what aquatic mites are doing on a non-aquatic host and how they get there in the first place. The story is quite fascinating and was worked out nearly forty years ago with painstaking thoroughness. The young mites hatch from their aquatic eggs and swimming in the water climb on to the backs of the mosquito pupae. Several larval mites may come to invade a single pupa where they settle down on each side of the line along which the skin will later split when the adult emerges. When this occurs, and the adult thorax pushes steadily upwards through the split, the tiny six-legged mite larvae step across like passengers boarding a moving staircase.

These mites belong to the Hydrobatidae, and several species may infest a single mosquito. During the short period when the newly emerged adult mosquito is standing on the surface of the water before taking to the air on its maiden flight, it may be invaded by a second wave of mites. These mites, which belong to the Thyasidae, are much smaller than the others and pass directly from the surface of the water to the resting mosquito [334].

The larval mites of both types remain on the host mosquito until it returns to the water for its first oviposition, when they drop off to continue their development as aquatic animals. The mites thus apparently derive a protein-rich food supply during their week or so aboard the flying host and are also carried to new waters; most important, perhaps, mites coming from one place are brought into contact with those from another—three important advantages for any parasite. Mosquitos that actually stand on the water surface for egg-laying (and this includes many *Anopheles* as well as those mosquitos that either scatter their eggs over dry ground or lay them above the water-level) are likely to pick up another load of the second type of mites, the small and much less conspicuous thyasids. Indeed, fresh instalments of thyasids are likely to be picked up at each oviposition.

When a mite of the first type is attached to the mosquito a long chitinous tube develops in the integument of the host at the head end of the mite. This structure apparently functions as a feeding tube, but it is not clear whether it is secreted by the parasite or whether it is formed as a reaction on the part of the host. Similar feeding tubes occur on slugs that are infested with slug mites. The special point of interest with these curious structures is that they remain behind when the parasite leaves the host. Thus it is possible to see the extent of previous occupation of a mosquito by mites by the presence of these tubes. No such tubes or scars are left by thyasid mites.

It is clear that if mites of the first type are present on mosquitos that come to bite, the mosquitos are likely to be nulliparous and coming to take their first blood meal. Indeed, the presence of mites on *Mansonia africana* was used in 1951 to sort out young mosquitos from their older companions [337]. In any study of mosquitos as carriers of disease it is important to know what proportion of the mosquito population is coming to feed for the first time. Mosquitos, unlike ticks and their allies, are not dangerous in nature unless they have fed as an adult on an infected person or other host. It follows that mosquitos coming to feed for the first time are harmless and if the bulk of the population is made up of these young mosquitos it does not constitute the threat that a population with an older age-profile would present. Nor will the picture necessarily change with the passage of time, since other parasites and predators may take such a heavy toll as to maintain an essentially young population.

Not all young mosquitos carry these mites; the proportion varies from species to species and from one season to another. Mosquitos that breed in small containers seem to be quite free from them at all times. Nor, it seems, are all mosquitos carrying these mites necessarily young, since, as Philip Corbet has recently shown, a small proportion of these parasites fail to leave the host at the first oviposition and others die *in situ* without dropping off. Nevertheless, provided one excludes the much less conspicuous thyasids from consideration and disregards any hydrobatids that have died while remaining attached to the host, the method can be used to separate young adults from their older companions with something like ninety-seven per cent accuracy.

After this slight digression in which we have seen how biological reasoning about the private life of a parasite may give information on the private life of the host, we may return to our main subject. The other common external parasites of mosquitos are certain midges which fasten on to recently fed females and take a blood meal second-hand. It seems that they are attracted by the warmth of the recently engorged mosquito. Whether there is then any exchange of parasites between the two insects is not known, but it is an interesting thought, for it opens up the theoretical possibility for the mosquito to acquire a blood parasite that is outside its normal repertoire. If we assume that the midge is not too fussy and will feed on more than one species of mosquito, then a primarily man- or monkey-feeding mosquito might pick up second-hand a pathogen from, say, a bird- or bat-feeding mosquito. Such a mechanism might serve not only to infect mosquitos from time to time with exotic parasites but to reintroduce familiar parasites from sources outside the recognised cycles of transmission. Where parasite life histories are concerned we have long ago learned to expect the unexpected.

Midges feeding on mosquitos form a convenient link between parasites on the one hand and predators on the other. Indeed, they are sometimes regarded as one and sometimes as the other. In this book, however, we restrict the term predators to those enemies that actually eat the whole mosquito or at least suck it dry.

This the midge does not do, for it apparently feeds on the freshly taken blood meal, not on the mosquito's own blood, and the mosquito survives to play host again and again. Spiders, of course, are a different matter, and immediately come to mind as predators of insects. Yet the position of spiders *vis-à-vis* mosquitos is not always a simple one of predator and prey. Marston Bates found *An. maculipennis* in southern Europe actually using spiders' webs as resting places. These large mosquitos were able to alight on and leave the web without difficulty. Indeed, he considered the webs actually to be favoured resting places by this mosquito, serving to protect them from the attacks of ants and other predators. Why the mosquitos were immune to attack by the owner of the web remained a mystery. Many spiders tend to disregard their catch until its struggling presence is signalled by the vibrations set up in the web. Perhaps *An. maculipennis,* enabled by their light weight to rest on the web without being ensnared, simply remained still. It hardly seems possible that the presence of the mosquito on the web confers any benefit to the spider unless, perhaps, it serves to attract more palatable prey to the snare.

Not all mosquitos and spiders have this apparent understanding; as long ago as 1529 we read, '–lyke unto cobwebbes in whych the lyttle Knattes, and Flyes stycke styll and hange fast'. Living in tree-holes in the forests of equatorial Africa is a species of spider that does not construct the familiar large vertical snare but sets out a miniature affair stretched horizontally like a tiny hammock across but a little way below the opening of the hole. It is not merely that this is the only way that the web could be supported inside a tree-hole, for if this spider is kept in a cage of ample proportions it still persists in making its web in this characteristic fashion. If a mosquito lands on this web it can in fact usually manage to fly off again without difficulty; presumably the relatively large surface area of its landing gear made up of its six long tarsi support its light weight without penetrating the surface film, or whatever is necessary for adhesion. Normally, however, the spider is there waiting on the underside of the hammock somewhere near the middle. Immediately a mosquito alights the spider shakes the web violently up and down with great rapidity. The legs of the mosquito penetrate and become ensnared; it struggles but this seems only to ensnare it even more securely. Almost at once the spider rushes in and stabs the struggling mosquito through the net, from below, and injects it with an immobilising poison. The contents of the mosquito are then partially digested and sucked dry. A frequent visitor to these tree-holes is our old friend, or rather enemy, *Ae. africanus,* and it may or may not be significant that these spiders bear a pair of crescent-shaped silver markings on a black background closely resembling those on the thorax of this mosquito.

Some spiders, although leaving a trailing safety line behind them like a climber's rope ready to give instant support in case of an emergency, do not actually construct a snare for their prey. Instead, they lie in wait and pounce with extreme accuracy on any suitable prey that comes into range. These jumping spiders are very common in many parts of the world and include mosquitos in their diet. They will spring upon and capture a mosquito in hovering flight as well as one walking near by. They have splendid vision and if a moving mosquito is too far away to capture the spider will stalk it till it comes into range for the final jump.

In warm tropical countries ants abound everywhere. One cannot sit down even in the open on a well-kept lawn, without being covered by them. The surface of every branch of a tree, even the surface of reeds emerging from water, is busy with streams of these animals hurrying to and fro. Many of these are carnivorous foragers and include resting mosquitos in their diet if they happen to be in the right place. Once the leg of a mosquito has been seized by an ant it is not long before reinforcements appear and the mosquito is torn limb from limb. It is as if the first ant to find the mosquito is able in some way to signal the other ants to the scene.

Also lurking on the trunks of trees in some parts of the world are the giant emesine bugs. These fantastic-looking creatures, with their grotesquely long legs and antennae, spend the day-time lying motionless against the bark of the tree, but when night falls they raise themselves on their incredible legs and start wandering over the face of the tree trunk like acrobats on stilts. Somewhat surprisingly, these insects are four-legged, functionally at any rate, for the front pair is highly modified as a jointed prey-capturing device –similar to the raptorial legs of a praying mantis. Almost as long as the four walking legs, and in fact looking very much like a third pair of legs, are the antennae. As with the legs, each of these geniculate appendages slopes upwards at about 45° as far as the first main joint and then elbows down again at about the same angle to end in a whip-like segment so fine as to be scarcely visible.

Looking like a four-legged daddy-long-legs each bug walks in a slow zig-zag path, rocking backwards and forwards between each step in the manner of a chameleon. As it pursues its hesitant course the long antennae, held up in front like a couple of fishing rods, move slowly from side to side, sweeping a twelve-centimetre-wide path in front of it. But if the tip of one antenna should come into contact with a resting mosquito the bug stops abruptly in its tracks and bringing the other antenna across, gently strokes the mosquito with the excessively fine tips. So fine are these organs and so gentle is the movement that the mosquito does not appear to notice them. The bug in the meantime, having judged the distance between it and the mosquito by touch, adjusts its body by pivoting backwards or forwards on its long legs without having to take another step. Brought thus into range, the raptorial front legs are very slowly raised and moved back over its head. Then suddenly, with incredible speed, the modified legs are shot forward to seize the prey, which is immediately drawn towards the predator's short stabbing proboscis and slowly sucked dry. A single bug can account for more than twenty mosquitos in a night [336].

A mosquito at rest often keeps its hind legs held raised off the tree-trunk, wall or other substrate. At times these legs undergo alternate up-and-down movements, one leg moving slowly up as the other moves slowly down. This movement is enough to attract the attention of certain lizards, particularly some of the geckos, which, with one sudden movement, seize the mosquito and eat it up. Many geckos take up their abode in houses where their antics on the wall each evening and their distinctive 'song' make them welcome tenants, particularly in Burma, Ceylon and other parts

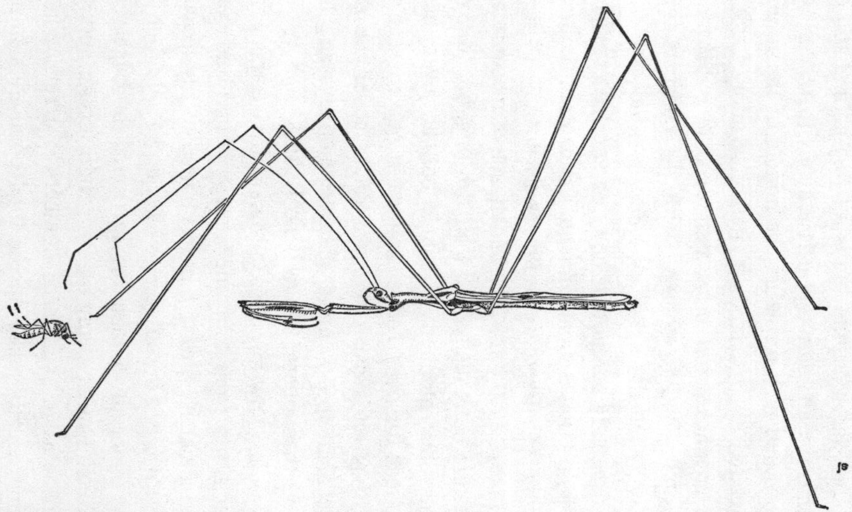

Figure 18. Giant Emesine bug, walking on its stilt-like legs, about to locate an adult mosquito by means of its long slender antennae. One of these bugs can capture and suck dry as many as twenty-four mosquitos in a day.

of Asia. There are day geckos as well as the more familiar night ones, and between them they play a significant part in keeping down the numbers of house-haunting mosquitos.

Thus far we have dealt with a few of the predators that attack adult mosquitos when at rest. But, like the bird threatened by a cat on the ground, the mosquito too is a flying animal and can always escape into the air, although it is unlikely to take more than an evasive flip round if disturbed in the middle of the compulsory rest period dictated by its circadian rhythm. Waiting for it in the air are other enemies, even more numerous than before. As a larva the mosquito had to contend with the constant threat of dragonfly larvae. Now the flying mosquito is faced with the vigilant, cruising dragonfly adults that hawk to and fro ready to seize anything else flying that is small enough. On capture the mosquito is eaten up while the dragonfly continues its flight.

Bats, too, take a heavy toll of evening-active mosquitos; these agile predators are not usually in the air for long, but they may be present in enormous numbers. A single bat, however, has been known to account for over 700 mosquitos in one evening's sortie [1]. Insectivorous birds carry this war on to cover the daylight hours and the night too. Lemurs, galagos, monkeys and man all combine to add their quota. There is no let-up.

One may wonder how any mosquito can survive this combined onslaught that starts from the time an egg is laid and continues unabated throughout its life. If I have omitted to say much about eggs and pupae it is not because these have no enemies but because less is known about them than the far more thoroughly studied larvae and adults. That so many mosquitos do survive and at times come to torment us in such large numbers is a reflection of the vast hordes that usually make up the total population in any one place. If this population is to remain stable, losses from whatever cause must be balanced by replacement. This means that for every mature male and female there must be a single viable egg-producing pair in the next generation—no more and no less. A species of *Culex* producing 400 eggs in a single raft can thus sustain an average loss of 99.5 per cent between the time of egg-laying and, say, mid-maturity of the next generation. It also means that a reduction in total losses during this period of only half of one per cent will result in a doubling of the mosquito population. In the tropics, where larval life is short such changes can occur in a week. The real life situation is, of course, vastly more complex than these oversimplified vital statistics indicate. Nevertheless, these simple figures serve to show how violent fluctuations in population size can arise with only slight changes in *total* parasite and predator activity, and with the general vagaries of weather.

12. MOSQUITOS AS NUISANCE

In this chapter we are going to deal with mosquitos in their most familiar role: the nuisance, the annoyance, the disfigurement and at times the torment and torture that they bring to us and to our domestic animals and pets. The discomfort they cause us introduces certain difficulties to the lives of the mosquitos themselves, and we shall try also to see the problem, as it were, from the mosquito's point of view. But it is mainly as a pest that the world knows the mosquito: the lone visitor that ruins sleep on a warm summer's night as it hums momentarily but repeatedly past one's ear; the vast hordes that during the mosquito season destroy all sense of solitude even in the arctic wastes; the bringers of bumps and blebs that, scratched and septic, disfigure face and limb; the tropical representatives of this ubiquitous pest that, present all the year round, are accepted as part of the normal night.

Until the discovery of the New World and the great expansion in travel that followed, references to these insects in the literature are rather rare. A pre-Columbian one given in the *Oxford English Dictionary* merely refers to 'Gnattes smale to sen and sarp on bite'. When, however, man started on his exploration of the world he found the torment brought by these creatures so much greater than anything in his previous experience that time and again they figure in the accounts of these journeys. In 1583 we read, 'We were also often times greatly annoyed with a kinde of flie . . . the Spanyards called them musketas.' Then, in 1600, Pedro Teixeira describing his Mexican journey wrote these tragic lines, 'Almost all along this road [Acapulco to San Juan de Uluâ] is a plague of mosquitos, so terrible and grievous that no defence avails against them, and they stung my best slave to death.'

Few other early references are as vivid or dramatic as this last one. None the less they draw our attention to some of the unexpected difficulties that were added to the dangers of travel in those days. Some of these references are worth quoting if only for the variations they show on the spelling of the word mosquito. John Davis, on his second voyage to find the northwest Passage, wrote in 1586 from Desolation, Greenland, 'We were troubled with a flie which is called Muskyto, for they did sting grievously.' In 1655, E. Terry, in an account of his voyage to 'East India', included the following: 'In the night we were very much disquieted with another sort [of fly] called musqueetoes', and ten years later we read, 'Howbeit the muskitto or gnats pestered us extreamly.' Many other references in similar vein can be found: 'They were grievously infested with moschetos', 'While these buzzing muscatos have been fluttering round their eyes and ears' and 'Nurseries of Muskettos and other Vermin'. One of these takes a rather unexpected turn: Josselyn writing on New England in 1674 reported that the country was strangely incommoded with flies, which the *English* called 'musketoes'.

Experiences of this sort are by no means confined to the past. There is a species of mosquito in tropical Africa known as *Coquillettidia fuscopennata* that seems to feed readily and avidly on any vertebrate that comes within its orbit of activity, whether warm-blooded or cold. Mention has already been made of the catholic taste for blood of this species. Its larvae, as we saw earlier, live out their lives hooked on to the roots, stems and leaves of aquatic plants and are thus largely immune to the usual methods of control. In certain years of unusually heavy and persistent rains, when new ground is inundated with the rising waters of lakes and pools, the size of the *Coq. fuscopennata* populations rises to prodigious heights; 1941–2 was such a year in Uganda and these large brown mosquitos with their conspicuous yellow-and-black-striped legs seemed to fill the air the moment the sun went down. At night they so covered the metal mosquito gauze that protected the windows of the houses that they not only obstructed the free passage of air but made sleep difficult by the continuous sound of their buzzing.

One evening during this period I visited a friend who owned a large black-backed dog, and the time came when we could no longer put off letting the animal out. Wrapping ourselves up as a protection against the expected onslaught we ventured out with the dog. I held a torch or flashlight and directed the beam to its back; within seconds its familiar black colour turned to a homogeneous brown-yellow, as every square centimetre was almost simultaneously and uniformly covered with these mosquitos. Before returning indoors I brought my hand down smartly on the poor animal's back and was appalled to find the browny-yellow colour turn dramatically to red; the dog's back and my hand were now covered with blood explosively squashed out from the engorging mosquitos.

A similar experience occurred some years later, this time in England. I was walking one evening along the bank of a canal when my attention was drawn to the peculiar appearance of a horse. While it is usual to find the underside of wild animals white or pale, whatever the coat colour on the upper parts, this is far from being the case with domestic animals. This particular horse was a very dark brown all over, except for an unusual contrasting paleness beneath. On investigation I found the whole belly and chest region to be coated with feeding *Anopheles,* massed one against another like people at some protest meeting. Once again I attempted to remove them with a smack of the hand (waving my hand near them had no effect except momentarily to disturb the flight of those still hovering in search of a space), and again the same bloody mess, this time on the horse's chest. At any one moment there must have been many thousands of these mosquitos sucking the blood from this unfortunate animal, and any vacant space was immediately occupied from the cloud of would-be feeders. This situation must be familiar to every river- and canal-side owner of horses, nor is it new for in 1471 we read, 'I would fayne my gray horse wer kept in mew for gnattys.'

In Chapter 1 I referred to the torment caused by mosquitos in the far north. Species of the largely temperate subgenus of *Aedes* known as *Ochlerotatus* have been recorded from the northern parts of the USSR, Scandinavia, Greenland and Canada. Detailed studies of the habits of northern mosquitos have been made in Labrador where the main culprits include *Aedes communis, Ae. pionips* and *Ae. punctor,* and in Manitoba where these species are joined by *Ae. hexodontus, Ae. impiger* and *Ae. nigripes.* These are all pest species, breeding over vast areas within the subarctic zone in which northern forest gives over gradually to tundra. They hibernate as eggs and after larval and pupal development emerge as adults during the summer to plague the people in numbers unimagined elsewhere; conversation is brought to a halt and even breathing becomes difficult or dangerous [353, 356, 363, 364].

The feeding habits of *Ae. impiger* and *Ae. nigripes* have recently been studied by Philip Corbet and co-workers in northern Ellesmere Island, nearly 82°N or only some 900 km from the North Pole. Larvae of these two species develop in the vast number of ponds that appear when the snows melt. The adult mosquitos are active mostly in mid-July but they may also be found during the early part of August. Anyone venturing into these regions of arctic wasteland during the very short summer season will be pounced upon by black clouds of these hungry mosquitos.

But man is only an occasional visitor to this land of treeless tundra and one may well ask what these mosquitos feed on most of the time. Corbet and Downe made a thorough study of *Ae. impiger* and *Ae. nigripes* on the shores of Lake Hazen. Inhospitable though this region may seem it provides a limited livelihood for no less than twenty-one species of bird and for seven species of mammal including herds of caribou (the western representative of the reindeer) and musk ox. Numerous clouds of these mosquitos hover around and over these animals like smoke from a row of chimney-pots. Not all, it seems, are successful in obtaining a blood meal. Many of the smaller and more active birds and most of the mammals avoid being bitten by sudden evasive movements; the arctic hare, for example, flicks its long ears, while the arctic fox rubs its muzzle in the dust. The best hosts from the mosquito's point of view are birds that are sitting on their nests and cannot leave their eggs for long periods, and the larger slow-moving mammals. Nests may be almost covered by probing mosquitos, and even the eggs, left warm by a temporarily absent parent, avidly but unsuccessfully attacked. Actual feeding is often restricted to certain regions of the bird's body, notably around the eyes [351].

The most successful hosts, again from the mosquito's point of view, are water-birds such as the eider duck that are nesting in and around the ponds, and the lumbering musk ox. Earlier accounts describe how herds of these animals are accompanied by dense masses of mosquitos. Again, feeding takes place around the eyes, lips and other regions of warm, exposed skin. The animals repeatedly shake their heads and roll over on their backs in what appear to be desperate attempts to rid themselves of this pest. Near Isersiutilik in West Greenland hares and reindeer move to higher ground during the mosquito season and so come largely to avoid the attacks of *Ae. nigripes* that plague the regions lower down [359]. Further north these mosquitos have even been seen attacking dead animals. It seems that when food is at a premium host specificity is less important; indeed, catholicity of taste becomes a prerequisite of survival.

Another adaptation for life in these high and inhospitable latitudes is a tendency towards autogeny; the female mosquitos are able to develop the first batch of eggs in the absence of a blood meal, provided that they have access to nectar. The number of autogenously produced eggs tends to be much fewer than otherwise; often only a single egg reaches maturity and is laid [350]. Autogeny has also been noted in other northern mosquitos: *Aedes communis, Ae. flavescens* and *Ae. punctor*.

Before leaving the subject of the suffering caused by the attacks of mosquitos I must at least mention what is probably one of the earliest accounts. Marcellinus, writing in the third century of our era and cited by F. Cowan in his book of 1885 on curious facts about insects,* relates how in Mesopotamia lions are so tormented by mosquitos that they seek the refuge of water, where they either drown or go mad. Some doubt has been cast on the true identity of these insects since the narrator states that they attack the eyes of these beasts. But we have already seen from the observations in the far north, both on birds and on musk oxen, that the eyes can be the favoured sites for attack.

Although a mosquito takes only some two to eight milligrammes of blood per meal, depending largely on the size of the mosquito, the total amount taken when thousands upon thousands are involved can be considerable. There are cases on record of dogs and even cattle dying following mass attack by mosquitos; reference has already been made to human death following persistent and wholesale attack. Brian Hocking has shown that the attack rate by north Canadian mosquitos on a single exposed forearm may be as high as 289 bites per minute. Thus a totally unprotected man could receive more than 9000 bites per minute, which would result in a loss of about half his blood in less than two hours. But death following mass mosquito attack is not always due to loss of blood *per se*.

People and other hosts develop an allergy to mosquito bites and this response can on occasion be very severe. Fortunately such cases are rare; usually the allergic response involves no more than the familiar itching red wheal. That the response on the part of the host is an allergic one and not merely a reaction to some poison injected by the mosquito is shown by our having first to become sensitised to the bites of mosquitos. When we encounter a species that is new in our experience we at first show no reaction at all following its bites. Later, however, if we are bitten regularly by members of this species we begin to exhibit a delayed reaction some hours or even a day or two after the mosquito has fed. If we persist in this exercise, or, as is more usually the case, if the mosquito persists, we become sensitised and develop an-

* I am indebted to Sir Rickard Christophers' classic book on *Aedes aegypti* for this reference, a book, by the way, of 700 pages devoted to a single species of mosquito.

other response to these bites—a reaction starts almost immediately after the mosquito has finished feeding. This is known as the 'immediate' reaction and it later comes to replace the delayed reaction altogether [357, 360, 362].

When we are the subject of constant mosquito biting it is usually the immediate reaction which we notice—the typical raised wheal accompanied by intense irritation. It has long been known, however, that if we are bitten regularly over a sufficiently long period we eventually lose the immediate reaction as well; we become desensitised—we develop a sort of immunity. A graphic description of this phenomenon appears in an account of the climate and diseases of America written by J. D. Schoepff in 1781. Other articles have followed at intervals ever since, notably by J. C. White in 1871 and by Carlos Finlay of yellow fever fame in 1886.

When a mosquito settles on the skin it spends a few seconds exploring the site before penetration of the skin begins in earnest. Penetration of the skin and tapping of the blood supply usually take about half a minute. During this period the mosquito's saliva flows into the wound and it is this that will produce the allergic reaction in a sensitised person. Two to three minutes after this initial penetration of the mouth fascicle into the skin and the accompanying outpouring of saliva the sensitised host is going to react and be made aware of what is going on, for the onset of the so-called immediate reaction is usually heralded by a sudden and more or less intense irritation. For the mosquito, then, the host's reaction introduces a serious hazard: the insect must tap the blood supply, feed and make its getaway, all before the host has been alerted by the onset of irritation.

It seems strange that the mosquito should carry with it its own built-in host-warning system, for surely the alerting of the host in this way must place a serious handicap in the way of successful feeding. It is rather as if a burglar were to wind up and preset his own burglar alarm before entering a house, giving himself a limited time in which to complete his job before the alarm goes off to wake his 'host'. Absurd? Yes, but if all burglars were compelled to do this, burglary would be much more efficient, for only those that finished and made good their escape within the time limit would survive as burglars; there would be a steady and rigorous selection in favour of fast workers. This is, in fact, precisely what happens with mosquitos—there is a natural selection for fast penetration and feeding, with the onset of irritation on the part of the host acting as the main selection force. Mosquitos that have not completed their meal and made good their escape by the time the host is alerted will be in danger of being crushed to death or at best forced to leave the site with only a partial load of blood. Since, as we have seen, egg development depends on the blood meal and since the number of eggs developed is largely dependent on the size of the blood meal, it follows that mosquitos disturbed before completing their meal will lay fewer eggs. It is, therefore, not necessary for the slower mosquitos to be killed for natural selection to play its relentless part; surviving slow feeders will leave less progeny than their faster companions [352].

Speed of penetration and feeding, then, are important for survival since the safe period between salivary injection by the mosquito (which may start only six seconds after the beginning of penetration) and the onset of irritation in a sensitised host is a matter of minutes only. Studies on wild mosquitos have shown that where the safe period lasts for two and a half minutes, feeding usually takes only a minute and a half, thus leaving a safety margin of a whole minute. These are, of course, average figures. Some individuals are slower, some quicker to complete the initial penetration; others are slower or quicker in the actual uptake of blood.

While the safe period for the mosquito is normally adequate, what of the individual that is just a little slower than the others but not slow enough to be caught out, the mosquito whose combined penetration and feeding periods bring it perilously near the time for the alerting of the host? Here is where the analogy of the burglar with his preset burglar alarm ceases. Since salivary injection occurs near the beginning of the penetration period and it is this that determines the time of onset of irritation in the sensitised host, you might think that the feeding mosquito that is a little slower than the others in penetrating the skin and tapping the blood supply will automatically have less time in which to complete the act, like our burglar who is slow in gaining entry. But no, in the case of the mosquito that is marginally slower in the initial tapping of the blood supply there is also a slight delay in the onset of the allergic response of the host, and the mosquito thus has a little extra time in which to complete the feed. It needs this extra time for not only has the penetration period already encroached on the feeding time but slow penetration tends to be followed by slower-than-usual feeding [352].

This rather complicated story can now be summarised in the following terms: it seems that properties of the saliva that affect the speed of tapping of the blood supply and its subsequent uptake affect also its antigenic activity and result in a compensatory delay in the onset of the host's reaction. Mosquitos that are slightly slower to penetrate and tap the blood supply tend also to be slower to feed, but these do not run into danger because they also have a longer time in which to complete the act before alerting the host. This remarkable safety mechanism rescues only those that are marginally slower—those mosquitos whose combined penetration and feeding periods take just a little longer than the usual three minutes; any mosquitos that are very slow will be caught in the act and destroyed, or at least disturbed, by the host. The timing of the various phases applies to wild *Ae. africanus* and man; the duration of each phase may well be very different with other species and indeed with other hosts, but the principles are likely to be the same.

The reaction of the host, then, which is I suppose what mainly concerns most of us, brings with it special problems for the mosquito. It may appear that I have given undue weight to this aspect of the subject, but it seems to me that when considering the role of mosquitos as pests and when discussing the torment which they can cause us, we should at least be made aware of the problem from the other side. Even man with his intellectual superiority is more often than not outsmarted by these long-legged slender creatures. But mosquitos were here long before man and they have had time enough to perfect their strategies of approach, uptake and getaway:

When did you start your tricks,
Monsieur?

What do you stand on such high legs for?
Why this length of shredded shank,
You exaltation?

Is it so that you shall lift your centre of gravity upwards
And weigh no more than air as you alight upon me,
Stand upon me weightless, you phantom?

I heard a woman call you the Winged Victory
In sluggish Venice.
You turn your head towards your tail, and smile.

How can you put so much devilry
Into that translucent phantom shred
Of a frail corpus?

Queer, with your thin wings and your streaming legs,
How you sail like a heron, or a dull clot of air,
A nothingness.

Yet what an aura surrounds you;
Your evil little aura, prowling, and casting numbness on my mind.
That is your trick, your bit of filthy magic:
Invisibility, and the anaesthetic power
To deaden my attention in your direction.

But I know your game now, streaky sorcerer.
Queer, how you stalk and prowl the air
In circles and evasions, enveloping me,
Ghoul on wings
Winged Victory.

Settle, and stand on long thin shanks
Eyeing me sideways, and cunningly conscious that I am aware,
You speck.

I hate the way you lurch off sideways into the air
Having read my thoughts against you.

Come then, let us play at unawares,
And see who wins in this sly game of bluff.
Man or mosquito.

You don't know that I exist, and I don't know that you exist.
Now then!

It is your trump,
It is your hateful little trump,
You pointed fiend,
Which shakes my sudden blood to hatred of you:
It is your small, high, hateful bugle in my ear.

Why do you do it?
Surely it is bad policy.
They say you can't help it.

If that is so, then I believe a little in Providence protecting the innocent.
But it sounds so amazingly like a slogan,
A yell of triumph as you snatch my scalp.

Blood, red blood
Super-magical
Forbidden liquor.

I behold you stand
For a second enspasmed in oblivion,
Obscenely ecstasied
Sucking live blood,
My blood.

Such silence, such suspended transport,
Such gorging,
Such obscenity of trespass.

You stagger
As well you may.
Only your accursed hairy frailty,
Your own imponderable weightlessness
Saves you, wafts you away on the very draught my anger makes in its snatching.

Away with a paean of derision,
You winged blood-drop.

Can I not overtake you?
Are you one too many for me,
Winged Victory?
Am I not mosquito enough to out-mosquito you?

Queer what a big stain my sucked blood makes
Beside the infinitesimal faint smear of you!
Queer, what a dim dark smudge you have disappeared into! [358]

13. MOSQUITOS AND DISEASE

We must now have a closer look at the part played by mosquitos in the transmission of disease, for in their capacity to act as vectors of parasitic worms, protozoa and viruses lies their importance to man. The suggestion that mosquitos might be responsible for the transmission of human disease is not new. It has in fact been put forward from time to time over the past 2500 years, usually to be unheeded, ridiculed or forgotten. The earliest reference seems to date from about 500 BC and comes from India: a far-sighted Brahmin priest named Suśruta declared that mosquitos were responsible for the spread of malaria. It is true that at about this time Empedocles successfully eliminated fever from Selinus in Sicily by draining the swamps that surrounded the town. But Empedocles seems to have been entirely unaware of the significance of the mosquitos arising from these swamps, as indeed were most others, who nevertheless associated swamps with malaria. The connection between swamps and mosquitos was, however, recognised by Columella, who in AD 100 advocated the avoidance of building near marshes, 'for,' as he says, 'they breed animals with mischievous stings which fly upon us in exceeding thick swarms'. Whether he also saw the connection between mosquitos and disease is not so apparent.

The first clear suggestion in more recent times that mosquitos might be responsible for the spread of disease comes from the writings of one Henry Hawks, who in 1572 sent back a report from Mexico that should have alerted all those concerned to this grisly possibility. Hawks was an Englishman who spent five years trading in Vera Cruz. He recorded the torment of those bitten by mosquitos and noted that it was mainly the newcomer who suffered. Most important, however, he observed that attack by mosquitos could be followed by disease and death:

> This towne is inclined to many kinde of diseases, by reason of the great heat, and a certeine gnat or flie which they call a musquito, which biteth both men and women in their sleepe; and as soone as they are bitten, incontinently the flesh swelleth as though they had bene bitten with some venimous worme. And this musquito or gnat doth most follow such as are newly come into the countrey. Many there are that die of this annoyance.

These important observations were quoted by Hakluyt and published between 1598 and 1600, but the book remained for the most part unread and apparently even uncut [439, 458].

FILARIA AND ELEPHANTIASIS

It is difficult for us to imagine the climate of belief and prejudice that still existed 300 years later when in 1878 Sir Patrick Manson showed for the first time that the filarial worm-parasite of man known as *Wuchereria bancrofti* underwent part of its development within the mosquito. The suggestion that mosquitos might be involved in the transmission of diseases to man had in fact only recently been put forward again by more than one person, but Manson's was the first direct evidence of mosquito participation. It was fashionable at the time to believe, as indeed it still is in some quarters, that all insects are ephemeral creatures. Manson apparently shared these beliefs to some extent and presumed that mosquitos fed only once in their lives. He reasoned, therefore, that after undergoing development in the mosquito the filarial worms must find their way into water at the time of egg-laying of their host, and that man became infected on drinking such water

[416]. We now know this to be wrong, and that infection is transmitted directly when the mosquito bites man for the second or third time in its life. Nevertheless, his epoch-making demonstration opened up a new dimension: infection could not only survive in another host species but could be transmitted over distances to people who had had no direct contact with human patients or with their belongings.

Human infections with *W. bancrofti* may lead to the distressing condition known as elephantiasis, which is still all too commonly found in many parts of the world from eastern Europe to Japan, from the Pacific islands and Australia to most of Africa south of the Sahara and the warmer parts of the western hemisphere. The adult worms reside in the lymph nodes and associated ducts of man and when present in large numbers following multiple infection tend to block the lymphatic canals. This eventually may lead to the characteristic gross enlargement of limbs, breasts and sex organs. Cases have been reported of elephantiasis in which the scrotum has weighed more than forty kilogrammes and has had to be carried on a wheeled cart in front of the patient. The region of enlargement depends to some extent on the original site of biting by the mosquitos; thus species tending to feed mainly on the lower parts of the body will usually give rise to elephantiasis in the lower extremities, since the adult filarial worms will block the associated lymphatics. While the number of people today infected with *W. bancrofti* is numbered in millions, luckily a relatively small proportion of these actually show symptoms of elephantiasis.

The adult female worm produces many thousands of larvae known as microfilariae. The larval worms exhibit a regular periodicity, appearing in the peripheral blood-flow of the host at night and returning to the lungs during the day-time. They are thus available to night-feeding mosquitos only, the most important of which are *Culex pipiens fatigans* and various species of *Anopheles,* and *Mansonia.* This periodic behaviour can be reversed if man changes his cycle of activity and rest by taking up night work and adopting the day for sleep. But nature is rarely straightforward and simple,

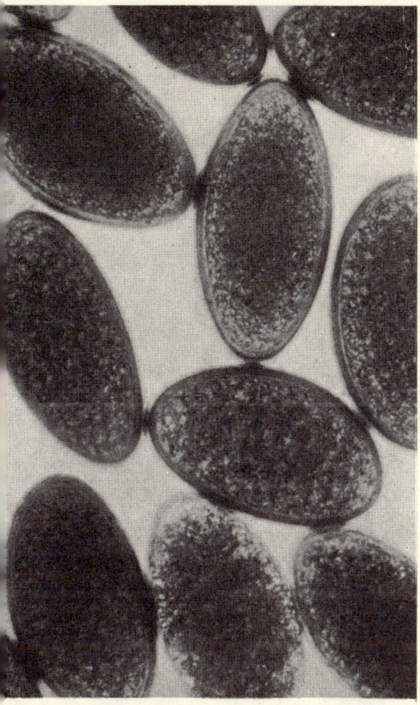

19. Fungal sporangia of *Coelomomyces psorophorae,* taken from infected larva of *Ades vexans.*

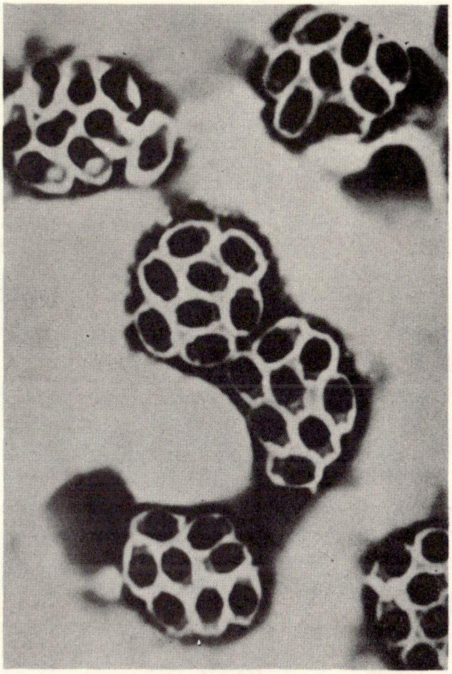

20. Sporoblasts (each containing eight spores) of the protozoan parasite, *Thelohania legeri,* taken from infected larva of *Anopheles maculipennis.*

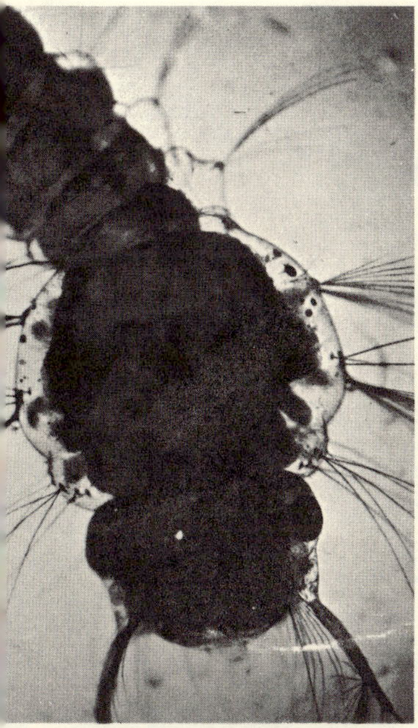

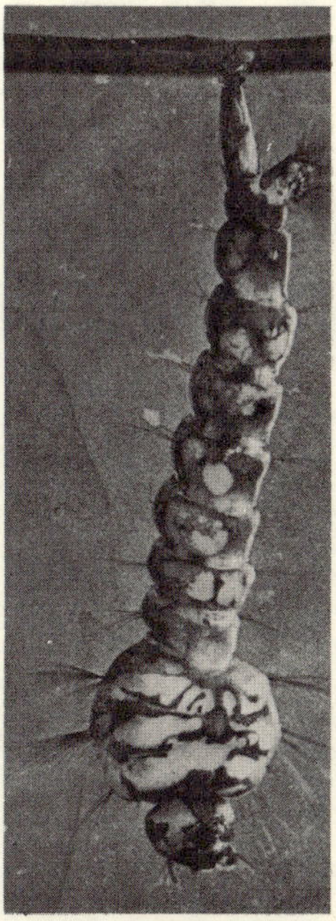

21. (far left) Larva of *Culex tarsalis* showing bloated thorax and dark spots associated with infection by inclusion body virus.

22. (near left) Larva of *Aedes squamiger* swollen and curiously opaque following infection by the protozoal parasite *Thelohania campbelli.*

23. Standing water and debris under a house in Thailand. *Culex fatigans* breeding in the water can sometimes be successfully controlled by the addition of larvivorous fish to the water.

24. Mosquitos settling on the caps of soldiers at Churchill, Manitoba.

25. A trapper in northern Canada, grotesquely garbed to protect himself against the hordes of hungry mosquitos that appear during the short summer season, vainly tries to protect his husky dog by the application of repellant; clouds of mosquitos can be seen against the black fur of the dog.

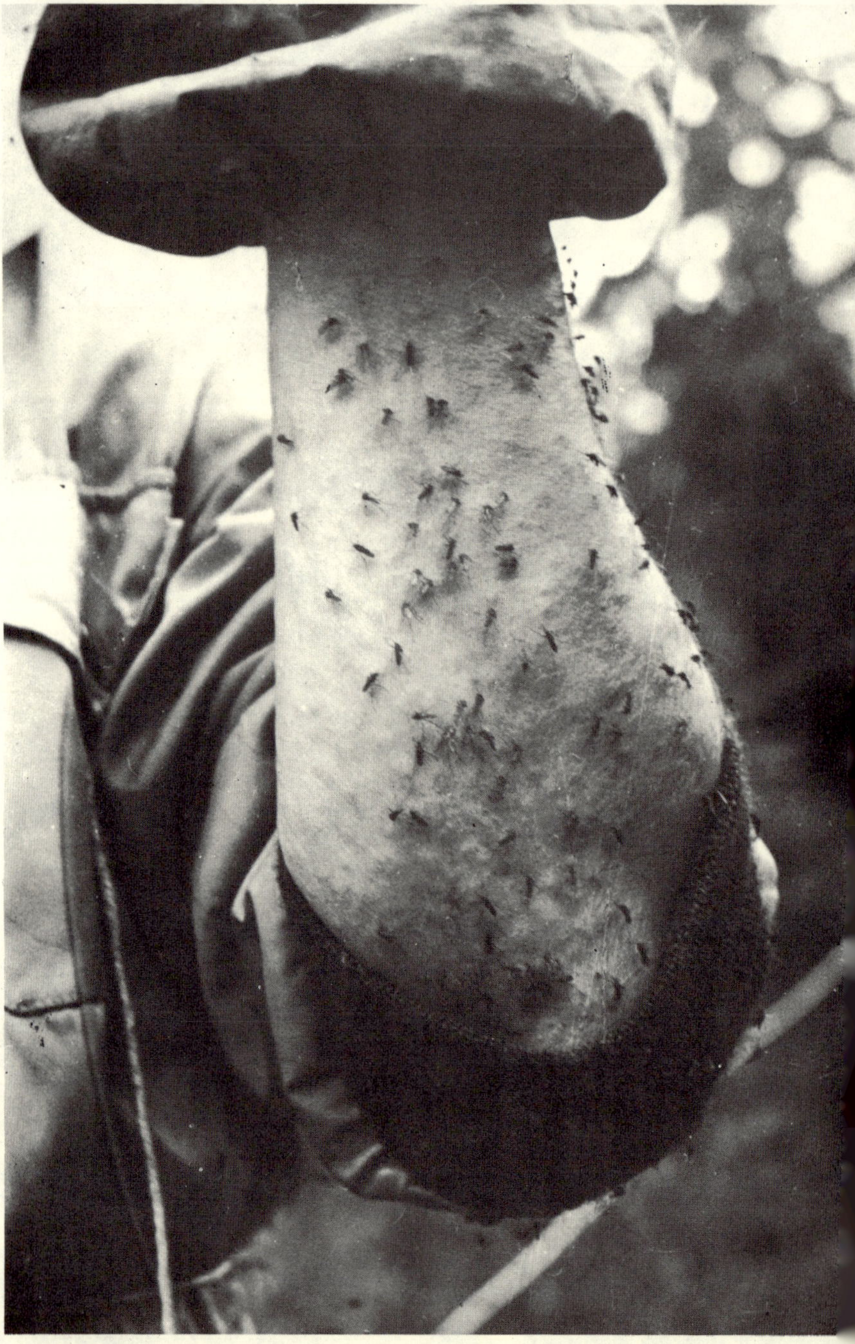

26. Carrying out a one-minute feeding-rate count of Arctic mosquitos settling on the forearm; as many as 289 mosquitos per minute have been recorded.

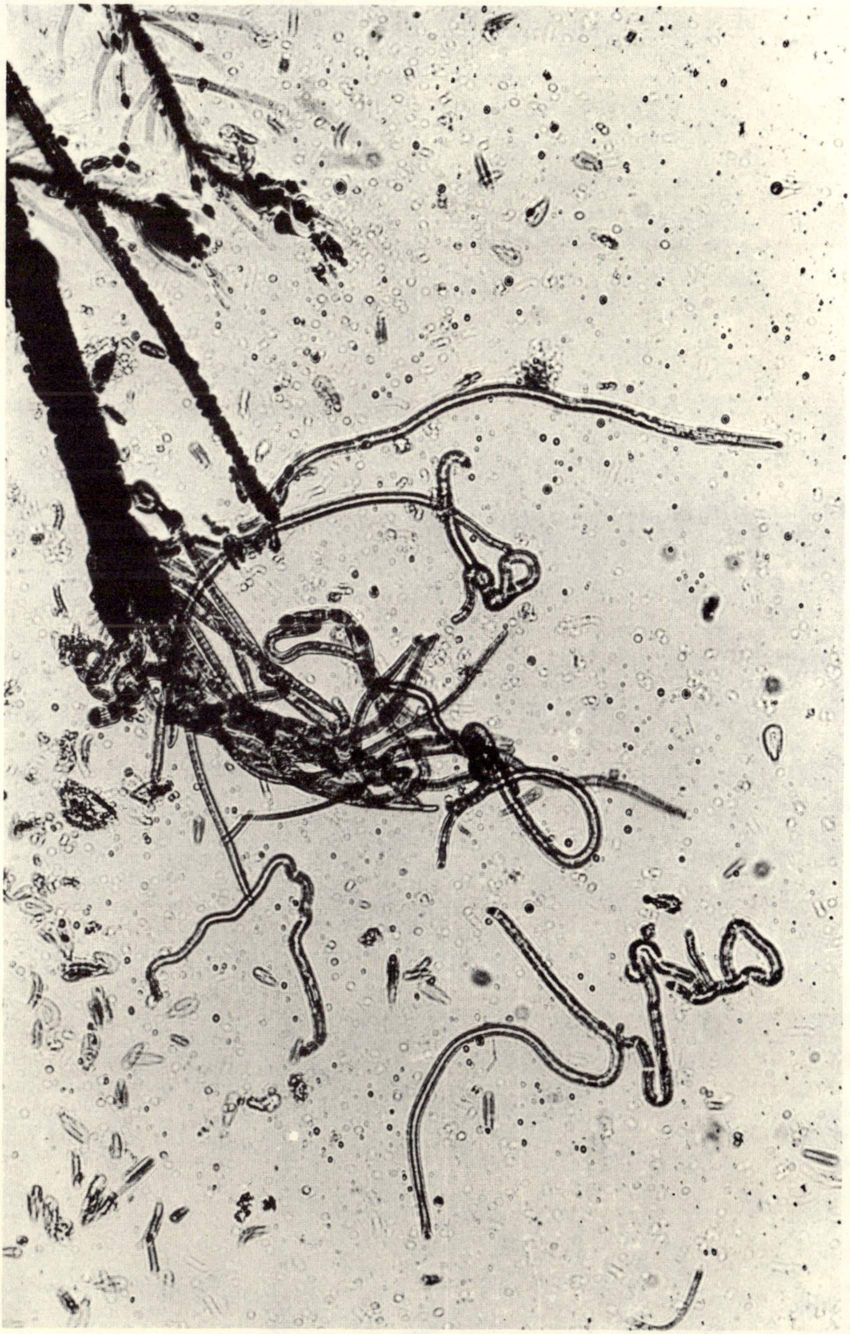

27. Mature larvae of the filarial worm, *Brugia pahangi,* emerging from the tip of the proboscis of a feeding mosquito.

28. Unfed, resting adult of *Anopheles balabacensis,* an important transmitter of malaria in Southeast Asia.

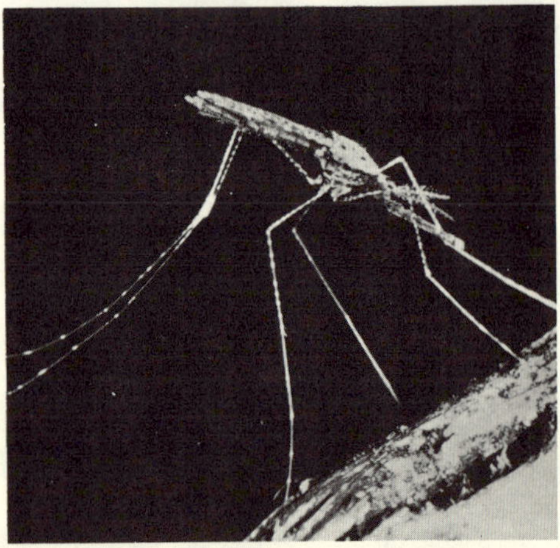

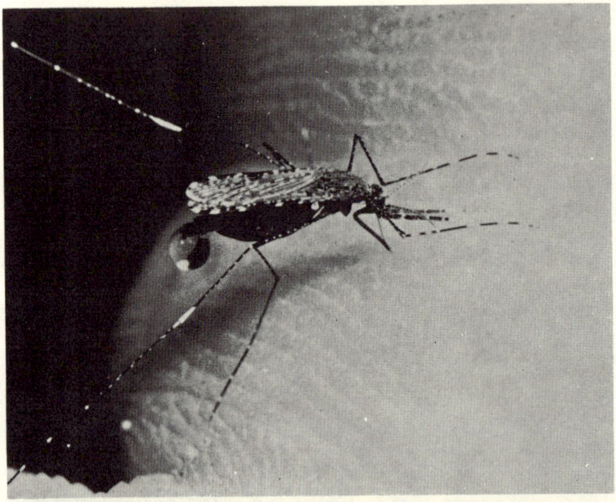

29. *An. balabacensis* in the act of feeding on man. Note the clear drop of fluid being passed from the anus (the production of one or two drops of clear fluid during blood-feeding is characteristic of mosquitos).

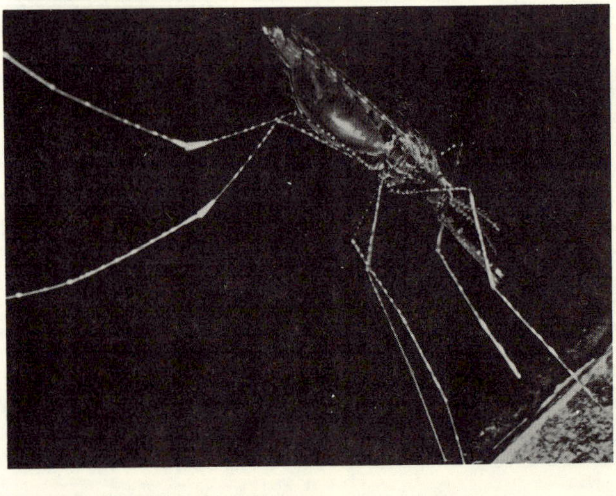

30. Recently fed adult of *An. balabacensis*. Note the greatly distended abdomen compared with the slender profile of the unfed female shown in Plate 28.

31. Dr. Louis Beauperthuy (1807–71)

32. Dr. Carlos Finlay (1833–1915)

33. Sir Patrick Manson (1844–1922)

34. Sir Ronald Ross (1857–1932)

35. Dr. Walter Reed (1851–1902)

36. General William Gorgas (1854–1920)

37. Dr. Jesse Lazear (1866–1900)

38. Dr. Adrian Stokes (1887–1927)

however much it would suit us to have her so; there is a Pacific strain of *W. bancrofti* that shows a naturally reversed periodicity, the microfilariae tending to concentrate in the peripheral blood in the day-time in a people who observe the normal pattern of sleep and activity. These parasites are picked up by day-feeding mosquitos, including *Ae. pseudoscutellaris.* Thus, whatever serves to trigger off this periodic behaviour, the circadian rhythms of parasite and vector in nature are perfectly adapted to each other.

Another remarkable thing about these microfilaria is that they are not always found in the same concentration throughout the peripheral flow. For example, a drop of blood taken from the ear may have a higher concentration of microfilaria in it than an equal quantity of blood taken from the finger-tip. We have already seen that some species of mosquitos have preferred biting sites on the human body (Chapter 6) and one may predict that some of the vectors of *W. bancrofti* show similar selectivity.

Less widespread in its distribution but also extremely important is another filarial worm, *Brugia malayi.* This parasite typically also shows a nocturnal periodicity and is transmitted mainly by *An. barbirostris, M. uniformis* and *M. annulifera.* There are also what are known as subperiodic strains of the parasite in which the density in the peripheral blood rises in the early afternoon and attains its peak level by the early evening. In West Malaysia these subperiodic strains of *B. malayi* are transmitted by evening-biting *M. dives* and *M. bonneae;* the latter species is also involved in the Philippines [408].

Mosquito-borne filarial infections are not confined to man. The subperiodic strain of *B. malayi* is also to be found in certain monkeys, domestic and wild cats, civets and in the pangolin [380]. Thus, even when the human population is treated and the direct transmission from man to man *via* mosquitos has been minimised, there is always the threat of reintroduction from wild animals to man. We shall come across this phenomenon again and again when we come to consider the mosquito transmission of viruses. To be effective, of course, transmission by mosquitos from animals to man demands the presence of mosquitos that will feed on more than one host species. There are filarial parasites of horses, dogs, cats, cattle, zebu, sheep and goats among domestic animals which are transmitted by mosquitos. There are others to be found in tigers, monkeys, bats, porcupines and racoons. *Dirofilaria tenuis* shows a nocturnal periodicity in racoons even though these animals, being of nocturnal habit themselves, are active during the night. Once again we see an association between the rhythm of the parasite in the vertebrate host and the habits of the night-feeding vectors, which appear to be *An. quadrimaculatus* and *Ae. taeniorhynchus*. The list of vertebrate hosts does not cease with the mammals, for birds, lizards, snakes and frogs are also involved. Filarial worms are also transmitted by arthropods other than mosquitos, but these need not concern us here [380].

It is evident that mosquitos provide a satisfactory environment for these worms, the association between the two being very widespread indeed. Many different species of mosquito belonging to at least six genera have been incriminated as vectors. One might be tempted to wonder if in fact most mosquitos are able to act as intermediary host to these worms, the main qualification being the readiness of the species to bite the right host in nature. But again things are not as simple as that. On the island of Pemba off the east coast of Africa, *Ae. pembaensis* is a vicious biter of man and other vertebrates. Although it serves as host to no less than five different species of filarial worm, it will not sustain development of *W. bancrofti*. We thus have host specificity. Moreover, there are also known to be genetic strains of the same species showing clear-cut differences in their capacity to act as hosts. For example, *Ae. aegypti* has a geographical race which is refractory to infection with subperiodic *B. malayi*. Thus in addition to host specificity we have intraspecific variation. Nevertheless, very many species of mosquito are involved, but they do not harbour these parasites without damage to themselves [409].

Before we consider the effect that these worms have on the mosquito host we must trace the fate of the parasites from the moment they are taken up with the blood of man or other vertebrate host until they enter the blood of another victim. The microfilariae of typical *W. bancrofti* are present in the blood of man in fantastic numbers. If one takes a single drop of blood during the hours of darkness by pricking the finger of someone suffering from filariasis (or, perhaps better, someone *with* rather than suffering, for he may or may not be aware that anything is wrong) and examines this drop while still fresh under the microscope, one finds the microfilariae in almost every field examined. Nor are they merely passive passengers; they may be seen as long slender creatures about 0.3 mm long, wriggling actively with regular and beautiful movements, knocking the blood cells this way and that. Each larva is sheathed in the elongated remains of its old egg-shell. We have already said something about the size of a mosquito's blood meal which could be described as being comparable in size to that of a large drop of blood. It is clear that any mosquito taking a blood meal at the time of maximum circulation or microfilariae in the host will suck up a number of these larval forms with the blood. The actual number varies greatly, but averages about ten microfilariae in every cubic millimetre of blood taken.

On reaching the stomach of the mosquito some of the microfilariae fail to survive or are digested along with the blood meal. Two to six hours later the surviving larval worms penetrate the wall of the midgut or stomach of the mosquito and migrate forward to the thorax. Once in the thorax they first move into the spaces between the individual fibres of the flight muscles and then actually penetrate the fibres themselves. Inside the muscle fibres the larvae shed their outer skin or moult. The second-stage larvae are shorter and more sausage-shaped. Later they undergo a second moult to produce greatly elongated filiform third stage larvae; these are 1.5 to 2 mm in length, and migrate first into the haemocoele and then into the head, where they take up positions within the proboscis sheath. The mature infective larval worms may be extremely active. Mature larvae of some species have been observed to move from the head of the mosquito and wander freely in the haemocoele. The infective larvae escape from the mosquito's proboscis, either from the tip of the labella or by rupturing the thin membranes at the base of the insect's mouth parts. Successful infection of a new vertebrate host depends on whether or not a potential male and a potential female worm reach the same lymph node or other site for adult development [409].

It seems that warmth and moisture from the vertebrate host of the mosquito stimulate migration of the infective larvae into the insect's labium, but that escape follows flexing of the labium and the entry of blood into the stomach. Infective larvae also sometimes leave when the mosquito is feeding on fruit juice. An important point to be remembered when we come to consider some of the other mosquito-borne diseases is that the mosquito may rid itself of all its filarial load and thus become non-infective again. This sometimes happens at a single feed. Mosquitos may also excrete their newly acquired microfilariae; *An. barbirostris* has been said to suffer from a bloody diarrhoea when fed on cats infected with *B. pahangi*. They have been known to pass more than 100 microfilariae through the anus during a single feed [212].

It is somewhat surprising that a mosquito is not only able to withstand invasion of its body by these relatively huge worms but can also live long enough to transmit them to a new vertebrate host. *Aedes aegypti* has been recorded in laboratory infections with as many as twenty larval *B. pahangi* in its thoracic muscles. One must remember that these worms at the infective stage may be two millimetres in length, and that the thoracic muscles form the 'engine' that drives the wings in flight. Little wonder, then, that heavy infection may reduce flying capacity and life expectancy. Pathogenicity of filarial worms to mosquitos has been studied in detail in the dog heart-worm, *D. immitis*. The larval worms of this species leave the gut and enter the malpighian tubules: eleven to twelve days later they migrate to the coelom and thence to the fat-bodies in the lower part of the thorax, before moving into the mouth parts. Damage by the larvae may lead to the premature death of the mosquito, invasion of the malpighian tubules leading to cytoplasmic degeneration of the lining cells, leaving the worms enclosed within a bag of basement membrane [406].

Despite these attendant risks no less than nineteen species of filarial worms are known to be transmitted by mosqui[illegible] more than thirty species of mosquitos have been shown to be involved. What, then, are the essential requirements for successful transmission? From our brief survey we may say at once that moderate levels of microfilaraemia in the vertebrate are more likely to lead to success than either very low or very high concentrations. If the mosquito ingests few microfilariae they may be killed or digested along with the blood meal. If on the other hand there are very many, the mosquito may die through gross damage to the wall of the gut. Now the adults of filarial worms are normally found in deep-seated organs within the vertebrate host. The larval worms too may spend much of the time out of reach of feeding mosquitos, but their circadian migration into the peripheral bloodstream makes them immediately accessible.

This sojourn in the peripheral flow for a part of each 24-hour period is clearly one of the requirements for successful transmission; if it coincides with the feeding cycle of large numbers of anthropophilic mosquitos contact between parasite and vector is assured. Deviant microfilariae will be unsuccessful in finding transport and harbourage for the next stages of development. We have here two independent circadian rhythms, one taking the microfilariae from the lungs to the peripheral blood once each twenty-four hours (whatever signal serves as a time-cue), the other bringing the mosquitos from their resting haunts to feed on the host. Natural selection will ensure that the circadian rhythm of the parasite is brought into phase with that of the vector [397].

MALARIA

Malaria and mosquitos–this association between insects and human disease is familiar to most of us. Indeed, to many the mere mention of mosquitos is enough to suggest malaria. It is, however, a sobering thought to recollect that it is not so long since medical dictionaries defined 'malaria' as a fever contracted from the miasma rising from swamps and prevented by sleeping with the windows closed. I have such a dictionary in my possession, dated

1900. Malaria—bad air. How easy it is to be right for the wrong reason. In nature malaria can only be contracted *via* the mosquito, and in theory is readily controlled by breaking the cycle of infection by attacking the insect at some stage. Yet it has been estimated as recently as 1957 that there are 250 million human cases each year, with more than two million deaths. Malaria is an incapacitating and debilitating disease and thus even without the mortality the economic losses are too great for us to be able to estimate. Nor are these losses evenly distributed throughout the world, for it is precisely those communities that can least afford this loss that suffer most from this scourge. Or, perhaps it would be more meaningful to put it the other way round—it is this sapping of energy and effort by malaria that holds down large areas of the warmer parts of the world to a low state of productivity.

It was Manson again, following close on his success with filariasis, who suggested to Ross that he should look in the mosquito for malaria parasites. In 1897 Ross duly found them in anopheline mosquitos and during the next three years much of the life cycle of these parasites was worked out. It was not, however, until nearly fifty years later that the complete life-history became clear. Through the work of Garnham and Shortt it is now known that the malaria parasites injected into the bloodstream by an infected mosquito first find their way to the liver, where they undergo what is called the exoerythrocytic cycle of multiplication. These protozoal parasites are thus faced with the same problem as that facing helminthic parasites causing filariasis dealt with in the previous section: how do parasites living in deep-seated organs of the human body provide a supply of infective forms in the peripheral blood flow, where they may be picked up by bloodsucking mosquitos?

Since our main concern is with mosquitos we shall not go too deeply into the complexities of the life-history of malaria parasites within the human host. We may note, however, that after a varying number of cycles of asexual reproduction and multiplication within the cells of the liver, the number of cycles depending largely on which of the four species of human malaria parasite is concerned, the resulting merozoites leave the liver and invade the red cells of the blood, and so initiate the erythrocytic phase. The period between the original infection of the human being by a mosquito and the appearance of parasites in the peripheral blood also varies with the species of parasite concerned, but is usually between six and ten days.

Once in the red blood cells a further cycle of asexual reproduction follows, each parasite dividing successively to give sixteen or so new merozoites. The red cell ruptures and releases these daughter parasites, every one of which in turn infects another red cell. Each of these erythrocytic cycles takes 1½ to 3 days (depending on species) and it needs little imagination to see that build-up of the number of parasites in the blood is prodigious. It is important to note that although there has been multiplication on a very large scale the merozoites released from the red cells are infective only to other red cells (or in some cases to cells of the liver) of the patient himself; if taken up in the blood by a mosquito they are merely digested. So also are the developing stages of the parasites within the red cells themselves. It is, however, interesting to note in passing that malaria parasites in man also exhibit a rhythm, resulting in synchronous division usually in the late morning. It is, in fact, the synchronous rupturing of countless red cells and release of the newly formed merozoites that is associated with the characteristic chill or rigour that so often heralds an attack of malaria.

Thus far, then, we have failed to answer the question of how the parasites become available to a feeding mosquito, even though there is an increasing number of parasites in the peripheral blood of the human host. The answer may at first seem strange. A relatively small number of the parasites that enter the red cells fail to divide into two, four, eight and so on, but develop into either male or female forms known as micro- and macro-gametocytes respectively. These are the infective forms. They are released at the time of the synchronous rupturing of the red cells, but unlike the merozoites they do no further harm to the host, but, so to speak, float around in the bloodstream in readiness to be picked up by a feeding mosquito—or so it used to be thought. But recent work by Frank Hawking and his team suggests that the gametocytes are infective for a limited period only, and that this is reached some seven hours after their release from the red cells. Thus they become mature after dark, just at the time when malaria-transmitting *Anopheles* begin to bite. Again we see how the circadian rhythm of the parasite is brought into phase with that of the vector [397, 404].

While in filariasis we saw multiplication leading to availability to the vector, in malaria multiplication and availability are independent of each other. Multiplication goes on by asexual reproduction on a huge scale within the human host, but development of sexual forms, the only forms available to the mosquito, involves really a decrease in numbers since each developing parasite produces either a male or a female cell, and these must eventually combine to form a single zygote if development is to proceed further.

If the blood meal of a mosquito contains male and female stages of one of the four species of human malaria parasites, one of two things can happen: they will be digested along with the blood meal or they initiate the next stage of development. Which of these happens depends on the mosquito concerned. Human malaria parasites have been found to develop only in anopheline mosquitos, but there are many species of *Anopheles* and we shall have more to say about this later on. What interests us now is the fate of, say, a single micro- and a single macro-gametocyte on entering the mosquito's stomach. This is a truly fascinating story and was described in detail soon after the discovery that mosquitos were responsible for the transmission of this disease. On arrival in the mosquito's stomach, the male or micro-gametocyte develops a number of whiplike tentacles known as flagella. These break away by a process known as exflagellation and become independently motile. If one flagellum finds the female or macro-gametocyte it enters the cell and sexual fusion takes place to form the zygote. The male cell that gave rise to the flagellum dies, but the fertilised zygote migrates to the wall of the insect host's stomach and penetrating the lining epithelium forms what is called the oöcyst stage.

Once settled in the stomach wall a series of nuclear divisions follows which eventually give rise to something like 10,000 daughter parasites known as sporozoites. The oöcyst bursts and releases the sporozoites into the body cavity of the mosquito, where they actively migrate to the salivary glands within the thorax. Here they remain in their thousands in readiness for introduction to a new human host. Unlike development in the human host, the interval between a mosquito's becoming *infected* with the male and female forms of the parasite to the time it becomes *infective* to another human being depends on the ambient temperature, since the mosquito provides the environment for this stage of the life cycle of the parasite and the mosquito's temperature fluctuates with that of the air in which it lives. It thus takes five to thirty days for the mosquito to reach the infective stage, when the sporozoites pass with the mosquito's saliva into the blood of the new host. Development of all four species of malaria parasite pathogenic to man is arrested in the mosquito at temperatures below 18°C and at 35°C or above. Although the time required at 30°C is about half that at 20°C, the latter is usually considered to be the optimum temperature, since the number of resulting sporozoites is said to be greater.

Summing up what we have said so far, we see that there are two distinct types of reproduction in the malaria parasite: a series of asexual cycles in the vertebrate host leading to the periodic release of merozoites in the blood, and a sexual cycle in the mosquito host leading to the formation of infective sporozoites in the insect's salivary glands. Both lead to prodigious numbers of parasites, the merozoites in the blood infecting further red cells, the sporozoites in the mosquito being held, as it were, in readiness to infect the liver and red cells of a new victim. In the example we took earlier we considered the case of a single male and a single female gametocyte entering the mosquito. In fact many gametocytes may be sucked up in a single meal and the mosquito's stomach wall may come to be dotted with oöcysts: if each produces some 10,000 sporozoites, the resulting numbers will be prodigious indeed: and yet, unlike the filarial infection considered earlier, the malaria parasites do not appear to affect the mosquitos adversely. The parasites are, of course, exceedingly small.

The four species of *Plasmodium* that cause malaria in man seem capable of undergoing their sexual cycle of reproduction within mosquitos of the genus *Anopheles* only. Why this should be so is not understood, particularly as there are many other species of *Plasmodium* infecting other vertebrates that develop satisfactorily in mosquitos of other genera. It must not, however, be thought that all species of *Anopheles* are automatically and equally implicated. Much depends on the feeding habits of the species concerned and also of the race within a species. In Africa there are some 100 species of *Anopheles,* and yet only two of these, *An. funestus* and *An. gambiae,* are dangerous vectors throughout their range. Even these include intra-specific forms or variants that for one reason or another play little part in the transmission of this disease, but this aspect is treated at greater length in Chapter 10. *An. coustani,* one of the most widespread anophelines in Africa, is usually of no importance whatever in the spread of this disease. One hesitates even to wonder how much money has been thrown away in attempts to control this species, the larvae of which are more commonly found than those of any other species of anopheline. Other species of African *Anopheles* are of local or temporary importance. Much the same applies in other parts of the world. Of the 300 or so species of *Anopheles* in the world list, about forty are of real importance as transmitters of human malaria [380].

In addition to the human malarias, *Plasmodium* parasites are found in apes, monkeys, rodents, birds, lizards and frogs. Bird malaria is transmitted by culicine mosquitos and has been much used in laboratory experiments on malaria and its treatment. Altogether some sixty or more species of malaria parasite are known, but while these are found in many different groups of vertebrates, the human malaria parasite appears to be confined to man (and certain species of *Anopheles*). It is this that lies behind the great campaigns of malaria eradication that have been and are now being carried out mainly by the World Health Organisation. If there are no animal reservoirs of the human malaria parasites then there is hope of breaking the link in the chain of infection somewhere, either by treating man, or by exterminating the vector. We shall take a further look at how successful this has been in Chapter 14. In the meantime we must record what may be a small cloud on the horizon. In 1960 one of the monkey malarias, a subspecies of *P. cynomolgi,* was found to be transmissible to man through the mosquito *An. freeborni.* Since, however, this species of *Plasmodium* is not normally found in man, the alarm is probably false. Human malarias have been recorded in chimpanzees—but these apes are probably not sufficiently numerous to constitute more than an occasional threat.

It must not be thought that every individual *An. gambiae* or other malaria vector species is infected with malaria parasites. Each species of mosquito is split up into intra-breeding populations that are to a varying extent separated from other populations of the same species. The proportion of females of *An. gambiae* with infective sporozoites in the salivary glands will vary from population to population and with the passage of time in any single population. This proportion may remain very low indeed during endemic conditions when the disease smoulders on with just a sufficient number of human cases to ensure a steady but low spread of infection in the mosquito population, and just sufficient of the mosquitos thus infected living long enough to allow the parasites to reach the salivary glands which, as we have seen, may take over three weeks when the mean temperature is low. The proportion may rise to as high as thirteen per cent during epidemic conditions [11]. We have seen that filarial infections in mosquitos may be self-limiting since if the parasites become too numerous the vectors stand less chance of survival. No such limiting factor appears to operate in malarial infection unless death in a significantly high proportion of the human host materially reduces the supply of infective gametocytes. We shall have much more to say about this aspect when we consider yellow fever and the other virus infections involving mosquitos.

YELLOW FEVER

In 1848 Josiah Nott, an American doctor living in Alabama, wrote suggesting that both malaria and yellow fever were spread by mosquitos. Six years later, Dr Louis Beauperthuy of Venezuela put forward a similar argument, but confined himself to yellow fever. He believed that a poison was responsible for this disease and that certain mosquitos picked this up from swamps and injected it into man at the time of feeding [366, 370]. But the world was still not ready for these original thinkers with their off-beat ideas and, as so often, they were dismissed by the current weight of medical opinion.

The views put forward by these two men were as bold as they were novel. They must be seen against a background of medical opinion still convinced of the importance of vapours and effluvias in the spread of disease, and against a background of lay opinion still obsessed by the power of evil and witchcraft. Louis Pasteur was still a young organic chemist. Robert Koch had not yet reached his teens. In the meantime countless numbers of people were to go on suffering and dying from these two scourges.

Yellow fever in the mid-nineteenth century lay seige to the whole Caribbean region; every island and coastal town was subject to devastating attacks of the disease. One army after another was totally defeated by the unseen enemy before the troops had had time for any engagement with the human armies of the opposing side. Work on the Panama Canal was brought to a complete standstill because of the ravages of yellow jack, as the disease was generally called by the seafaring folk who had to run its gauntlet. Nor was the disease restricted to the Caribbean, for its ravages were largely responsible for the title of 'white man's grave' given to parts of West Africa.

Some thirty years after Nott and Beauperthuy independently put forward their original but still premature suggestions that mosquitos were responsible for the spread of yellow fever, Carlos Finlay not only made a similar suggestion but went further and put his hypothesis to the test in a series of carefully planned experiments. Finlay was a medical man practising in Havana. He had already demonstrated to his own satisfaction that an outbreak of cholera that visited the city in 1867 was water-borne, even though his views, running counter to accepted opinion of the time, were refused publication. He turned his attention to yellow fever in 1872 and produced a long series of papers on this terrible disease. He published his mosquito theory in 1881, even accurately naming the species of mosquito responsible, the one now known as *Ae. aegypti* [389, 390].

Finlay noted that epidemics did not occur in the absence of a previous case. He also noted that survivors were immune to further attacks and that epidemics did not occur unless non-immune subjects were present. He even established the concept of inapparent infections, showing that yellow fever in the indigenous population was sometimes of a very mild nature as far as the symptoms were concerned; it was mainly in new immigrants that the typical fulminating symptoms occurred, so often leading to death. He postulated the presence of a necessary transmitting agent which was both independent of the disease and of the patient. He had already noted the prevalence of *Ae. aegypti* in houses during epidemic periods, including those of patients, and that the season of mosquito prevalence and outbreaks of the disease coincided. It is tragic that although this keen and accurate observer spent the following twenty years trying to demonstrate the validity of his hypothesis, he was never actually able to transmit the disease by mosquito bite. He carried out no less than 103 transmission trials and altogether published about ninety papers on yellow fever. But it was left to Walter Reed in 1900 to make the experiments that finally convinced the world, experiments really of desperation when all else had failed.

Reed was well grounded in the new germ theory of disease that had begun to revolutionise medicine since 1866. He went to Cuba as head of the Fourth American Commission for the Study of

Yellow Fever in 1900. The three earlier commissions had ruled out Finlay's claims in favour of bacterial contamination through direct and indirect contact. Volunteers slept in huts previously occupied by yellow fever victims. Others slept elsewhere but wrapped themselves in the blankets that had previously covered those who had died of the disease. Reed and his three companions, Carroll, Lazear and Agramonte, found themselves continuing this work and started off by studying the gut flora of yellow fever cases and other similar bacteriological work. Finally it was suggested to Reed that he should put Finlay's claim to the test, and in a series of experiments his team was able to show that the spread of this terrible disease depended solely on the mosquito *Ae. aegypti.* But this great triumph was bought dearly, for Lazear died of yellow fever in the course of these very experiments [421, 432].

What had gone wrong with Finlay's experiments? On reading through his original papers one is instantly struck by the careful planning that went into his work. One thing is certain, he was no mere fanatic obsessed with an idea but lacking in scientific method and discipline. It seems probable that he did not know of Manson's work on the development of filariae in the mosquito. His hypothesis was based on careful observation; his experiments were everything that good experiments should be. Why, then, was he dogged with consistent failure for twenty years? Again, one must try to see the problem in its context. He postulated an agent in the blood that could be picked up by a blood-feeding mosquito and then inoculated by it into a susceptible person. No one could have foreseen that the agent or pathogen causing yellow fever was to turn out to be a virus, so small that it could never be seen by the ordinary light microscope. Nor could one have predicted that the virus on entering the mosquito virtually 'disappears' or becomes undetectable during what is now known as the eclipse period, and that only later does it 'reappear', or rather becomes detectable again. Virus particles that are not digested by the mosquito invade and multiply within the actual cells lining the stomach. It is during this intra-cellular phase that they become undetectable. After a varying period depending on the outside temperature, they break out into the haemocoele and invade the whole body including the salivary glands. At average temperatures of around 23°C it takes as long as eighteen days before the virus reaches the salivary glands and the mosquito becomes infective.

Finlay did not keep his mosquitos long enough to allow for this extrinsic incubation period. But how could he have known: at least Manson and Ross could see their parasites. Not only can yellow fever virus not be seen under the ordinary microscope but it causes no visible response whatever on the part of the mosquito's tissues even when the virus has multiplied within them by a factor of a hundred thousand or so. We know now that a mosquito infected with yellow fever virus lives as long as its uninfected companions, and once the eclipse period is over the mosquito remains both infected and infective for life. Reed's team kept their mosquitos long enough; this seems to have been the only difference.

The early history of what seemed then to be the conquest of yellow fever is a long and exciting one, and I have only touched on those aspects here that illustrate certain points which we shall need to bear in mind as we skip the years and attempt to bring this fascinating story up-to-date. I shall not say much of the seven others who gave their lives in fitting the various pieces into the jigsaw, for we are really concerned mainly with the mosquito's role. Nevertheless before I can begin to bring matters up-to-date I shall have to mention certain key events that have led us to our present understanding of this complex story. The Reed Commission had demonstrated that yellow fever was transmitted by the mosquito, *Ae. aegypti.* Now this species is essentially a domestic mosquito; throughout most of its range the larvae are found almost exclusively in man-made containers: discarded cans, water-storage tanks, broken bottles, blocked gutters and so on. It was not surprising, therefore, that the disease rapidly disappeared when sanitary measures were enforced in towns and villages first in Cuba and then in Central and South America. It should be noted, however, that while the disease was eradicated in Cuba by

1901, only a year after the Reed report, mysterious outbreaks kept on appearing in various parts of South America.

In Africa the situation was somewhat different. Although the same vector species was concerned, most of the cases that occurred in the African population were very mild, resembling the inapparent infections that Finlay had found in Cuba. When, however, the virus was transmitted to the previously unexposed European population, the disease was in every way similar to the yellow fever found in the Americas—fulminating fever, bloody-black or coffee-ground vomit, jaundice and, often, death. Little wonder that West Africa became known as white man's grave.

The first of the key events that will help us through this story is clearly the discovery of the role of the mosquito *Ae. aegypti,* including the period of development of the virus in the mosquito. This extrinsic incubation period varies from five to thirty days, depending on temperature. The second is the presence of mild inapparent infections in a high proportion of indigenous peoples. The third is the immunity in man that follows recovery from an attack of the disease, whether the attack had been mild or severe. The fourth key event that was later to prove of far greater significance than was at first supposed was the demonstration that primates other than man are not only susceptible to the virus but may show symptoms almost identical with those exhibited by man, leading to death or to the development of a similar lasting immunity. A chimpanzee was infected in 1907 and rhesus monkeys in 1927. Adrian Stokes, a member of the team connected with this important landmark in the history of yellow fever, died of the disease later that year; *Ae. stokesi,* a proven yellow fever transmitter in the laboratory bears his name [432].

Manson saw the implications even before the successful transfer of virus to monkeys was achieved. While it was true that it would be extremely useful to have a laboratory animal that was susceptible to yellow fever, and certainly the lack of one had seriously hampered research thus far, there was another more ominous side to the story. As early as 1909 he drew attention to the possibility that if monkeys could be infected by mosquitos that had fed on infected human beings, the reverse could also be true. Up till now, however, everything had fitted neatly into the simple story of transmission from man to man by an easily controlled species of mosquito. Then suddenly in 1932 cases of yellow fever came to be recognised in a rural area of South America and in the complete absence of *Ae. aegypti*. This, then, was the fifth event and one that rocked the whole story. It seemed that there were two entities: urban and jungle yellow fever. Epidemiologically this is indeed the case. The virus is, however, the same. It is the vector and the events leading up to human infections that differ. In South and Central America we know now that yellow fever can really be considered as a disease of forest monkeys. Indeed, the progress of an epidemic can be followed in space and time by recording the presence of dead and dying monkeys in the forest [372].

Although jungle yellow fever was not finally recognised until 1933 and the forest cycle in monkeys was not worked out until ten years or so later by John Bugher, Jorge Boshell and their collaborators, certain events which had previously been recorded but promptly swept under the mat as of no relevance suddenly became clear. The South American monkeys involved include the howlers, *Alouatta,* so called on account of the very loud call which they make. There is a report of a sinister silence in the jungles of Verapaces when, in 1901, the howlers were wiped out by a wave of disease; this was followed by a severe human epidemic of yellow fever. In 1914 it was reported that the people of Trinidad held firmly to a 'superstition' that epidemics of yellow fever could be foretold by the numbers of howlers that died in the forest.

The monkey or forest cycle in South America is transmitted by certain forest mosquitos in the total absence of *Ae. aegypti* and man. The most important of these are *Haemagogus spegazzinii falco* and other species of this genus, and to a lesser extent *Ae. leucocelaenus*. These mosquitos, like the New World monkeys on which they prey, are for the most part arboreal. Both mosquito and monkey spend most of their time in the forest canopy and it is usually only the people who cut down large forest trees that expose themselves to attacks by these mosquitos. Indeed, so regular is this pattern that jungle yellow fever in South America has been described as an occupational disease of woodcutters. Jungle yellow fever occurs when the independent forest cycle spills over in this way to a human being; what has been called the tangential spread to man.

The incubation period of yellow fever, or the intrinsic incubation period as it is called, is about four to five days in man and usually somewhat less in monkeys. At this time virus 'appears' in the peripheral blood whether the symptoms of the disease are overt or inapparent. The virus may be present in enormous quantities, one one-hundred-millionth of one millilitre of blood being enough to infect a non-immune monkey. The important thing, however, is that the virus then 'disappears' after about two more days, never to 'appear' in the blood again. Indeed, as the virus level drops antibody appears in the blood and gives rise to an immunity that usually lasts for life. Thus, and this may be taken as point number six, virus is accessible to a feeding mosquito for about two days only in the life of any one man, or for that matter of any one monkey. This is quite unlike anything with which we have so far dealt. In filariasis we found a nightly invasion of microfilariae into the peripheral blood; in untreated malaria we saw a near permanent parasitaemia. In yellow fever we have a single transient viraemia of about two days, followed by permanent immunity. If, then, we refer to monkeys as reservoirs of virus we must be careful to emphasise that we are referring to populations and not to individual monkeys.

Event number seven in this story is the finding in 1931 by Max Theiler that immunity in man can be put to the test at any time. It thus became possible to carry out immunity surveys in human populations; anyone who had had the disease would show up as an immune, even if the event occurred in childhood and the test was not made till twenty or forty years later. For this and his work on developing a yellow fever vaccine Theiler won the Nobel Prize. When the immunity test was applied to Africa in an effort to map out the distribution of yellow fever in that vast continent some surprising things emerged. The Rockefeller Foundation, which had virtually organised all the research on yellow fever since 1916, extended its activities to Africa in the twenties, first to West Africa where yellow fever was almost continuously present and then to the area known politically as East Africa where no case of yellow fever or anything even remotely resembling it had ever been reported. Not unnaturally this work came in for some pretty severe criticism from people in the country who could not be expected to appreciate the full significance of the presence of immunes and the concept of inapparent infections. This came not only from the administration but even from members of the medical profession: why spend all this money on yellow fever research; why study a disease that is not there, when people are actually suffering and dying from sleeping sickness, plague, leprosy and malaria?

The work in East Africa, starting as an immunity survey in an attempt to define the eastern limit of the disease, turned up scattered immunes among the human population of Uganda [405]. This led to full-scale investigations which partly followed and partly overlapped the work going on in South America that was later to reveal the forest cycle. Virus was eventually isolated from man and from mosquitos in a cultivated rural area in a clearing near the edge of the Semliki Forest in Bwamba, a small region lying on the Congo–Uganda border in the western rift valley at the foot of the Ruwenzori mountains [415]. Now the extraordinary thing about this region is that although *Ae. aegypti* is present, it does not feed on man. One may find the larvae in tree-holes in significant numbers, but water-storage pots are rarely occupied. The adults themselves are hard to find. The role of domestic mosquito seems to have been taken over by *Ae. simpsoni,* a species, as we have seen in Chapter 3, breeding in the leaf axils of certain plants: colocasias, dracaenas, pineapples, *Sansevieria* and two of the many species and varieties of plantains or bananas. These are all in one way or another economic or subsistence crop plants, although the dracaena's role is merely hedging and shelter. Thus the plants are grown in concentrations of no great size around every house in Bwamba, almost every area being alongside or cut out of the forest itself. Each axil, as we have seen, does not hold more than a few millilitres of water, but what they lack in size they more than make up for in number. *Ae. simpsoni* feeds as an adult in the day-time, often in bright sunlight. It is present in enormous numbers in each cultivated area around each homestead. It is responsible for the human cycle of yellow fever in these rural areas, a cycle that otherwise resembles the classical urban cycle with *Ae. aegypti.*

The focus of attention now moved to the forest in what may be described as a gigantic ecological study of mosquito, monkey and virus. Rather, it started on a very wide front embracing most of the forest animals [428]. When, however, antibody studies indicated that of all the vertebrates only monkeys were involved, and some of these lived their lives exclusively in the forest canopy [403], the work narrowed down to a study of the possible vectors in the canopy. The picture of the stratification of mosquitos in rain-forest that emerged and the separation of their feeding by level or by time or both has been dealt with in a previous chapter [229]. Not only was this forest cycle found to occur in every lowland rain-forest studied in East Africa, but an identical cycle involving monkeys and *Ae. africanus* was also shown to occur in West Africa, some 3000 km away across this vast continent [235]. In Africa, then, as in South and Central America, yellow fever turns out to be essentially a disease of forest monkeys. It is transmitted directly from monkey to monkey by *Ae. africanus,* a species of mosquito that not only breeds in tree-holes but feeds as an adult mainly in the forest canopy. This mosquito, as we have already seen, feeds during the first half hour following sundown, at a time when monkeys, being diurnal creatures like ourselves, have taken up their positions for the night in the crowns of their sleeping trees [411, 412, 430]. As in tropical America this forest cycle or extrahuman cycle as it is sometimes called goes on in the absence of man and *Ae. aegypti.*

We saw that in tropical America man becomes infected by disturbance of the tree-top mosquito, *Haemagogus spegazzinii,* during tree-felling. While occasional contact is made in Africa between man and *Ae. africanus* on the ground, these are relatively isolated occurrences involving a few individual mosquitos only, not clouds of them as in South America. The statistical chances of virus transfer at such contacts are clearly low for not only is the mosquito population for the most part twenty to thirty metres above the ground but feeding mainly takes place after dark. While man may occasionally find himself in the forest at sundown or he may conceivably climb to the leafy crown of one of the large forest trees, it is extremely unlikely that he will do both of these things at the same time. Nevertheless, we have seen that *Ae. africanus* does feed in ones and twos away from its typical feeding site and there is a steady trickle of isolated cases as shown by subsequent antibody studies that can only be accounted for by such contact [394, 395]. It must be borne in mind that once immune, always immune, and a person showing antibodies to yellow fever today may in fact have become immune following infection many years ago. These isolated cases are of persistent immunity not of virus activity and thus the temporal element must be taken into account when considering them. It may well be that as the timber industry grows in tropical Africa direct contact between *Ae. africanus* and man will become more frequent and thus forge a direct link between the forest and the human cycle similar to that now found in tropical America.

There is another and more important difference which, although seemingly small in itself, is of great biological interest and has the effect of making the epidemiology of yellow fever in Africa very different from that in tropical America. For the most part African monkeys are relatively insusceptible to the symptoms of yellow fever. While a particular strain of virus transmitted by *Ae. aegypti* in the laboratory to American monkeys will cause severe illness and probably death, the same strain of virus transmitted by the same species of mosquito to African monkeys very often leads to an inapparent infection or only mild illness for one or two days. In tropical America the monkey cycle can only spill over to start a human cycle by the agency of a mosquito that will bite both hosts, this part being played mainly by *Haemagogus*. In Africa a monkey circulating the virus in its blood is usually well enough to travel about in normal fashion. While many of the tree-top monkeys live almost exclusively in the canopy itself, rarely coming down to ground level and probably never leaving the forest, there is one species, the red-tailed guenon, that, while spending much of its time in the canopy, makes daily excursions outside the forest [403]. Bananas are a great attraction to the red-tail, and since, as we have seen, these plants are usually grown in clearings cut out of the edge of the forest itself, no great journey is involved to reach them. A red-tail with a high-level yellow fever viraemia and highly infectious to a feeding mosquito but otherwise perfectly normal, may thus come to be fed on by *Ae. simpsoni,* the banana-axil mosquito responsible for rural yellow fever. If this, or as is more likely, these mosquitos survive long enough to allow the completion of the extrinsic incubation period of the virus, they will transmit the virus to man and on occasion to another visiting red-tail, provided that both man and visiting monkey are non-immune.

We thus have all the requirements both for the maintenance of the virus in the *Ae. africanus* transmitted tree-top forest cycle in monkeys and for the establishment of the *Ae. simpsoni* transmitted rural cycle in man, with the red-tail monkey serving as the mobile link between the two. Why, then, do we not find evidence of widespread human yellow fever in most of the African villages situated alongside or even cut out of the forest itself. We know that virus activity is going on in the forest canopy from the evidence provided by immunity in the monkeys and by isolations of virus from *Ae. africanus.* We know that red-tails come in from the forest to raid the banana gardens where *Ae. simpsoni* abounds. We have seen in Bwamba that this results in the establishment of rural cycles in man. Elsewhere, for the most part, human beings remain free from infection, even in areas where *Ae. simpsoni* is as plentiful as in Bwamba [386].

The explanation of this apparent enigma was to be found not so much in the ways of man and monkeys, as was at one time supposed, but in the behaviour of the mosquitos. Just as *Ae. aegypti* occurs in mainly anthropophilic and mainly non-anthropophilic populations, so also does *Ae. simpsoni.* Throughout much of its range this species simply does not feed on man; it prefers other hosts and the human cycle of yellow fever is broken or, rather, it is never completed. In most of Uganda the distribution of rural yellow fever fits closely the known distribution of the man-biting populations of this species. Where *Ae. simpsoni* does feed on man then, provided that raiding monkeys from nearby lowland forest are also present, a risk of rural yellow fever exists [394, 395].

On the cultivated lower slopes of Kilimanjaro man-biting *Ae. simpsoni* are present in fantastic numbers but the risk of yellow fever would appear to be nil. Below the cultivated region there is no forest to speak of; above, in the mountain-forest where monkeys are certainly present, the mean temperature is too low to allow sufficiently rapid multiplication of any virus within a potential vector. Once again we see the value and importance of considering each species of mosquito as a collection of partially isolated and largely non-interbreeding populations [222].

The difference between yellow fever epidemiology in tropical America where the mosquito serves as the link connecting the forest and the human cycles, and in Africa where the monkey itself forms this link, depends mostly on one thing: in tropical America the monkeys are sick and die, whereas in Africa they are largely unaffected. Survivors in both continents, however, are solidly immune and can play no further part. It is true that the first experiments on transmitting yellow fever virus to monkeys by mosquito bite were carried out in Africa, and these monkeys, as we have seen, showed all the severe symptoms one normally associates with yellow fever. But the monkeys used were *Rhesus* monkeys shipped from India.

Now this difference between the response of African and New World monkeys has been the subject of much argument and speculation but it has always appeared to me to have a simple and beautiful explanation, biologically speaking. Monkeys are slow-developing animals and do not reach the breeding age until they are four or five years old. Hence there is ample time for contact to be made with infected mosquitos before offspring are produced. Those monkeys showing a violent response to the virus will be

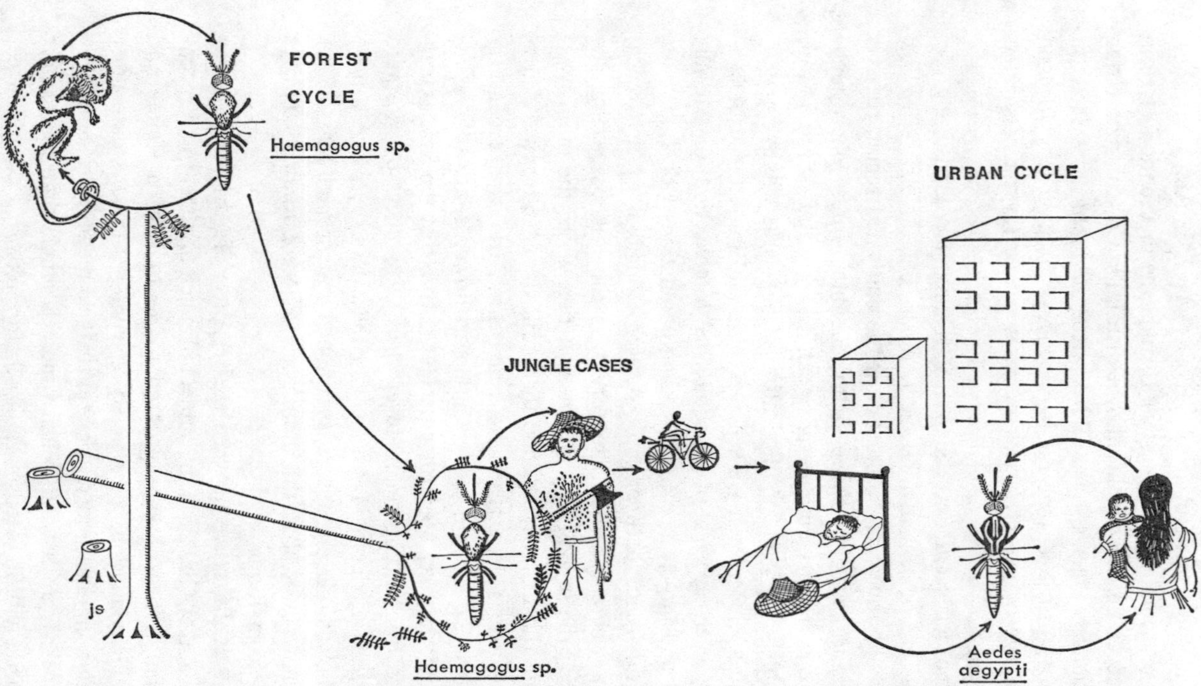

Figure 19. The forest cycle of yellow fever transmission in Central and South America, leading to isolated cases of jungle yellow fever among tree-cutters and, indirectly, to epidemics of urban yellow fever. The infected monkeys usually die of the disease and it is the *infected mosquitos* that act as the *link-host* between the tree-top monkey cycle and human cases below, while man himself acts as the link-host between jungle cases and urban epidemics.

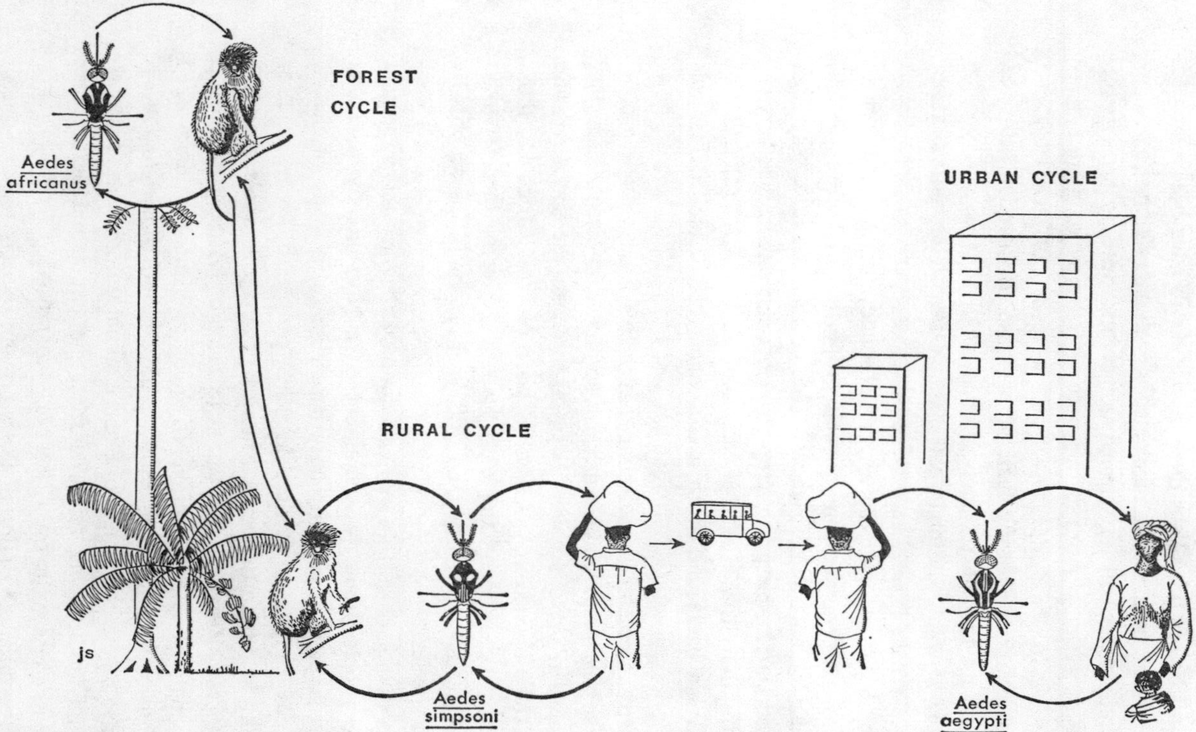

Figure 20. The forest cycle of yellow fever transmission in tropical Africa, leading to outbreaks of rural yellow fever in nearby villages and, indirectly, to epidemics of urban yellow fever. The *infected monkeys are rarely adversely affected* and, remaining active, themselves act as the *link-host* between the tree-top cycle and man-feeding mosquitos below, while man acts as the link-host between rural outbreak and urban epidemic.

killed off before this. Thus most monkeys will have been parented by relatively insusceptible monkeys; in each generation the susceptible individuals will be selected out before breeding age. In the course of every generation this natural selection of monkeys insusceptible to the symptoms of yellow fever can have only one result: monkeys showing lack of response to infection with the virus except for the production of antibodies. It is a very clear example of what one might call classical Darwinian natural selection.

The absence of symptoms is not absolute; African monkeys do show violent symptoms in a small proportion of the population, whereas Indian monkeys show violent symptoms in a very high proportion. The picture in tropical American monkeys is very interesting indeed. We have already seen that a high proportion of these react badly and death is a frequent result, and at first this may appear to contradict what has just been said. There is, however, strong reason for believing that yellow fever virus is a comparative newcomer to the American continent, infected *Ae. aegypti* and infected persons having been taken across the South Atlantic when trade opened up between the ports of West Africa and those of South America and the Caribbean. If this is so then it would seem that it takes more than 300 years to build insusceptibility into a species.

We cannot close this section without extending this argument to the people in the two continents. We have already seen that yellow fever as a clinical entity hardly exists in many of the indigenous peoples of Africa. In other people, however, whether residing in Africa or in the Americas, there is a whole spectrum of symptoms from the inapparent to those of typical yellow fever, but people of African origin living in the Americas make up a high proportion of those showing little or no symptoms. Again we have a perfect example of Darwinian natural selection, susceptible individuals being removed from the population before they can pass these traits on to the next generation.

The gradual unfolding of this story must surely rank as one of the great feats of detective epidemiology of this century. It is a story of imagination and grandeur of design. So many new concepts were established during this work: transient viraemia in the vertebrate host followed by lasting and testable immunity in the survivors; intrinsic incubation period in the vertebrate host of more or less fixed duration; extrinsic incubation period including an eclipse period in the mosquito host of variable duration, depending on temperature; intracellular virus multiplication in both hosts; extrahuman cycles of transmission involving mosquitos and other vertebrate species; mosquito- and monkey-links between the two cycles.

OTHER DISEASES

It has been necessary to deal at some length with yellow fever since it was the first of very many mosquito-borne diseases to be shown to be of viral origin, and the methods developed for its study and the concepts that have grown out of these have now become the model for all such studies. It will help us take stock of the situation if we remember that at the turn of the century yellow fever was the only arthropod-borne virus disease known: there are now more than 200 different viruses known to be so transmitted, over 100 of them involving mosquitos. Clearly it would be beyond the scope of this book to deal with all of these, and I have selected only those that are of more importance to man, or those that pinpoint certain special features in the part the mosquito plays in these complex relations of insect, pathogen and vertebrate host. It will not be necessary to deal even with these at such length since the main concepts which have been established with yellow fever apply to most of them. More than a hundred species of mosquitos belonging to no less than sixteen of the thirty-three genera have so far been incriminated one way or another in the transmission of viruses; at least forty of these belong to the genus *Aedes* and some twenty to *Culex* (see appendices).

During the 1930s severe epidemics of what is now known as western equine encephalomyelitis occurred in the enormous horse populations of the western United States. These frequently involved death of horses on a large scale and the economic loss to a country in which the horse played such a key role was considerable. But while all this was going on there occurred also cases of mysterious sickness in man with symptoms not unlike those of poliomyelitis. Indeed, the disease was at first diagnosed as poliomyelitis, so often did it leave the survivors with a lasting paralysis in one part of the body or another. When, however, the new virological techniques developed by the Rockefeller Foundation in their work on yellow fever were applied to this problem it was found that a single virus was responsible both for the disease in horses and for the polio-like disease in man. Western equine encephalomyelitis virus, or WEE virus as it is usually called, could be isolated from the blood of horses and from the blood of man during epidemics of the disease. Survivors of both species developed lasting antibodies to the virus, even those in which the infection had been inapparent or symptomless. The parallel with yellow fever seemed at first to be so close that the assumption was made that all that had now to be done was to find the vector linking the cycles in horse and man.

Work on this important disease, carried out mainly by William Hammon and William Reeves led eventually to the recovery of WEE virus from the mosquito *Culex tarsalis*. All seemed set for the completion of the story: here were what appeared to be the cycles in animals and man, and here was the vector which subsequent study showed fed on both hosts. But once again it is dangerous to jump to conclusions where nature is concerned; although WEE virus duly appears in the blood of horses during the early stages of the disease, it is present at such low levels that mosquitos feeding on these horses rarely become infected. Moreover, the same applies in human infections. Thus neither horse nor man can play the part of maintenance hosts to the virus; infection in both species has to be regarded as a dead-end infection and not part of any cycle. Where, then, does *C. tarsalis* pick up its virus if not from horse or man?

Since the general plan of attack followed very closely the work already done or then being done on yellow fever, I will have to limit myself here to the main conclusions and omit the exciting steps that led to them. *Culex* as we have seen in Chapter 8 is, with notable exceptions, a bird-feeding genus of mosquitos. While *C. tarsalis* is essentially a bird-feeder too, it will also attack any available vertebrate that comes its way, including reptiles and man. It picks up its virus load from birds and after an extrinsic incubation period during which the virus may multiply within the mosquito a million-fold, passes it on to other birds in an avian cycle and to horses and man as dead-end tangential spin-offs [371, 422].

In the central valley of California, one of the great crop-growing areas in the world, food plants of all sorts, both temperate and subtropical, are grown on a scale almost unequalled in any other part of the world. Rice, oranges, corn and olives grow in such abundance that one is apt to overlook the rather extraordinary fact that the rainfall is only some 150 mm (6 in.) per year. In other words, the valley is desert. The secret of course is irrigation; the whole valley, which is 900 km in length and some 200 km in width, is surrounded by mountain ranges and water has been tapped and piped or channelled to all the main growing areas. Water is so plentiful that great stretches of land are at times lying under the water. It is in these man-made mosquito nurseries that *C. tarsalis* breeds in unimaginable numbers. It seems probable that without the irrigation that supports the crops WEE would have remained an undetected virus infection of birds—one of many.

In 1953 an earthquake of considerable magnitude, greater in fact than the famous one that resulted in the destruction of San Francisco in 1906, occurred in the region of Bakersfield, an oil town in the middle of the San Joachin Valley, to give the central valley its proper name. Bill Reeves has told me how the earthquake was followed by a human epidemic of WEE, also of considerable magnitude. It is estimated that the epidemic infected thousands of people, of whom 375 were severely ill and nine died. Now it is not unusual for earthquakes and other major upheavals, both natural and man-induced, to be followed by pestilence of one sort or another. Usually, however, these plagues can be attributed directly or indirectly to the breakdown of water supplies or to the breakdown of public services and to the appearance of hordes of rats. But why should damage to property lead to a mosquito-borne virus disease? The answer, as so often, was really obvious in the clear light of hindsight. When buildings began to crumble and fall many people left the city and went to sleep out in the fields. Houses in Bakersfield, as in many other parts of the United States, are protected from mosquitos by metal gauze over all the windows and doors. Outside, the people slept unprotected from the onslaught of vast hordes of *C. tarsalis*.

Laboratory experiments have shown that although as many as sixteen different species of mosquitos belonging to five separate genera can be infected with WEE virus, only eleven of these actually transmit the virus to another host when given the opportunity for a subsequent blood meal. While some of these species may be concerned in the avian cycle, only *C. tarsalis* seems to be important in transmitting the virus from birds to horses and from birds to man [371, 374, 378].

Another disease producing similar symptoms and also occurring both in horses and man is found in other parts of the United States, notably in the south-eastern seaboard. It is known as eastern equine encephalomyelitis, or EEE. As in WEE, horses and man appear generally to be dead-end infections, the natural cycles occurring mainly in waders and other water-birds and in the swamp-breeding mosquito *Culiseta melanura,* and to a lesser extent in *Coquillettidia perturbans*. This swamp cycle goes on at a low level independent of man or horse, only about one per cent of the water-birds becoming involved. From time to time virus activity appears to increase, and the infection rate in the water-birds rises; when it reaches about eleven per cent the virus tends to 'spill over' into other birds, including sparrows and pigeons. At such times other species of mosquito also become involved, species that bite man and horses as well as birds. Characteristic outbreaks then occur in man and horses, but these outbreaks are rarely so extensive or on such a large scale as those of the western virus. No less than twenty-one different species of mosquitos have been infected when fed on viraemic hosts but only thirteen

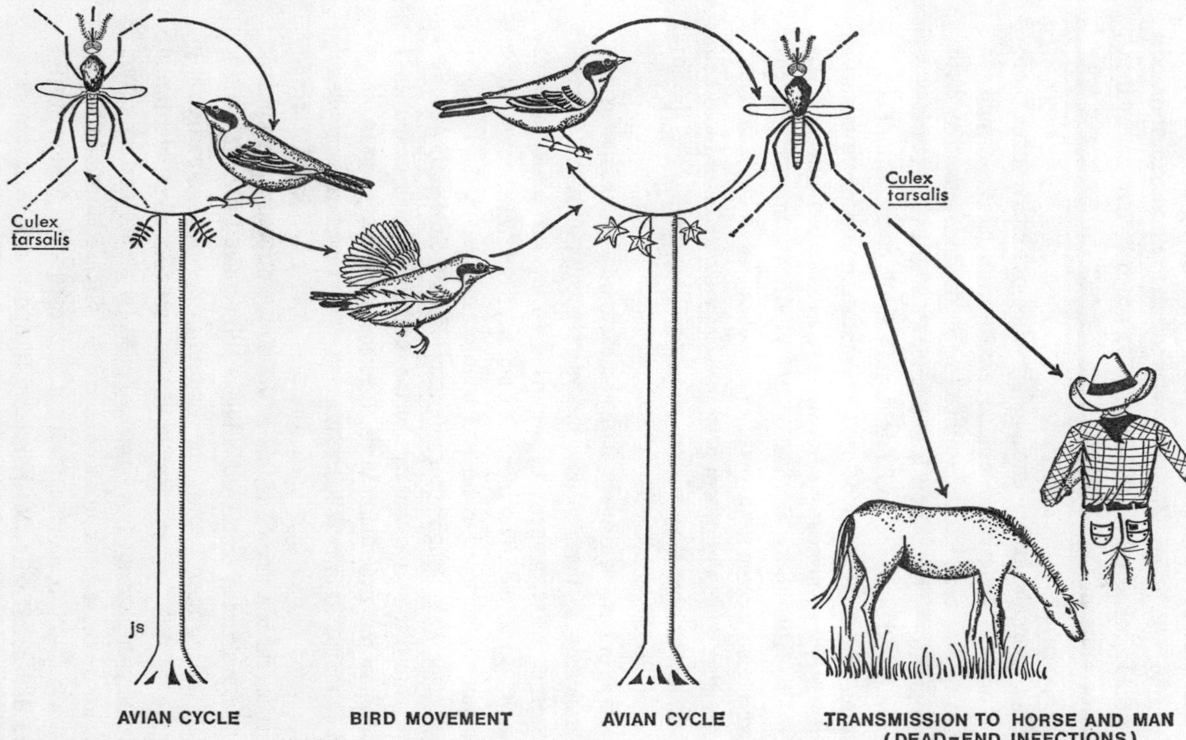

Figure 21. The tree-top avian cycle of western equine encephalomyelitis and tangential spread to man and horses as 'dead-end' infections.

of these were able actually to transmit the virus by bite under experimental conditions. Some species of mosquitos are infected more readily than others, *Culiseta melanura* becoming infected at levels of virus in the blood too low to infect most other species [375–7].

Other mosquito-borne encephalomyelitis-producing viruses are also known: Venezuelan equine encephalomyelitis or VEE is related to and resembles the other two in its involvement of both horses and man. Like them it is essentially a bird virus but is confined to South America and Trinidad. St Louis encephalitis (SLE) virus on the other hand belongs to another group of viruses, although it is transmitted in rural parts of the western United States by *C. tarsalis,* the main vector of WEE. At times *C. pipiens fatigans* also becomes involved, particularly in urban outbreaks further east. Like the others it is essentially a disease of birds and outbreaks in domestic fowls are common. Elsewhere other species of mosquito are concerned.

In one important respect, with the exception of VEE virus, the epidemiology of these viruses differs markedly from that of yellow fever. What happens to the virus during the winter months when it is too cold for mosquito activity? Several theories have been put to the test but no real solution has yet been found. At one time it was thought that migrating birds from South America reintroduced the virus, but this seems unlikely because of the short duration of the viraemia in birds. It seems more likely that the virus is maintained in mosquitos resting in cellars and similar places. WEE virus has in fact been found in over-wintering *C. tarsalis* but it is still not known if this is the only mechanism involved. It has also been recovered from a hibernating snake [407, 422, 435].

It has long been known that visitors to the tropics, particularly to parts of Asia and the Pacific, are subject to a short, sharp and extremely unpleasant fever accompanied by a rash and by severe pain in the back, neck or elsewhere. The disease is usually known as dengue, or break-bone fever. It is caused by infection with a virus that is not only closely related to that of yellow fever, but is transmitted to man mainly by *Ae. aegypti.* This much has been known since 1916. It seems that *Ae. aegypti* picks up the virus from unprotected patients suffering from dengue, or from persons with inapparent infections, and following an extrinsic incubation period, passes it on to susceptible people who, as might be expected, are usually newcomers to the area. From time to time dengue also breaks out in areas north and south of the tropical endemic zone. These epidemics, which have sometimes been of enormous size, are confined to the summer-time when *Ae. aegypti* is able temporarily to spread its range, and they often involve most of the human population in the area. In the tropics *Ae. albopictus* seems to be concerned in the more rural areas. But while in all the other mosquito-borne viruses that we have so far considered, and in many others besides, careful and patient search has revealed other vertebrate hosts and cycles of infection that are independent of man, no such extra-human cycle has been found with dengue virus although there is mounting evidence that monkeys and other primates may be involved.

There is another very curious thing about dengue virus, or rather dengue viruses, for there are several different types of the virus. I have already pointed out that it is related to yellow fever virus and that it is transmitted, at least in urban areas, by the classical yellow fever mosquito, *Ae. aegypti*. But the two viruses seem in some way to be mutually exclusive; there are yellow fever areas and there are dengue areas, and where these do overlap they seem to be separated in time. Yet, it seems probable that yellow fever virus must have been taken across the Indian Ocean at some time over the centuries, just as it was taken across the South Atlantic when trade opened up between West Africa and the Caribbean and South American ports. Until rigid mosquito-control measures were introduced at Mombasa and at the other East African ports in 1941 following a yellow fever epidemic in the Sudan and cases in Western Uganda, almost every Arab dhow that left the coast had its own population of *Ae. aegypti* aboard. These were self-perpetuating populations, the adult mosquitos feeding on the crew of the dhow and breeding in the stored water. It only needed a single human case to develop aboard to infect these adult mosquitos and so produce further cases some two weeks later. It is one of the mysteries remaining to be solved. There is no known case of yellow fever recorded from India, Ceylon, Burma, Malaysia, Indonesia or from any part of Asia, although *Ae. aegypti* abounds in the urban areas. Neither are there any signs of previous activity as shown by the absence of immunity. Dengue appears to take its place.

We have already seen that different strains of a single species of mosquito may vary in their capacity to act as hosts to a particular species of parasite and as a working hypothesis it has been suggested that *Ae. aegypti* in Asia differs from those in Africa and tropical America. But this hypothesis has had to be rejected for Asian strains of *Ae. aegypti* have been tested and found to be perfectly capable of transmitting yellow fever virus from one monkey to another. The mystery remains unsolved [400].

Equally one wonders why dengue does not occur in East Africa, for human cases and infected mosquitos could just as easily have travelled from the Asian ports as to them. Actually there are reports of the disease in the clinical literature but recent events make one wonder at their accuracy. At the end of 1952 and during the first months of 1953 an epidemic occurred among the people of the Makonde Plateau in the southeastern corner of Tanzania. The disease resembled dengue in almost every way and was in fact so diagnosed. There was the fever, the rash and the crippling pain. The pain in the joints was so severe that the affected people called the affliction *chikungunya,* which may be translated as 'bonebreaker'. *Ae. aegypti* was present in enormous numbers; water is at a premium on the Makonde Plateau and large water-storage vessels were found in every house. Almost every one of these bred its own *Ae. aegypti.* The adult mosquitos fed on the human occupants both within the house itself and on the verandah that protects each house. Everything seemed to support the original diagnosis of dengue [413].

About 40,000 cases of this dengue-like disease occurred during this epidemic and a virus was duly recovered both from human cases and from the *Ae. aegypti.* Tests on the virus, however, showed it to be unrelated to dengue virus. It was a virus belonging to the same group as WEE and EEE viruses and was named chikungunya virus after the name given to the disease by the people themselves. Chikungunya virus was also isolated from some other species of mosquitos including *Anopheles,* but the significance of this was not fully appreciated at the time. Although many isolations of the virus were made from *Ae. aegypti* all attempts at transmitting the virus at the time by mosquito bite failed, even to a human volunteer. A curious thing about this whole episode was that a second virus was also isolated from some of the patients. It was named Makonde virus, after the plateau which was the scene of the main epidemic. Its significance was not understood and it was suggested that it must be a contaminant that had crept in during the complex procedures that follow any virus isolation [424].

Six years passed with no further incident except for some scattered isolations of chikungunya virus from mosquitos 1500 kilometres from the site of the original epidemic. Several isolations were made from *Ae. africanus* and some from *Mansonia* and *Coquillettidia* spp. Then in 1959 began a most extraordinary epidemic of gigantic size in the north-west corner of Uganda where this country borders both Congo and Sudan. The afflicted people were smitten with fever, a rash and such severe pain in the joints that they named the disease *o'nyong-nyong,* meaning bone-breaker in their language, a language quite different from that of the people of the Makonde Plateau. The medical profession pronounced dengue, which it was for all intents and purposes. O'nyong-nyong struck at men and women of all ages and at children with equal ferocity. So many people were laid out with the crippling pain in the joints that there was almost complete disruption of work. Schools in some areas were without children or teachers or both.

Once again the agent isolated from patients proved to be unrelated to dengue virus. Nor was it chikungunya virus, although it resembled it very closely indeed. The agent was given the name o'nyong-nyong virus. Strangely, *Ae. aegypti* hardly existed as a man-biting mosquito in the epidemic areas, nor did any other culicine mosquito in sufficient numbers to account for the tremendous size of the epidemic. In fact the only man-biting mosquitos present in numbers that made sense were the two important malaria mosquitos, *An. funestus* and *An. gambiae*. Although isolations of viruses of one sort or another had been made from *Anopheles* mosquitos from time to time in various parts of the world, including chikungunya virus from Tanzania, the genus had more or less been written off as unimportant as far as virus transmission was concerned. Indeed, more than one famous textbook on the subject states categorically that *Anopheles* simply do not transmit viruses. It was, then, surprising indeed when isolation after isolation of o'nyong-nyong virus was made from these two species of *Anopheles,* the decision to concentrate on them being dictated by all the evidence that presented itself. It should be recorded that other modes of transmission were also considered, including the agency of lice and bed bugs and direct contact. All had to be discarded in favour of the two malaria mosquitos, *An. funestus* and *An. gambiae*.

O'nyong-nyong saturated the area and moved on. It became possible to predict its progress by studying the distribution of endemic malaria, which coincided, for the most part, with the distribution of *An. funestus*. This mosquito, as we have seen in Chapter 3, usually breeds in more permanent waters than its partner *An. gambiae*. The epidemic moved in a southeasterly direction on about a 100 kilometre front and struck down successively first the Acholi people and then the Teso, who called it *o'bili-bili* also meaning 'bone-breaker'. It moved on to the region of Mount Elgon on the borderland between Uganda and Kenya. Here it divided into two arms, one sweeping round the north side of the mountain, where it came to be known by the people as *kemujunjon,* the other moving southwards to be called *nyanginya*. Both these names and the many others which it received as it passed into different language groups, had roughly the same meaning, 'bone-breaker', which is not so unlike break-bone fever, our own name for dengue.

Two other names given to this disease deserve special mention. When o'nyong-nyong, as it officially came to be called, arrived in Budama, immediately west of Mount Elgon, it came to be called simply, *nylon!* Everyone, man, woman and child knew what nylon meant. One had only to mention the word and the local people would grip their elbows or their knees and indicate pain by the expression on their faces. But why nylon and why not the local equivalent of bone-breaker? It seems that nylon clothing had arrived in the local shops at about the same time as the first sudden onslaught of the epidemic. Delightedly the Badama people donned their new nylon shirts and blouses only to find that that night or the next, for o'nyong-nyong usually struck at night, they were transfixed in their beds with unbendable arms or legs. The association was clear. Curiously, no ill-feeling seems to have been engendered against this technological innovation. The other name of special note was given to the disease when it struck the Jaluo people in the Kavirondo Gulf of Lake Victoria. These people all called it Johnny, a rather unexpected and improbable-sounding name to come out of central Africa [382]. About a week before the epidemic struck the Jaluo people the British Army had carried out a house-to-house survey in the area. British Tommies fresh out from England were not expected to know much of the local customs or the elaborate etiquette of greeting, nor were they expected to speak the local language. For reasons of their own they pursued their inquiry addressing every male as Johnny; 'Hey you, Johnny!' 'How long have you lived here, Johnny?' 'Is this your wife, Johnny?' and so on. The Jaluo people, somewhat understandably perhaps, were confused but intrigued. All became clear a week later when many of those visited found themselves unable to straighten their legs or their arms–the magic word had worked and the homely name of Johnny took on somewhat sinister overtones.

O'nyong-nyong in the meantime continued on its course, receiving a new name in each new area invaded. These names helped considerably with the untangling of the problem. Malaria, leprosy, plague, sleeping-sickness and other common diseases in tropical Africa all have their vernacular names. That a new word was coined for o'nyong-nyong in each area was itself evidence that we were dealing with something new—a disease, with its characteristic joint pains, the like of which was unknown in living memory. That all ages were equally affected supports this view; had the disease visited these areas before, the older people would still be immune and o'nyong-nyong would have shown itself as a children's disease, and would already have received a name.

Where did o'nyong-nyong come from, or for that matter, chikungunya? No evidence has been produced of other vertebrate hosts, or extra-human cycles of transmission with other species of mosquitos. Chikungunya virus has turned up in recent years in parts of India where numerous isolations have been made both from human patients and from *Ae. aegypti*. These two viruses not only produce extremely similar clinical symptoms in man but they are very closely related to each other. It seems more than likely that o'nyong-nyong virus is in fact an *Anopheles* adapted strain of chikungunya virus. If this is so we have recently witnessed what can happen when a virus, through what was probably a single mutation, invades a new mosquito host-species. This incredible epidemic of new disease was the result. Although not fatal, several million people were affected during the passage of this epidemic across Uganda and into parts of Kenya and Tanzania. No other outbreak of mosquito-borne virus disease has involved so many people and the role of *Anopheles* in the transmission of viruses to man has had to be reassessed.

Could it be that what was sometimes called dengue in East Africa, before developments in virus technology allowed unequivocal diagnoses to be made, was in fact chikungunya? And what about some of the so-called dengue cases elsewhere? We have seen that chikungunya virus occurs in India and, as in East Africa, the disease in man very closely resembles dengue clinically. The whole situation is rather confused and certain new developments have not helped to clarify it [419]. In recent years both dengue and chikungunya virus have been recovered from patients in Thailand, suffering from the serious and often fatal symptoms of haemorrhagic fever. These two viruses belong to entirely different serological groups, and one now begins to wonder again about the significance of Makonde virus that was apparently isolated along with chikungunya virus in the original epidemic of 1952–3.

Chikungunya virus is also closely related to Semliki Forest virus, an agent isolated in 1942 from mosquitos belonging to the *Ae. abnormalis* group. Although man becomes infected from time to time, as shown by the presence of neutralising antibodies in his blood, the virus is not known to produce any symptoms of disease. Antibodies have also been found in monkeys [432]. The observant reader may have noticed how some of these mosquito-borne viruses are named after the diseases they cause while others bear the names of places. When the original isolation of a virus is from a patient suffering from recognisable symptoms, the virus is named after the disease; for example, yellow fever virus, western equine encephalomyelitis virus and, of course, chikungunya and o'nyong-nyong viruses. When, however, the original isolation of virus is made in ignorance of any specific disease that it may cause in man or in other vertebrates, it is customary to name the virus after the place in which the isolation was made; for example, we have Zika, Bunyamwera and Semliki Forest viruses.

It is interesting to reflect on how the Semliki Forest itself received its name. It seems that when Henry M. Stanley passed along the western foothills of the Ruwenzori mountains he could see in the distance some of the loops and bends of a great river which meandered through the black forest below. Now Stanley conscientiously inquired about the name of any geographical feature that he encountered on his travels and duly entered it in his diary. Even today one comes across many hills and lakes still bearing these names. When he asked his African companions what was over there in the forest, they replied, *'Semliki'!* But apparently the word simply means 'nothing' and what his companions were really trying to tell him was, that there was in fact 'damn-all' over there. But Stanley wrote it down, 'S-E-M-L-I-K-I' and added the word 'River', and Semliki it has been ever since. Later the forest too came to be known as the Semliki Forest and now the virus that was recovered by Ken Smithburn from mosquitos captured in the forest by Alec Haddow helps perpetuate the name.

We come then to a completely new situation in the study of mosquito-borne or indeed of any other type of disease. The classical method, following a description of the disease itself, has always been a search for the micro-organisms or pathogens responsible for the disease, followed by an investigation into the mode of transmission. In 1937 a virus was isolated from a woman in the West Nile District of Uganda as a by-product of yellow fever investigations. Nothing was known about any specific disease caused by this new virus; it was simply named West Nile virus and remained somewhat of a curiosity for the next thirteen years. Other viruses similarly turned up in Uganda during the 1940s, most of them being recovered from mosquitos [386, 432]. We thus had a new situation–the discovery and description of pathogens before any diseases they may cause were recognised–viruses in search of a disease. The phenomenon is now well known in virology and has led to the designation 'orphan viruses'!

West Nile virus differed from the other so-called orphan viruses in not being recovered from mosquitos. Then in 1950, and the years that followed, the virus was isolated from human patients in Israel and Egypt. Concentrated study revealed that the virus occurs in an avian cycle; hooded crows and rock pigeons seem to be the birds that are mostly concerned with bird-feeding *Culex* transmitting the virus from bird to bird, from bird to man and from man to man. *Culex univittatus* is the mosquito most concerned, but *C. antennatus* and probably *C. pipiens* also play a part [388, 422]. Thus, thirteen years after its discovery and some 3000 km from the site of the original isolation of virus the full story came to be worked out. West Nile virus has recently been isolated again in Uganda, this time from the bird-feeding mosquito *Coquillettidia metallica* [438].

Nearly as dramatic is the case of Zika virus, which was first isolated from *Ae. africanus* in the Zika Forest, Uganda, in 1947. No disease was known to be associated with this virus until 1954 when it was recovered from human cases of jaundice in Nigeria, also some 3000 km away from the site of the original isolation.

Other parts of the world also produced their quota of mosquito-borne viruses. The list from South and Central America exceeds even that from Africa in length and we have already mentioned several from the United States. In the Far East there is another bird virus that is related to West Nile and St Louis encephalitis viruses. As in the American encephalitides, the virus spills over from time to time into both horses and man, the disease in man being called Japanese B encephalitis. The virus, as might be expected, is transmitted by *Culex* mosquitos, the main vector varying from country to country; in Japan, where most of the work on this virus has been done, it is *C. tritaeniorhynchus*. The study of Jap B, or simply JE virus as it is now often called, has been particularly interesting for three reasons: it has shown how a mosquito-borne virus builds up in several hosts before spilling over into horses and man; it has indicated how mosquito-borne viruses in the more temperate regions may survive the winter; and it has shed considerable light on the fate of viruses within the mosquito itself.

In the springtime *C. tritaeniorhynchus* that have survived the winter in a state of rest lay their eggs to give rise to the first new brood of adults. These feed mainly on black-crowned night-herons and egrets. The mosquito numbers build up and virus transmission takes place between mosquito and bird, both species of bird filling the role of maintenance host. In the meantime the bird population also builds up and by July the heronries are full of nestling birds. The night-herons are nocturnally active and leave their chicks unprotected and exposed to the attacks of the increasing numbers of *C. tritaeniorhynchus*. Exchange of virus between night-heron chick and the July population of mosquitos reaches such a pitch that infection rates in both hosts rise to a very high level, the night-heron chicks acting as what has been described as an amplifier host. The virus becomes dispersed by bird and mosquito and new foci are built up. *C. tritaeniorhynchus* also feeds on mammals, one of its preferred hosts being the pig. During August pigs become infected on a very large scale indeed and act as what have been called link hosts between bird and man or between bird and horse [367].

Insect-eating bats may also be involved in this complex story. Not only do they exhibit a viraemia following the bite of an infected mosquito, but they may become infected by swallowing an infected mosquito. The viraemia may last for six days or more during which time the animals are normally active. Moreover, the viraemia has been shown to be high enough in level to infect other mosquitos that feed on them. What is particularly interesting and may well prove to be of the highest importance is the demonstration that latent infections may survive for more than 100 days during simulated hibernation. On return to warm conditions the virus appears in the blood three days later, and again the titre of the virus is high enough to infect mosquitos. This, then, could be the over-wintering mechanism for JE virus [417]. An increasing number of viruses is being recovered from these animals and their role in mosquito-borne virus epidemiology may turn out to be extremely important.

JE virus has been studied in detail in the mosquito, from the time of ingestion with the blood meal to the time of its build-up in the salivary glands. Initial multiplication of the virus occurs within the cells of the midgut during the first three weeks following the infective meal. From the second to the fourth week, virus appears in the haemolymph and is transported to every part of the mosquito's body. Virus in the salivary glands reaches high titres by the end of the second week and continues at a high level. Virus also builds up in the nervous tissue, the head and thoracic ganglia becoming especially involved. Other organs also become infected including the ovaries. The points of special interest are that the viraemia is of limited duration and that multiplication of virus then proceeds within the cells of the salivary glands without further reinforcement from the haemolymph. It is interesting also to note that this virus which has an affinity for nervous tissue in man and thus produces the symptoms of encephalitis, reaches higher levels in the mosquito's nerve cells than in any other tissue, including the salivary glands [418]. Finally, while it reaches detectable levels in the ovaries about three weeks after the infective blood meal, so far there has been no evidence of transovarial transmission of this

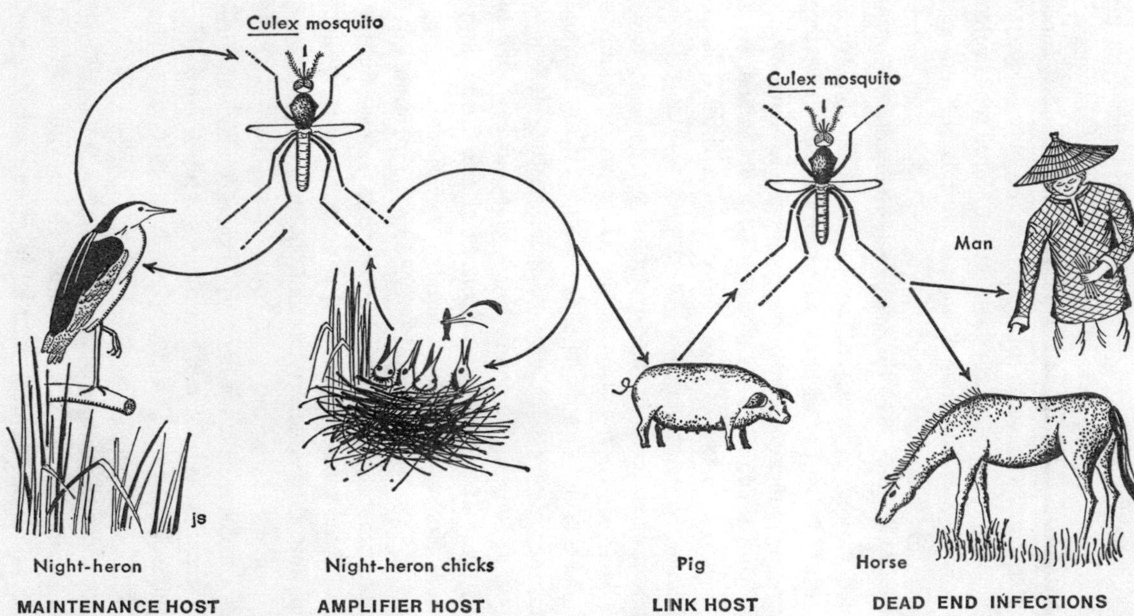

Figure 22. The ground-level avian cycle of Japanese B encephalitis and build-up of virus levels in nestling birds acting as *amplifier hosts* and eventual spread through pigs acting as *link-hosts,* to man and horses.

virus, or indeed of any other virus as far as mosquitos are concerned, although it is a common enough phenomenon in ticks [401].

In Australia JE virus appears to be replaced by Murray Valley encephalitis virus, the cause of epidemics of encephalitis from time to time, with a particularly high rate of mortality. The vector is not known for certain, but *C. annulirostris* is well up on the list of suspects [388, 422].

But what of Europe, does it not also have its quota of mosquito transmitted viruses? It does, but not to the extent that we have seen elsewhere. A virus has been recovered from *Ae. caspius* and *Ae. vexans* on numerous occasions by Czechoslovakian workers. Most of these isolations were made in or near Ťahiňa, in East Slovakia, and the virus has been named Ťahiňa virus. It is transmitted mainly by *Ae. vexans* but *Ae. caspius* also plays its part. Both these species breed in flood-waters and very large numbers of *Ae. vexans* can be found resting in cow-sheds. Immunity surveys in man indicate that the virus is widespread from Finland in the north, through Poland, Czechoslovakia, Austria, Hungary and Yugoslavia to Italy in the south. But by far and away the most important areas as far as man is concerned are Czechoslovakia, Austria and Hungary, where in certain areas antibodies may be found in half the population [368, 369, 383]. Nothing is yet known about the disease caused by this virus. It may well prove that many of the summer illnesses that occur in the warmer parts of Europe owe their origin to mosquito-borne viruses.

That so many of the mosquito-borne viruses have come from Africa and tropical America is partly because to begin with these are both gigantic regions and together make up a large proportion of the tropical world, particularly in terms of continuous land masses. But both regions also still have large areas covered with rain-forest, those of South and Central America being particularly extensive. Most of these viruses are associated with reservoirs in the forests and it is only when man infringes on these that leakage occurs resulting in human cases or, rarely, human cycles. But there is another reason for these two regions producing between them about sixty per cent of all the known arthropod-borne viruses. The Rockefeller Foundation set up yellow fever research stations in both continents and once the story of this disease had been written, attention was turned to the many new viruses that were unearthed as by-products of this work. Study of these viruses and their mosquito hosts was then continued in its own right. Thus it is, as so often is the case, as much a distribution of virus- and mosquito-workers as of the pathogens and vectors themselves. A third reason for this unequal treatment in this book is that I spent more than twenty years in Africa myself, taking a part in this exciting work, and as I have stated in the Preface, I have deliberately set out to draw on personal experience whenever possible. We have come a long way since *Ae. aegypti* was declared to be *the* yellow fever mosquito. We now have no less than sixteen different genera of mosquitos known to be involved with viruses.

During this brief survey we have considered the part played by mosquitos in filaria, malaria and some of the virus diseases. We have seen how the filarial worms undergo a cycle of development within the mosquito host, but do not actually multiply in number; multiplication occurs only in the vertebrate host. In the malaria parasite, on the other hand, cyclical development is accompanied by multiplication, a single zygote giving rise to several thousand infective sporozoites in the mosquito. The arthropod-borne viruses, or arboviruses, also multiply within the mosquito, cases being reported of million-fold increases, but there is apparently no cyclical development from one stage to another. We must now consider another type of transmission altogether, one in which the mosquito plays merely a passive part, carrying the pathogens from one host to another without either development or multiplication. Myxomatosis is a virus disease of rabbits that has received considerable publicity in recent years. In Australia it is transmitted from rabbit to rabbit by mosquitos, notably by *C. annulirostris* and *An. annulipes*. The insects pick up the virus on their mouthparts and carry it directly to other rabbits, the vector merely acting as what Frank Fenner has called a flying pin; the proboscis may remain infective for two to three weeks. In France the disease was spread in 1952 by *An. atroparvus*. When it reached Britain, however, in the following year, transmission was mainly by rabbit fleas, although *An. atroparvus* seems to have been implicated in the early stages of the epidemic when the disease first arrived [367], and in 1970 isolations have been made from *An. claviger*.

In tropical America the larvae of the human botfly, *Dermatobia hominis,* undergo their development in the skin of man, cattle, dogs and occasionally birds. Each larva lives separated from its fellows in a boil-like pouch communicating with the atmospheric air through a small opening. Heavy infestations have been known to kill cattle and dogs. Now it is a curious fact that the female fly does not lay her eggs directly on the potential host, but attaches them to the abdominal integument of a feeding mosquito or other blood-sucking insect; the mosquitos most often involved are *Aedes serratus* and *Psorophora ferox* [1]. The fly settles on the cow, horse or man and lies in wait. If a mosquito settles near by it is seized and carried off. Without harming the mosquito the fly then proceeds to attach a series of eggs side by side in a neat row along the sides of the abdomen. All this is done while hovering near the host, after which the mosquito is released into the air. In due course the eggs hatch and when the mosquito carrier again visits a mammalian host the young larvae drop off and at once burrow into the skin. One may marvel and wonder and carefully observe 'when' and 'how'. One may hazard a guess or two about selective advantage, but one does not ask 'why'.

14. MOSQUITOS AND HISTORY

It would be especially satisfying if, as we near the end of our survey of the world of mosquitos, we could attribute this historical turning point or that to the ravages of malaria or yellow fever or to any of the other epidemic diseases which as we have seen are transmitted to man by these insects. In fact, quite an impressive number of historical events, some of them dramatic in the extreme, can be so explained. Most of these, however, are comparatively short-term affairs lasting a few years or at most a decade or two, such as the outcome of various battles and the débâcle over the building of the Panama Canal. When we come to look at some of the long-term changes that have shaped the course of history we cannot be so confident; it is all too easy to confuse cause with effect, particularly when we go back far in time. Nevertheless our story would be incomplete without some brief survey of malaria in ancient times and of its possible effects on history.

Among the early inhabitants of Greece about whom we have record, malaria was apparently scarce or absent even though many of the inhabitants settled in low-lying marshy places. Indeed, it seems that they deliberately chose these places because they were safe; either the local population of marsh-breeding man-biting anophelines were free from infection or, possibly, the right man-biting species or subspecies had not yet been introduced to the region. A parallel situation has occurred in more recent times on the island of Mauritius. Occasional cases of malaria used to occur on the island, but these were almost entirely restricted to the immigrant Indian population, many of whom doubtless arrived already harbouring latent infections from India. Until 1866 there were apparently no dangerous anophelines on the island to spread the disease. In that year, however, *Anopheles gambiae* from the mainland of Africa was accidentally introduced and started to breed on the island. Even if none of these was infected on arrival it was not long before they picked up infection from the sporadic Indian cases [439]. A huge epidemic followed and malaria has been a problem almost ever since. Of course, no one associated these two events in 1866.

To return to ancient Greece, references to malaria become fairly frequent during the fifth century BC and it has been suggested that the disease was first introduced by the invading Persians. References to malaria become even more frequent during the following century; this was the period of the Peloponnesian War and it may be that returning soldiers carried the parasites with them in their blood and so spread the disease to new regions by infecting previously harmless local anophelines. Or perhaps it was more like the story of Mauritius and man-biting anophelines were accidentally introduced. At any rate during this period, when for a generation Greek fought Greek, the disease spread rapidly throughout Attica, judging by the accounts that have come down to us [439]. By the first century BC malaria was very widespread indeed and from then onwards seems to have been constantly present; even up to recent times it is reported that Greek peasants valued quinine as highly as bread!

More than one historian has put forward the view that the constant malaria that now hung over the land led eventually and inevitably to the fall of Greece. Disorder replaced order. But how far cultural decline followed widespread disease and the disability and apathy that are its constant companions, and how far malaria spread as a result of breakdown in authority is really a matter of conjecture. Malaria must certainly have contributed to the downfall of Greece, just as in more recent times it has been one of the key factors in holding back development in certain regions of the world.

From all accounts the story of Rome is extraordinarily similar. There seems to have been an absence of malaria during the early years of the Republic; the city itself was built around land that was previously swampy. Malaria seems, however, to have been widespread by the fifth century BC, and during the century to follow it hit the peasants so hard that there was wholesale migration from the low-lying malarious regions into the cities where the swamps had at least been drained. This disruption of the peasant way of life had an important effect on the Roman economy, since it was from these people that the army was mainly drawn. The same arguments apply to Rome as they do to Greece: malaria almost certainly played some part in the eventual eclipse of the Empire, but how much it is not possible to say [439].

Many questions remain unanswered: is it, for example, only coincidence that the records of widespread malaria both from Greece and from Rome date from the fifth century? Perhaps the explanations about returning troops and invasions are altogether too facile. Could the fifth and fourth centuries BC have seen great pandemics of the disease such as those of plague that struck the world in the sixth, fourteenth and nineteenth centuries of our own era, or of influenza in 1919? There have been pandemics of malaria in more recent times: the first well-known one struck Europe in 1557; others occurred in 1657 and 1669; a fourth in 1806–7 during the Napoleonic Wars, and a fifth, that laid low the people of Europe from southern Norway to Greece, in 1826. Such matters seldom lend themselves to simple explanations, especially when they occurred so long ago. More often than not the causes are multiple and highly complex and we have as much hope of finding a simple explanation as we have of finding a single cause of cancer or of yesterday's weather.

We are on much firmer ground when we come to look at the effects of mosquito-borne disease on man's short-term activities, particularly those that have influenced the outcome of battles, on which, alas, so much of history depends. In 1387 the Earl of

Arundel's fleet was forced to abandon its campaign at Sluys because of malaria. The sickness was attributed to 'the unhealthy water lying on the surrounding country'. In 1585 Drake left Plymouth with twenty-nine ships carrying 1500 seamen and 800 soldiers, with Frobisher as vice-admiral. The men went ashore for a time on the Cape Verde Islands, where they were attacked by mosquitos. Drake later wrote, 'wee were not many dayes at sea but there beganne among our people such mortalitie, as in a fewe dayes there were dead above two and three hundred men', and 'until some seven or eight dayes after our coming from St Iago, there had not died one man of sicknesse in all the fleete: the sicknesse shewed not her infection wherewith so many were stroken, untill we were departed thence, and then seazed our people with extreme hot burning and continual agues' [458].

Drake lost a further 500 men when he reached the Caribbean three years later and attributed this disaster to the early evening air ('first night ayre'), 'who so is then abroad in the open ayre, shall certainly be infected to death'. We know now, of course, that the incubation period of malaria is usually around ten days, but seven or eight days is not uncommon. And we know a little more about the circadian rhythm controlling the mosquito's activity that makes the first night air so dangerous.

The fate of many another military expedition was determined more by malaria than by enemy action, the best known in more recent times being the catastrophe that overtook the British at Walcheren in 1809. The outcome of various French and British invasions in Indo-China, Madagascar, West Africa and elsewhere was similarly decided [458]. Fighting in Macedonia in the First World War was virtually brought to a standstill by the disease. The Caribbean epidemic that wrought disaster in Drake's fleet in 1588 might conceivably have been yellow fever and not malaria, but from all accounts malaria seems much the more likely. It is clear from his account of the affair that Drake was a very keen and accurate observer and such highly dramatic symptoms as jaundice and coffee-ground vomit that characterise a proportion of yellow fever cases would be hardly likely to have gone unrecorded.

The Guinea slave trade operated by the Spaniards started in 1501; British participation in this grisly business apparently dates from 1593 with slave ships running mainly from Angola to the River Plate. There is now no doubt whatever that yellow fever, which as we have seen is endemic in parts of tropical Africa, was carried across the South Atlantic in these ships. There are accounts of European crew dying of yellow fever on these voyages while the African 'cargo' below remained free from the disease (only to die of dysentery instead). This is exactly what we would expect since, as we have seen in the last chapter, the African from yellow fever endemic zones would either be wholly immune as a result of a previous infection with the virus, or would be likely to react mildly to the disease; he might run a slightly raised temperature, but would usually be free of yellow jaundice and black vomit. Every ship, of course, would provide ideal breeding conditions for *Ae. aegypti* and, like the Arab dhows plying the Indian Ocean today, would have its own breeding colony of this mosquito. Since an African with an inapparent infection would nevertheless be highly infectious to the mosquitos that fed on him, the long slow voyage would provide ample time for the ship to become infested with large numbers of infected mosquitos. Little wonder then that overt yellow fever would manifest itself among the susceptible, non-immune European crew.

The first clearly documented epidemics of yellow fever in the Caribbean were those of 1647 and 1648. Barbados was the first, and recorded some 6000 deaths; it is noteworthy that it was referred to at the time as '*nova pestis*'. Almost as many deaths occurred in St Christopher, and Guadaloupe was hit the following year. To the invading forces of Britain and France and to the Spaniards who occupied much of this region, yellow fever struck like an unseen enemy many times more powerful than anything their opponents could muster–an enemy that defeated expedition after expedition over a period of 250 years. But to others who were largely immune to the disease it must have appeared as a deliverer, as did the epidemic that smote the Assyrian hordes poised ready to invade the Israelites of old. Time and time again over this long period the little black and silver mosquito delivered its deadly load, often doing so before any troops could be landed or a shot fired. Troops left behind to garrison some of the islands suffered the same fate.

Admiral Sedgwick attacking Jamaica in 1655 lost 140 of his soldiers per week from yellow fever; the symptoms were described as three to four days' fever accompanied by jaundice. He died of the disease himself in the following year. In 1664 yellow fever accounted for all but eighty-nine out of 1500 men left to garrison Santa Lucia [458, 460].

In 1689 Captain Lawrence Wright set sail with eleven ships for the West Indies, carrying the Duke of Bolton's Regiment. He returned to Britain two years later having achieved precisely nothing; he had been defeated by yellow fever. In the following year Admiral Sir Francis Wheeler set off with no less than nineteen ships, planning first to attack Martinique and then to sail northward to attack the French in Canada. Even this illustrious gentleman with his impressive array of force was no match for *Ae. aegypti;* yellow fever struck while they rode at anchor off Barbados where he was compelled to stay a whole month. His fleet was utterly destroyed as a fighting force before he could even start his attack on Martinique and the Canadian venture had to be abandoned. His losses were so great that he had constantly to replace them by local recruitment and altogether lost 3100 men from yellow fever. He finally returned to England with scarcely sufficient men for safe navigation [458].

One of the most decisive defeats at the hand of this unseen enemy occurred in 1741 when Admiral Vernon, having set off from Southampton with 27,000 men with the intention of conquering Mexico and Peru, lost 20,000 of his men during an epidemic of black vomit, as the Spaniards called the disease. Similar fates awaited the military expeditions of Admiral Graydon in 1703, Gray in Martinique and Abercrombie, who lost most of his 13,000 men from yellow fever. Havana experienced several huge epidemics during the eighteenth century. One of these occurred in 1762 after the Earl of Albemarle had successfully captured the city; 8000 of his 15,000 men were killed by the disease. The Spanish forces on the island under General Bernado Galvez *en route* for Louisiana were almost wiped out and so was the Spanish fleet under General Aristizabal after its return from Santo Domingo [460].

Worse than any of these by virtue of the numbers concerned was the fate that awaited the expedition sent by Napoleon in 1802 to conquer Haiti and the Mississippi Valley. This massive army, led by Napoleon's brother-in-law, General Charles Leclerc, numbered 33,000 all told. It is said that 29,000 of them died of yellow fever, an ignominious and hideous defeat that led directly to the sale in 1803 of Louisiana to the United States of America. It seems we are always ready to let the other fellow have the land when we find or think that it is no good; if we can sell it so much the better [439].

Apart from the effect on invading forces in the Caribbean, yellow fever struck again and again at the civilian populations of the coastal towns both of South America and of the United States. New York was hit in the summer of 1668, Philadelphia and Charleston in 1690. In the year 1793 the population of Philadelphia was 55,000. That summer there were 5000 deaths from the disease out of some 17,000 cases, but to make matters worse a further 12,000 fled the city. Thus out of the original 55,000 only 11,000 were unaffected one way or another. New Orleans was overcome by successive epidemics between 1793 and 1853. It is difficult to imagine now the disruption these epidemics caused. During the last of these, when 11,000 died out of 40,000 cases, a further 50,000 or so fled the town; some estimates put the figure as low as 30,000, others at 75,000. This fleeing from the populated areas during times of pestilence allowed many to escape, but it also helped to spread the disease. Most important, however, it aggravated an already desperate situation in the cities themselves: there were not enough people left to carry on normal services let alone to care for the sick or to bury the dead [439, 460].

The attacks in the southern United States reached another climax in the year 1878, during the very time that Carlos Finlay was preaching his lonely gospel in Cuba; in that year the disease appeared in epidemic form in more than 100 American towns, killing some 20,000 US citizens. Economic loss was put at 100 million dollars.

Ships carried infected mosquitos to many parts of the world where they became temporarily established on land during the summer months. Spain was particularly hard hit, outbreaks occurring in most of the major cities: Barcelona, Cadiz, Cartagena and Jerez alone recording over 50,000 deaths. Gibraltar too was attacked. In 1856 the disease invaded Lisbon, Brest, Marseille, Saint Nazaire and Livorno, Lisbon alone losing 18,000 people. Three times it swept through the Canary Islands and the disease even reached as far north as Southampton and Dublin [432, 439].

Perhaps the best known of these events and one that made a great impact at the time was the tragedy and triumph of the Panama Canal. Ferdinand de Lesseps, fresh from his victory over the sands of Suez, was called upon to cut a similar canal across the isthmus of Panama. But whereas at Suez the canal had to be cut through arid country, in Panama the route chosen lay through swamps and mountains. Moreover, the high rainfall from the mountains ensured that any man-made containers left lying around to litter the labour camps and even the hospitals along the way would be water-filled for much of the time. Conditions were thus ideal for the breeding of vast numbers of *Anopheles* in the low-lying swampy areas and of *Ae. aegypti* in the many artificial containers. Little wonder that the isthmus came to be referred to in 1885 in the following terms: 'so much foul disease, such a hideous dungheap of physical and moral abomination'.

The canal was to be about 120 km long. De Lesseps brought in 500 young engineers to supervise the work, none of whom were to draw their first month's pay. Malaria and yellow fever between them incapacitated or killed off engineers, administrators and imported labour almost as fast as they could be recruited [454, 455]. Yellow fever alone accounted for 5627 deaths, 654 of them being buried during September 1884. (In that year also the entire crew of a visiting British ship died of the disease.) Altogether about one-third of the entire force of European stock, or 20,000 people, lost their lives in this disastrous undertaking, so similar to Suez in the eyes of the armchair planners, so utterly different in hideous reality.

It was not only sickness and death that affected the progress of this otherwise imaginative project, but the accompanying breakdown of administration and all ancillary services. Work on the canal came to a complete standstill and the whole project was abandoned in 1889. Not for the first time man capitulated before the mosquito.

The Panama Canal project was later taken up by the United States, but by this time Finlay, Manson, Ross and Reed had made their historic contributions to life for man on this planet. None the less in 1904 there was a sharp outbreak of yellow fever during which it is said that the coffins accumulated on the railway stations faster than they could be cleared away. Not unnaturally panic began to seize the men who had left their safe homes in the United States. Indeed, fear of yellow fever was so great that work on the canal was again severely hindered. John F. Stevens, the chief engineer in charge during the early part of the American venture, exclaimed in a speech intended to boost the morale of his men, 'There are only three diseases on the Isthmus, yellow fever, malaria and cold feet; and the greatest of these is cold feet' [455].

But the way was now open for a renewal of the conflict against these two mosquito-borne diseases on more favourable terms. William Gorgas, a major (later general) in the United States Army, was the man called upon to direct operations. Having cleared up Havana in 1901 after its 250-year history of yellow fever, he turned his attention to Panama and the canal. He was well fitted for this gigantic task; he was a man of exceptional energy and ability and both he and his wife had previously survived an overt attack of yellow fever. Indeed, Gorgas met his future wife while she was actually ill with the disease. Thus both were immune to further infection and could safely work in the worst affected areas. Gorgas' methods were simple in the extreme, but having military backing they were extremely effective. With a force of several thousand he killed off the swamp vegetation with powerful blasts of burning petrol. He filled the thousands of water-filled crab-holes with sand. He destroyed all unnecessary objects capable of holding water [454].

During the French venture the hospital at Ancon had become one of the main foci of infection with yellow fever; it was well known that anyone entering the hospital stood a high chance of contracting the disease within four days of his arrival there, even if he was admitted in the first place as a surgical case. The hospital became a place of foul repute; in the short span of nine years 1200 people died of yellow fever in the hospital. The French authorities had made every effort to make this hospital a place of rest and tranquillity. To help them in this illusion they planted flowering shrubs and orange trees in the grounds surrounding the building, but found that they must fight a losing battle against the umbrella-ants that were apt to swarm up these plants to collect leaves for their fungus gardens; a valuable plant could be completely defoliated over night. The French found that they could successfully protect these plants by encircling each with a trough of water, similar in principle to the ant-traps protecting the legs of meat-safes familiar to everyone in tropical countries.

When Gorgas arrived on the scene in 1904 he found these troughs filled with larvae of *Ae. aegypti*. He also found many of these pottery rings discarded and lying around the extensive grounds. It seems unbelievable now, but he collected and destroyed several thousand of these rings from the hospital grounds alone. He also uprooted all those lovely shrubs and trees. Yellow fever virtually disappeared from the hospital; it was as simple as that. It should be added that once the mosquito population had been so reduced as to make yellow fever no longer a threat he reintroduced the flowering shrubs and the fruit trees, but not before he had solved the problem of the leaf-cutting ants. He did this also in characteristic fashion: he traced each ant run back from the plant to the nest, where he poured in carbon bisulphide and exploded it [454].

In the meantime the United States got on with building the canal and with a force of some 50,000 to 60,000 men deaths from yellow fever dropped from seven to only two per cent per year; most of the cases of malaria were those that arrived in the area already infected. The general story is so well known that we need not elaborate on it here. Gorgas' methods were applied right down the coast of tropical South America. Between 1895 and 1908 there had been 28,078 deaths from yellow fever in Rio de Janeiro. Mosquito control on the Gorgas model was introduced in 1909 and the disease virtually disappeared.

In 1920 General Gorgas visited London on his way to advise on the eradication of yellow fever on the west coast of Africa. During his visit he became seriously ill and such was his standing that he was visited in hospital by the King. When, shortly afterwards, he died he was given a state funeral in St Paul's Cathedral. If his name does not figure in our history books along with other generals, admirals and shapers of history then it looks as though we may not have our priorities right.

The victory over yellow fever was not, as supposed at the time (and is still often so represented) an end of the story but only a resounding victory of one particular battle or series of battles. As we have seen in the last chapter, yellow fever goes on in the forest cycles of tropical America and Africa and may jump out and involve man at any time, as witness the recent epidemics that have occurred in Central America and Ethiopia; it is variously estimated that between 15,000 and 30,000 people died in the Ethiopian affair. Nor was it the end of malaria for although Gorgas' methods were widely adopted and put into practice the more we learned about the ways of mosquitos the more complicated everything seemed to become.

In 1934–5 Ceylon was hit by a devastating epidemic of malaria that is of particular interest. Unlike yellow fever, the death rate from malaria is not high, except in children and in untreated cases.

It is usually more a debilitating disease than a killer. It will, therefore, give some idea of the gigantic proportions of this epidemic when we consider that deaths alone amounted to about 100,000 in seven months, an average of nearly 500 per day. To some it seemed that all the hard-won facts about malaria transmission and its control would have to be thought out again for, surprisingly, the epidemic occurred during a period of exceptional drought when one might reasonably have expected some respite from the disease. But we now know that it is precisely at times of little rainfall that residual pools form in the dried-up river beds, and in these countless small pools *An. culicifacies* breeds in prodigious numbers [465, 466].

The drier zones of Ceylon are subject to malaria in most years. This particular epidemic hit what are normally the wetter parts of the island where the rains come twice a year: in May, June and July and at the turn of the year. In 1934 the early rains failed altogether, but to make matters worse the drought was broken in October by some gentle rain, not enough to flush out the pools but sufficient to replenish all those that had in the meantime dried out. Sir Vincent Wigglesworth, who toured the stricken areas at the height of the epidemic relates how the roads were almost blocked with the continuous stream of people struggling or being carried to the local treatment centres. The numbers receiving treatment in those small dispensaries rose from some 400 to over 4000 per week. The immense increase in the number of *An. culicifacies* ensured that few people escaped being bitten and the large number of infected people ensured in turn that a high proportion of these mosquitos became infected. Indeed, the infection rate in the mosquitos rose to fourteen per cent in the epidemic areas, and in one particular area to twenty-one per cent: one in every five female *An. culicifacies* was capable of transmitting the disease, a very high proportion indeed.

It seems that this year had been preceded by five exceptionally malaria-free years and it has been suggested that the people had in consequence lost some of their tolerance to the parasite. Thus, not only did conditions favour an unprecedented increase in the number of mosquitos, but the people themselves were in a particularly susceptible state–in short the malaria parasite must rarely have had it so good. In the past, however, such epidemics must have occurred from time to time, judging by the ruins of ancient cities that are to be found in the drier parts of the island. Indeed, these drier parts, which were once the site of the ancient Sinhalese civilisations, seem to have become largely depopulated by malaria by the mid-seventeenth century. Wigglesworth has drawn attention to the possibility that the Tamil conquerors so changed the ancient systems of agriculture and irrigation that malaria, formerly more or less in balance with the population or at least kept down to manageable proportions, suddenly became out of hand.

We have instances all around us, both in the past and in the present day, of one species of animal preying on another. But the relation between man and mosquito has been much more in the nature of all-out war, war between species, the small mosquito certainly giving as good as it got. Indeed, until the beginning of the present century it had things mostly its own way, for we lost almost every battle. Nor is this war over yet. Our casualties from malaria alone still number somewhere around one and a half million a year. These are deaths, human deaths. Nevertheless, more and more battles are beginning to be won against this elusive enemy, and almost everywhere it is in fact in retreat. Twenty years ago the number of human deaths each year from malaria was probably more than double the present figure–mostly children. Much has been accomplished, particularly by the World Health Organisation, but successes have for the most part been local and very much remains to be done.

The enemy, then, is being pushed back, but it is still far from being defeated. Like other persistent enemies it retreats here but counter-attacks there. And even where it retreats, it, so to speak, digs in. Each individual mosquito seems such a fragile thing and is so easily dispatched, but each species made up of many millions of such individuals is a very different matter, for it is tough and almost infinitely resilient. Attack the species with one weapon, say with one of the new insecticides, and we find that some of the individuals do not succumb. These do not respond to a particular poison or are able, by bringing into play unsuspected powers of detoxication, to come through the ordeal unscathed. Once their companions, who are at the same time rivals for space, mates and food, have been killed off by man's clumsy attempt at annihilation, these non-responders are able to multiply and fill the gap. Thus it is that a resistant strain appears and if it also attacks man the result is worse than before. A new insecticide must be found, one that will be effective against the enemy in its new guise.

It seems that nature's armoury is almost limitless. But, even if there should one day come a time when this widespread and thorny problem of resistance is conquered, we may be sure that nature has reserved other dodges up its ample sleeve, dodges that cannot now be guessed at, any more than resistance was widely predicted when the so-called new insecticides first began to be splashed around this planet with gay abandon a quarter of a century or more ago.

Despite this seemingly boundless resilience on the part of each species, there is ample record that species of animals and plants can and do come to an end; extinction has been a widespread phenomenon in the past. Species that thrived for long ages have disappeared from the face of the earth leaving by some accident of fate their fossil remains for us to record. But for every one so preserved many others we may be sure have disappeared without trace. Indeed, whole groups of species have gone for ever, groups that took their place in the general economy of living things for a 100 or even 200 million years, groups that were highly successful by whatever yardstick we use, save that of presence in today's scene. Even if 100,000 species have disappeared forever, this only means on average one every 6000 years, hardly frequent enough to make much of a stir. The rate, of course, has not been evenly distributed over the ages and may well now be increasing as man with his exponential rise in numbers begins to extend his frontiers and fouls up his environment.

Species of mosquitos too, and at least one genus, have become extinct during the past sixty million years as is shown by the fossil mosquitos that have so far been discovered. Nevertheless, with all man's advances in science and technology, not a single species of mosquito has become extinct since war was declared in earnest some three-score years and ten ago. *Ae. aegypti,* which in the Americas breeds almost exclusively in man-made containers, was at one time thought to be on its way out. Indeed in 1968 the United States Government saw fit to appropriate fifty-three million dollars to help it on its way. But *Ae. aegypti* has been entrenched on the west side of the Atlantic too long now to be easily dislodged. Its reappearance in recent years in Rio de Janeiro, a city long thought to be free, has caused surprise and alarm, and one of President Nixon's first acts after his accession to power was to withdraw all further support for the campaign. What was needed, of course, was a diversion of this sum from the eradication programme to more basic research.

To all that I have said there is one outstanding exception. In 1930 *An. gambiae,* the most important transmitter of malaria in the continent of Africa, was accidentally transported across the South Atlantic in a French mail-plane that used to make the flight across the narrowest sector of the ocean where the great bulge of West Africa reaches out towards the bulge of Brazil on the other side. *An. gambiae* found conditions in Brazil much to its liking and quickly established itself around the port of entry where it was discovered one Sunday morning by R. C. Shannon. Any increase in numbers was now dependent solely on breeding and not on reinforcement. Since, however, a female mosquito is fertile for life following a single mating it needed only one female to make the crossing; she would then lay fertile eggs after each blood meal. This part of the story is, of course, largely speculation, but the fact remains that *An. gambiae* soon spread out from this centre and by the following year had penetrated the South American continent to a depth of 150 km or so. In 1938 it reached the heavily populated areas of the Assu, Mossoro and Jaguaribe valleys, resulting in an epidemic of malaria with 14,000 deaths [463].

There followed a success story comparable with those of Gorgas in Havana and Panama. The matter was taken up by the International Health Division of the Rockefeller Foundation, the organisation that has done such splendid work in unravelling the complex story of yellow fever. Under the leadership of Fred Soper the anti-*gambiae* campaign was conducted on semi-military lines and with ruthless efficiency. Within two years it was possible to say that *An. gambiae* had been eradicated from Brazil, an astonishing feat that cost two million dollars. That this statement is justified is borne out by the thirty years or so that have followed: *An. gambiae* has not reappeared, it was well and truly defeated in South America.

Why then, it may justifiably be asked, was the campaign not carried on in Africa where the species remains virtually untouched except in a few local areas? Would not the same success follow if the attack were to be pressed home with the same vigour and imagination? The answer is, quite emphatically, no. Quite apart from the enormous size of tropical Africa, even compared with Brazil, and considerations of political frontiers and the like, there is one basic difference between the enemy abroad in Brazil and the same enemy at home in Africa. We have seen that it was likely that the crossing of the ocean was made by a single female mosquito or at any rate by a small number of individuals. The colonisers of the new continent, then, carried with them a very restricted selection of genes compared with the immense variety available in Africa and the almost limitless scope for recombination. It is, of course, this total gene pool in a species throughout its normal range that gives it its plasticity and resilience.

An. gambiae in South America was cut off from this mainstream of genetic potential and the thorny question of resistance was hardly likely to arise. Further, being new to the territory the species was genetically unprepared to face the natural enemies and predators that were entirely new to it and these must have gone a long way towards keeping the numbers down. Nevertheless it is a success story of great historical importance; *An. gambiae* would not only have spread throughout much of the continent, but in the fullness of time would, by mutation, have developed its own powers of defence and resilience bringing disaster and misery to millions.

History does not record whether the men responsible for the defeat of this invasion played bowls or not. In fact, history does not record the event at all.

Except in special situations the war between man and mosquito will not be won by total attrition but by understanding, understanding of the ways of man as well as those of the mosquito, for ignorance, prejudice and tradition do much to ensure continual contact between the two. *Aedes simpsoni,* for example, might cease to be a constant threat in many parts of rural Africa if the people could be persuaded to change their customs only slightly. On the fertile and heavily populated slopes of Kilimanjaro this species breeds in the leaf axils of certain plants including those of *Dracaena,* which as we have seen is used extensively for hedging and boundary marking. This is not a crop plant and although its use is deep-rooted in social tradition, heads of plant being broken off and used as peace offerings both in domestic and larger squabbles, a substitute could surely be found at least for its more utilitarian uses.

It is true that *Ae. simpsoni* also breeds in the axils of bananas. But of the many kinds grown in the area only a few are used extensively by *Ae. simpsoni.* It might mean giving up four or five of the seventeen or so varieties of banana now grown there. In parts of Uganda only two varieties from an even larger total are important breeders of this species. At present in both regions growers seem to have as many varieties of banana as possible and the few offending kinds are scattered indiscriminately or in small concentrations over vast seas of bananas. If these particular kinds are so desirable either for beer-making or as food, would they not be better grown in larger concentrations, making some sort of control measures possible? Colocasias or cocoyams and pineapples also constitute a threat and here the problems might be more difficult, but the point I am making is that it is of little use providing ideal conditions for *Ae. simpsoni* and then complaining when they come out in clouds to bite.

Much the same could be said for *Ae. aegypti, Culex fatigans, C. tarsalis* and *Anopheles gambiae,* four other important biters and transmitters of disease that depend largely on the conditions we provide for them for their continued close contact with us on a large enough scale to be dangerous. There are other species, to be sure, that present very difficult problems indeed, as for example, the various man-biting members of *Mansonia,* and *Coquillettidia* and the swamp-breeding species of *Anopheles*. Fortunately these are for the most part night-feeders and there are well-tried, if old-fashioned, methods for reducing contact with night-feeders to a minimum, if only man could be taught to use properly the knowledge he already has. *An. culicifacies* presents a very special problem. It has been successfully tackled by damming the rivers and periodically releasing the water by means of automatic siphons built into the dams; larvae in the residual pools below are flushed out and carried further downstream.

I am not advocating a policy of live and let live; we already know that this does not work. But unless some entirely new and at present totally unforeseen weapon becomes available, attempts at eradication, even on a small scale, are not likely to succeed. On small islands and other small areas with natural frontiers the numbers may be so reduced as effectively to break contact between mosquito and man. Indeed malaria has become a disease of history in a few such areas. Even in the vastness of India the number of cases of malaria has been reduced to about 300,000 per year; before the eradication schemes were put into operation the number stood somewhere between 50 and 100 million cases per year. But eradication of malaria does not mean eradication of the mosquitos and the laborious work has to be continued if the newly achieved absence of malaria is to have anything more than ephemeral success. In Ceylon, an island of modest proportions, the malaria eradication scheme was within sight of its goal when funds for its support were withdrawn; the disease has recently returned in almost epidemic proportions.

A new approach to these problems is now called for, an approach leading to a much deeper understanding of the ways of mosquitos. But even this will be of little avail unless at the same time a new effort is made at educating the people, who, after all, must be considered an integral part of many of the cycles of transmission. Until recently there was little hope of reaching the people on anything but a superficial level. But with today's mass media the situation is rapidly changing and a golden opportunity offers itself if only these powerful new aids can be employed for true education in place of the usual political indoctrination into someone or other's particular pet theory. The problems are many and immense and I would not wish to give the impression that they will be easily solved, only that new approaches are called for in what we may call the second phase of the war between man and mosquito: approaches that must be made with knowledge and imagination.

If, as seems likely, man shortly succeeds in killing off many of his larger fellow-tetrapods, and the elephant, gorilla, whale and crocodile become almost legendary figures of the past, he is not so easily going to dispose of the mosquitos; it seems he is going to have to share this planet with them and with many other six-legged creatures for a long time to come. How long may, perhaps, depend on whether in the meantime he also succeeds in disposing of himself, in which case the mosquitos will carry on regardless as indeed they did for the fifty million years or more before man first began to provide a change of diet and extra breeding places for them. I think the present time may well go down in history as the period when man sought to conquer and control nature as if he himself were something apart, when the ignorance of the nineteenth century was replaced by the arrogance of the twentieth. But I question the whole present-day attitude that defines progress and civilisation as conquests over nature. Progress can come only from an understanding of and an attempt to co-operate with the forces of nature in the full realisation that we ourselves as well as the mosquitos that bite us are moulded by these very forces and are as inseparable from them as the clouds from the sky.

APPENDICES

The nine appendices list most of the known species of mosquitos but do not include any of the subspecies or named varieties. *Culex pipiens pipiens, C. pipiens molestus* and *C. pipiens fatigans,* for example, appear merely as *C. pipiens*. This can be misleading, since *C. p. pipiens* feeds mainly on birds and is never troublesome, while *C. p. molestus* and *C. p. fatigans* feed on man in temperate and the warmer regions respectively and can be real pests. But, more important, it is only the subspecies *C. p. fatigans* that is a transmitter of filarial worms and viruses to man.

Species of mosquitos that have been implicated in the transmission of human disease, whether in nature or in the laboratory, are printed in heavy type and one or more of the letters *f, m* and *v* appear before the name; these indicate actual or potential transmitters of filaria, malaria or viruses, respectively, at least in some parts of the world. Caution is also needed here, since, for example, **C. pipiens** appears in heavy type with the letters *f* and *v* before it, even though it is only the subspecies *fatigans* that is involved. The letters placed before the names of disease transmitters should be interpreted as follows:

f = This species, or a subspecies of it, is capable of transmitting filarial worms to man in the warmer parts of the world.

m = This species, or a subspecies of it, is capable of transmitting human malaria in the warmer parts of the world (or during the warmer months of the year).

v = This species, or a subspecies of it, is capable of transmitting viruses to man in the warmer parts of the world.

APPENDIX 1 MOSQUITOS OF THE AUSTRALIAN REGION

(*Australia and Pacific Islands east of Borneo, excluding New Guinea*)

Bironella (*Bironella*) *gracilis* Theobald
B. (*B.*) *papuae* (Swellengrebel & Swellengrebel de Graaf)
B. (*Brugella*) *hollandi* Taylor
B. (*B.*) *travestita* (Brug)
Anopheles (*Anopheles*) *atratipes* Skuse
fmv **A. (A.) bancroftii** Giles
A. (*A.*) *colledgei* Marks
A. (*A.*) *powelli* Lee
A. (*A.*) *pseudostigmaticus* Dobrotworsky
A. (*A.*) *stigmaticus* Skuse
A. (*A.*) *tasmaniensis* Dobrotworsky
fv **A. (Cellia) amictus** Edwards
mv **A. (C.) annulipes** Walker
fmv **A. (C.) farauti** Laveran
fm **A. (C.) koliensis** Owen
A. (*C.*) *longirostris* Brug
A. (*C.*) *lungae* Belkin & Schlosser
v **A. (C.) meraukensis** Venhuis
A. (*C.*) *nataliae* Belkin
A. (*C.*) *novaguinensis* Venhuis
m **A. (C.) punctulatus** Dönitz
A. (*C.*) *solomonis* Belkin, Knight & Rozeboom
fm **A. (C.) tessellatus** Theobald
A. (*C.*) *vagus* Dönitz
Toxorhynchites (*Toxorhynchites*) *amboinensis* (Doleschall)
T. (*T.*) *inornatus* (Walker)
T. (*T.*) *speciosus* (Skuse)
T. (*T.*) *splendens* (Wiedemann)
Tripteroides (*Maorigoeldia*) *argyropus* (Walker)
T. (*Tripteroides*) *bimaculipes* (Theobald)
T. (*T.*) *binotatus* Belkin
T. (*T.*) *brevipalpis* Brug
T. (*T.*) *distigma* (Edwards)
T. (*T.*) *lipovskyi* Belkin
T. (*T.*) *magnesianus* (Edwards)
T. (*T.*) *purpuratus* (Edwards)
T. (*T.*) *quasiornatus* (Taylor)
T. (*Rachisoura*) *brevirhynchus* Brug
T. (*R.*) *filipes* (Walker)
T. (*R.*) *mathesoni* Belkin
T. (*R.*) *stonei* Belkin
T. (*R.*) *torokinae* Belkin
T. (*Polylepidomyia*) *atripes* (Skuse)
T. (*P.*) *caledonicus* (Edwards)
T. (*P.*) *coheni* Belkin
T. (*P.*) *collessi* Lee
T. (*P.*) *folicola* Belkin
T. (*P.*) *melanesiensis* Belkin
T. (*P.*) *punctolateralis* (Theobald)
T. (*P.*) *rotumanus* (Edwards)
T. (*P.*) *solomonis* (Edwards)
T. (*P.*) *subobscurus* Lee
T. (*P.*) *tasmaniensis* (Strickland)

Malaya genurostris Leicester
M. solomonis (Wharton)
Ficalbia (*Etorleptiomyia*) *elegans* (Taylor)
Coquillettidia (*Austromansonia*) *tenuipalpis* (Edwards)
C. (*Coquillettidia*) *crassipes* (van der Wulp)
C. (*C.*) *giblini* (Taylor)
C. (*C.*) *iracunda* (Walker)
C. (*C.*) *linealis* (Skuse)
C. (*C.*) *samoaensis* Stone
C. (*C.*) *xanthogaster* (Edwards)
f **Mansonia (Mansonioides) dives** (Schiner)
fv **M. (M.) uniformis** (Theobald)
Uranotaenia albescens Taylor
U. antennalis Taylor
U. argyrotarsis Leicester
U. atra Theobald
U. barnesi Belkin
U. civinskii Belkin
U. clara Dyar & Shannon
U. colocasiae Edwards
U. lateralis Ludlow
U. nivipes (Theobald)
U. painei Edwards
U. pygmaea Theobald
U. quadrimaculata Edwards
U. sexaueri Belkin
U. solomonis Belkin
U. tibialis Taylor
U. wysockii Belkin
Hodgesia cairnsensis Taylor
H. malayi Leicester
H. quasisanguinae Leicester
Orthopodomyia andamanensis Barraud
O. papuensis Zavortink
v **Aedeomyia catasticta** Knab
A. venustipes (Skuse)
Heizmannia aurea Brug
Aedes (*Mucidus*) *alternans* (Westwood)
A. (*M.*) *aurantius* (Theobald)
A. (*Ochlerotatus*) *aculeatus* (Theobald)
A. (*O.*) *albirostris* (Macquart)
A. (*O.*) *andersoni* Edwards
A. (*O.*) *antipodeus* (Edwards)
A. (*O.*) *burpengaryensis* (Theobald)
A. (*O.*) *calcariae* Marks
A. (*O.*) *camptorhynchus* (Thomson)
A. (*O.*) *clelandi* (Taylor)
A. (*O.*) *cunabulanus* Edwards
A. (*O.*) *edgari* Stone & Rosen
A. (*O.*) *flavifrons* (Skuse)
A. (*O.*) *luteifemur* Edwards
A. (*O.*) *macleayanus* Mackerras
A. (*O.*) *nigrithorax* (Macquart)
A. (*O.*) *nivalis* Edwards
v **A. (O.) normanensis** (Taylor)
A. (*O.*) *perkinsi* Marks
A. (*O.*) *pseudonormanensis* Marks
A. (*O.*) *purpuriventris* Edwards
A. (*O.*) *rubrithorax* (Macquart)
A. (*O.*) *sagax* (Skuse)
A. (*O.*) *stricklandi* (Edwards)
A. (*O.*) *theobaldi* (Taylor)
fv **A. (O.) vigilax** (Skuse)
A. (*O.*) *vittiger* (Skuse)
A. (*Finlaya*) *albilabris* Edwards
A. (*F.*) *alboannulatus* (Macquart)
A. (*F.*) *alocasicola* Marks
A. (*F.*) *auridorsum* Edwards
A. (*F.*) *australiensis* (Theobald)

A. (*F.*) *avistylus* Brug
A. (*F.*) *biocellatus* (Taylor)
A. (*F.*) *bougainvillensis* Marks
A. (*F.*) *britteni* Marks & Hodgkin
A. (*F.*) *candidoscutellum* Marks
A. (*F.*) *derooki* Brug
A. (*F.*) *dobrotworskyi* Marks
A. (*F.*) *dorseyi* Knight
A. (*F.*) *fijiensis* Marks
A. (*F.*) *freycinetiae* Laird
A. (*F.*) *gahnicola* Marks
A. (*F.*) *gressitti* Bohart
A. (*F.*) *hui* Bohart
A. (*F.*) *iwi* Marks
A. (*F.*) *josephinae* Marks
A. (*F.*) *keefei* King & Hoogstraal
A. (*F.*) *knighti* Stone & Bohart
f **A. (F.) kochi** (Dönitz)
A. (*F.*) *lauriei* (Carter)
A. (*F.*) *lewelleni* Starkey & Webb
A. (*F.*) *mackerrasi* Taylor
A. (*F.*) *mallochi* Taylor
A. (*F.*) *monocellatus* Marks
v **A. (F.) notoscriptus** (Skuse)
v **A. (F.) occidentalis** (Skuse)
A. (*F.*) *palmarum* Edwards
A. (*F.*) *papuensis* (Taylor)
A. (*F.*) *purpureus* (Theobald)
A. (*F.*) *quasirubithorax* (Theobald)
A. (*F.*) *quinquelineatus* Edwards
A. (*F.*) *samoanus* (Grünberg)
A. (*F.*) *savoryi* Bohart
A. (*F.*) *solomonis* Stone & Bohart
A. (*F.*) *subauridorsum* Marks
A. (*F.*) *wallacei* Edwards
A. (*F.*) *wasselli* Marks
A. (*Macleaya*) *tremulus* (Theobald)
A. (*Chaetocruiomyia*) *elchoensis* Taylor
A. (*C.*) *humeralis* Edwards
A. (*C.*) *moloiensis* Taylor
A. (*C.*) *spinosipes* Edwards
A. (*C.*) *tulliae* Taylor
A. (*C.*) *wattensis* Taylor
A. (*Pseudoskusea*) *ashworthi* Edwards
A. (*P.*) *australis* (Erichson)
A. (*P.*) *bancroftianus* Edwards
A. (*P.*) *culiciformis* (Theobald)
A. (*P.*) *multiplex* (Theobald)
A. (*Skusea*) *dasyorrhus* King & Hoogstraal
A. (*S.*) *lamelliferus* Bohart & Ingram
A. (*Geoskusea*) *daggyi* Stone & Bohart
A. (*G.*) *daliensis* (Taylor)
A. (*G.*) *longiforceps* Edwards
A. (*G.*) *tonsus* Edwards
A. (*Levua*) *geoskusea* Amos
fv **A. (Stegomyia) aegypti** (Linnaeus)
A. (*S.*) *agrihanensis* Bohart
A. (*S.*) *albolineatus* (Theobald)
v **A. (S.) albopictus** (Skuse)
A. (*S.*) *alorensis* Bonne-Wepster & Brug
A. (*S.*) *andrewsi* Edwards
A. (*S.*) *dybasi* Bohart
A. (*S.*) *edwardsi* (Barraud)
A. (*S.*) *guamensis* Farner & Bohart
A. (*S.*) *gurneyi* Stone & Bohart

A. (*S.*) *hakanssoni* Knight & Hurlbut
A. (*S.*) *hensilli* Farner
A. (*S.*) *horrescens* Edwards
A. (*S.*) *maehleri* Bohart
A. (*S.*) *marshallensis* Stone & Bohart
A. (*S.*) *neopandani* Bohart
A. (*S.*) *palauensis* Bohart
A. (*S.*) *pandani* Stone
A. (*S.*) *pernotatus* Farner & Bohart
fv **A. (S.) polynesiensis** Marks
A. (*S.*) *pseudalbopictus* (Borel)
fv **A. (S.) pseudoscutellaris** (Theobald)
A. (*S.*) *quasiscutellaris* Farner & Bohart
A. (*S.*) *rotanus* Bohart & Ingram
A. (*S.*) *saipanensis* Stone
f **A. (S.) scutellaris** (Walker)
A. (*S.*) *scutoscriptus* Bohart & Ingram
f **A. (S.) tongae** Edwards
A. (*S.*) *upolensis* Marks
A. (*Aedimorphus*) *alboscutellatus* (Theobald)
A. (*A.*) *caecus* (Theobald)
A. (*A.*) *lowisii* (Theobald)
A. (*A.*) *oakleyi* Stone
A. (*A.*) *senyavinensis* Knight & Hurlbut
A. (*A.*) *trukensis* Bohart
v **A. (A.) vexans** (Meigen)
A. (*Neomelaniconion*) *imprimens* (Walker)
fv **A. (N.) lineatopennis** (Ludlow)
A. (*Aedes*) *cunninghami* Taylor
A. (*A.*) *panayensis* Ludlow
A. (*A.*) *pipkini* Bohart
A. (*Verrallina*) *carmenti* Edwards
A. (*V.*) *funereus* (Theobald)
A. (*V.*) *lineatus* (Taylor)
A. (*V.*) *neomacrodixoa* King & Hoogstraal
A. (*V.*) *similis* (Theobald)
A. (*Leptosomatomyia*) *aurimargo* Edwards
Armigeres (*Armigeres*) *breinli* (Taylor)
A. (*A.*) *denbesteni* Brug
Opifex fuscus Hutton
Culiseta (*Austrotheobaldia*) *littleri* (Taylor)
C. (*Culicella*) *atra* (Lee)
C. (*C.*) *inconspicua* (Lee)
C. (*C.*) *victoriensis* (Dobrotworsky)
C. (*C.*) *weindorferi* (Edwards)
C. (*Climacura*) *novaezealandiae* Pillai
v **C. (C.) tonnoiri** (Edwards)
C. (*Neotheobaldia*) *frenchii* (Theobald)
C. (*N.*) *hilli* (Edwards)
f **Culex (Lutzia) fuscanus** Wiedemann
C. (*L.*) *halifaxii* Theobald
C. (*Neoculex*) *chaetoventralis* (Theobald)
C. (*N.*) *cheesmanae* Mattingly & Marks
C. (*N.*) *douglasi* Dobrotworsky
C. (*N.*) *fergusoni* (Taylor)
C. (*N.*) *latus* Dobrotworsky
C. (*N.*) *postspiraculosus* Lee
C. (*N.*) *pseudomelanoconia* Theobald
C. (*N.*) *tricuspis* Edwards
C. (*Mochthogenes*) *femineus* Edwards
C. (*Lophoceraomyia*) *atracus* Colless
C. (*L.*) *carolinensis* Bohart & Ingram

C. (*L.*) *christiani* Colless
C. (*L.*) *collessi* Sirivanakarn
C. (*L.*) *cottlei* Sirivanakarn
C. (*L.*) *cylindricus* Theobald
C. (*L.*) *fraudatrix* (Theobald)
C. (*L.*) *gossi* Bohart
C. (*L.*) *hilli* Edwards
C. (*L.*) *hurlbuti* Belkin
C. (*L.*) *insularis* Sirivanakarn
C. (*L.*) *kaviengensis* Sirivanakarn
C. (*L.*) *kusaiensis* Bohart
C. (*L.*) *orbostiensis* Dobrotworsky
C. (*L.*) *rajaneeae* Sirivanakarn
C. (*Culiciomyia*) *fragilis* Ludlow
C. (*C.*) *muplei* Knight & Hurlbut
C. (*C.*) *nigropunctatus* Edwards
f **C. (C.) pallidothorax** Theobald
C. (*C.*) *papuensis* (Taylor)
C. (*C.*) *pullus* Theobald
C. (*Acallyntrum*) *belkini* Stone & Penn
C. (*A.*) *pallidiceps* (Theobald)

C. (*Culex*) *albinervis* Edwards
C. (*C.*) *alis* Theobald
fv **C. (C.) annulirostris** Skuse
C. (*C.*) *asteliae* Belkin
C. (*C.*) *atriceps* Edwards
C. (*C.*) *bihamatus* Edwards
fv **C. (C.) bitaeniorhynchus** Giles
C. (*C.*) *boninensis* Bohart
C. (*C.*) *crinicauda* Edwards
C. (*C.*) *globocoxitus* Dobrotworsky
C. (*C.*) *iyengari* Mattingly & Rageau
C. (*C.*) *litoralis* Bohart
C. (*C.*) *marquesensis* Stone & Rosen
C. (*C.*) *mimulus* Edwards
C. (*C.*) *pacificus* Edwards
v **C. (C.) pervigilans** Bergroth
fv **C. (C.) pipiens** Linnaeus
C. (*C.*) *rotoruae* Belkin
C. (*C.*) *samoaensis* (Theobald)
f **C. (C.) sitiens** Wiedemann
v **C. (C.) squamosus** (Taylor)
C. (*C.*) *starckeae* Stone & Knight
C. (*C.*) *vicinus* (Taylor)
fv **C. (C.) vishnui** Theobald

APPENDIX 2 MOSQUITOS OF NEW GUINEA

Bironella (*Bironella*) *gracilis* Theobald
B. (*B.*) *brugi* Soesilo & van Slooten
B. (*B.*) *confusa* Bonne-Wepster
B. (*B.*) *papuae* (Swellengrebel & Swellengrebel de Graaf)
B. (*B.*) *soesiloi* (Strickland & Choudhury)

B. (*Brugella*) *hollandi* Taylor
B. (*B.*) *travestita* (Brug)
fmv **Anopheles (Anopheles) bancroftii** Giles
A. (*A.*) *papuensis* Dobrotworsky
A. (*Cellia*) *annulatus* Rook
mv **A. (C.) annulipes** Walker
A. (*C.*) *clowi* Rozeboom & Knight
fmv **A. (C.) farauti** Laveran

A. (*C.*) *incognitus* Brug
A. (*C.*) *karwari* (James)
fm **A. (C.) koliensis** Owen
A. (*C.*) *longirostris* Brug
A. (*C.*) *lungae* Belkin & Schlosser
v **A. (C.) meraukensis** Venhuis
A. (*C.*) *novaguinensis* Venhuis
fm **A. (C.) punctulatus** Dönitz
fm **A. (C.) tessellatus** Theobald
A. (*C.*) *vagus* Dönitz
Toxorhynchites (*Toxorhynchites*) *inornatus* (Walker)
T. (*T.*) *speciosus* (Skuse)
T. (*T.*) *splendens* (Wiedemann)
Tripteroides (*Tripteroides*) *alboscutellatus* Lee
T. (*T.*) *bimaculipes* (Theobald)
T. (*T.*) *brevipalpis* Brug
T. (*T.*) *elegans* Brug
T. (*T.*) *littlechildi* (Edwards)
T. (*T.*) *nissanensis* Lee
T. (*T.*) *purpuratus* (Edwards)
T. (*T.*) *quasiornatus* (Taylor)
T. (*T.*) *splendens* Lee
T. (*Rachisoura*) *bisquamatus* Lee
T. (*R.*) *brevirhynchus* Brug
T. (*R.*) *confusus* Lee
T. (*R.*) *exnebulis* Bonne-Wepster
T. (*R.*) *felicitatis* Bonne-Wepster
T. (*R.*) *filipes* (Walker)
T. (*R.*) *flabelliger* Bonne-Wepster
T. (*R.*) *fuliginosus* Lee
T. (*R.*) *fuscipleura* Lee
T. (*R.*) *kingi* Lee
T. (*R.*) *latisquama* (Edwards)
T. (*R.*) *longipalpatus* Lee
T. (*R.*) *pallidus* Lee
T. (*R.*) *papua* Brug
T. (*R.*) *pilosus* Lee
T. (*R.*) *plumiger* Bonne-Wepster
T. (*R.*) *simplex* Brug
T. (*R.*) *vanleeuweni* (Edwards)
T. (*Polylepidomyia*) *altivallis* Bonne-Wepster
T. (*P.*) *argenteiventris* (Theobald)
T. (*P.*) *ater* (Taylor)
T. (*P.*) *digoelensis* Brug
T. (*P.*) *microlepis* (Edwards)
T. (*P.*) *obscurus* Brug
T. (*P.*) *punctolateralis* (Theobald)
T. (*P.*) *subobscurus* Lee
Malaya leei (Wharton)
Ficalbia (*Mimomyia*) *flavens* King & Hoogstraal
F. (*M.*) *modesta* King & Hoogstraal
F. (*Etorleptiomyia*) *elegans* (Taylor)
Coquillettidia (*Coquillettidia*) *crassipes* (van der Wulp)
C. (*C.*) *memorans* (Bonne-Wepster)
C. (*C.*) *nigrochracea* (Bonne-Wepster)
C. (*C.*) *ochracea* (Theobald)
f **Mansonia (Mansonioides) annulifera** (Theobald)
M. (*M.*) *bonnewepsterae* Assem
f **M. (M.) indiana** Edwards
M. (*M.*) *papuensis* (Taylor)
v **M. (M.) septempunctata** Theobald
Uranotaenia diagonalis Brug
U. *fimbriata* King & Hoogstraal

U. neotibialis King & Hoogstraal
U. papua Brug
U. setosa King & Hoogstraal
U. subtibioclada King & Hoogstraal
U. tibioclada King & Hoogstraal
Hodgesia spoliata Edwards
Orthopodomyia papuensis Zavortink
Aedes (*Ochlerotatus*) *purpuraceus* Brug
A. (*Finlaya*) *alticola* Bonne-Wepster
A. (*F.*) *anggiensis* Bonne-Wepster
A. (*F.*) *argenteitarsis* Brug
A. (*F.*) *aureostriatus* (Doleschall)
A. (*F.*) *candidoscutellum* Marks
A. (*F.*) *clintoni* Taylor
A. (*F.*) *dobodurus* King & Hoogstraal
A. (*F.*) *gani* Bonne-Wepster
A. (*F.*) *gracilelineatus* Bonne-Wepster
A. (*F.*) *hollandius* King & Hoogstraal
A. (*F.*) *keefei* King & Hoogstraal
A. (*F.*) *novalbitarsis* King & Hoogstraal
A. (*F.*) *papuensis* (Taylor)
A. (*F.*) *plumiferus* King & Hoogstraal
f **A. (F.) poecilus** (Theobald)
A. (*F.*) *quasirubithorax* (Theobald)
A. (*F.*) *subalbitarsis* King & Hoogstraal
A. (*F.*) *toxopeusi* Bonne-Wepster
A. (*F.*) *tsiliensis* King & Hoogstraal
A. (*F.*) *wallacei* Edwards
A. (*Macleaya*) *littlechildi* Taylor
A. (*M.*) *trumulus* (Theobald)
A. (*Pseudoskusea*) *culiciformis* (Theobald)
A. (*P.*) *lunulatus* King & Hoogstraal
A. (*Skusea*) *dasyorrhus* King & Hoogstraal
A. (*Geoskusea*) *fimbripes* Edwards
fv **A. (Stegomyia) aegypti** (Linnaeus)
A. (*Aedimorphus*) *inexpectatus* Bonne-Wepster
v **A. (A.) vexans** (Meigen)
A. (*Neomelaniconion*) *brugi* Edwards
A. (*N.*) *imprimens* (Walker)
A. (*Verrallina*) *azureosquamatus* Bonne-Wepster
A. (*V.*) *bifoliatus* King & Hoogstraal
A. (*V.*) *carmenti* Edwards
A. (*V.*) *embiensis* Huang
A. (*V.*) *foliformis* King & Hoogstraal
A. (*V.*) *funereus* (Theobald)
A. (*V.*) *killertonis* Huang
A. (*V.*) *leilae* King & Hoogstraal
A. (*V.*) *lineatus* (Taylor)
A. (*V.*) *milnensis* King & Hoogstraal
A. (*V.*) *multifolium* King & Hoogstraal
A. (*V.*) *neomacrodixoa* King & Hoogstraal
A. (*V.*) *obsoletus* Huang
A. (*V.*) *panayensis* Ludlow
A. (*V.*) *parasimilis* King & Hoogstraal
A. (*V.*) *quadrifolium* Brug

A. (*V.*) *quadrispinatus* King & Hoogstraal
A. (*V.*) *reesi* King & Hoogstraal
A. (*V.*) *sentanius* King & Hoogstraal
A. (*V.*) *similis* (Theobald)
A. (*V.*) *simplus* King & Hoogstraal
A. (*V.*) *trispinatus* King & Hoogstraal
A. (*V.*) *vanapus* Huang
A. (*V.*) *variabilis* Huang
A. (*Leptosomatomyia*) *aurimargo* Edwards
A. (*Huaedes*) *medialis* Brug
A. (*H.*) *variepictus* King & Hoogstraal
A. (*H.*) *wauensis* Huang
Armigeres (*Armigeres*) *breinli* (Taylor)
A. (*A.*) *denbesteni* Brug
A. (*A.*) *lacuum* Edwards
A. (*A.*) *malayi* (Theobald)
A. (*A.*) *milnensis* Lee
Culex (*Lutzia*) *halifaxii* Theobald
C. (*Neoculex*) *brevipalpis* (Giles)
C. (*N.*) *caeruleus* King & Hoogstraal
C. (*N.*) *crassistylus* Brug
C. (*N.*) *pedicellus* King & Hoogstraal
C. (*Mochthogenes*) *cataractarum* Edwards
C. (*M.*) *malayi* (Leicester)
C. (*Lophoceraomyia*) *bolii* Sirivanakarn
C. (*L.*) *collessi* Sirivanakarn
C. (*L.*) *crowei* Sirivanakarn
C. (*L.*) *digoelensis* Brug
C. (*L.*) *durhami* Sirivanakarn
C. (*L.*) *fraudatrix* (Theobald)
C. (*L.*) *gressitti* Sirivanakarn
C. (*L.*) *kowiroensis* Sirivanakarn
C. (*L.*) *kuhnsi* King & Hoogstraal
C. (*L.*) *lakei* Sirivanakarn
C. (*L.*) *leei* King & Hoogstraal
C. (*L.*) *marksae* King & Hoogstraal
C. (*L.*) *minjensis* Sirivanakarn
C. (*L.*) *muruae* Sirivanakarn
C. (*L.*) *ornatus* (Theobald)
C. (*L.*) *petersi* Colless
C. (*L.*) *pseudornatus* Colless
C. (*L.*) *pseudorubithoracis* Sirivanakarn
C. (*L.*) *schilfgaardei* Sirivanakarn
C. (*L.*) *sedlacekae* Sirivanakarn
C. (*L.*) *shanahani* Sirivanakarn
C. (*L.*) *singuawaensis* Sirivanakarn
C. (*L.*) *steffani* Sirivanakarn
C. (*L.*) *uniformis* (Theobald)
C. (*L.*) *versabilis* Sirivanakarn
C. (*L.*) *wamanguae* Sirivanakarn
C. (*Culiciomyia*) *fragilis* Ludlow
C. (*C.*) *fuscicinctus* King & Hoogstraal
C. (*C.*) *nailoni* King & Hoogstraal
f **C. (C.) pallidothorax** Theobald
C. (*C.*) *papuensis* (Taylor)
C. (*C.*) *pullus* Theobald
C. (*C.*) *ruthi* Peters
C. (*Acallyntrum*) *bicki* Stone & Penn

C. (*A.*) *binigrolineatus* Knight & Rozeboom
C. (*A.*) *pallidiceps* (Theobald)
C. (*Culex*) *alis* Theobald
fv **C. (C.) bitaeniorhynchus** Giles
C. (*C.*) *litoralis* Bohart
C. (*C.*) *mimulus* Edwards
C. (*C.*) *miraculosus* Bonne-Wepster
fv **C. (C.) pipiens** Linnaeus
C. (*C.*) *solitarius* Bonne-Wepster
C. (*C.*) *squamosus* (Taylor)
C. (*C.*) *vicinus* (Taylor)
fv **C. (C.) vishnui** Theobald
f **C. (C.) whitmorei** (Giles)

APPENDIX 3 MOSQUITOS OF THE ETHIOPIAN REGION

(Africa and neighbouring islands south of the Sahara, but excluding Madagascar)

Anopheles (*Anopheles*) *caliginosus* De Meillon
A. (*A.*) *concolor* Edwards
A. (*A.*) *coustani* Laveran
A. (*A.*) *implexus* (Theobald)
A. (*A.*) *obscurus* (Grünberg)
m **A. (A.) paludis** Theobald
A. (*A.*) *symesi* Edwards
A. (*A.*) *tebebrosus* Dönitz
A. (*A.*) *ziemanni* Grünberg
A. (*Cellia*) *ardensis* (Theobald)
A. (*C.*) *argenteolobatus* (Gough)
A. (*C.*) *aruni* Sobti
A. (*C.*) *austenii* (Theobald)
A. (*C.*) *azaniae* Bailly-Choumara
A. (*C.*) *barberellus* Evans
A. (*C.*) *berghei* Vincke & Leleup
A. (*C.*) *bervoetsi* D'Haenens
A. (*C.*) *brohieri* Edwards
A. (*C.*) *brucei* Service
A. (*C.*) *brumpti* Hamon & Rickenbach
m **A. (C.) brunnipes** (Theobald)
A. (*C.*) *buxtoni* Service
A. (*C.*) *cameroni* De Meillon & Evans
A. (*C.*) *caroni* Adam
A. (*C.*) *carteri* Evans & De Meillon
A. (*C.*) *christyi* (Newstead & Carter)
A. (*C.*) *cinctus* (Newstead & Carter)
A. (*C.*) *cinereus* Theobald
A. (*C.*) *confusus* Evans & Leeson
A. (*C.*) *cydippis* De Meillon
A. (*C.*) *dancalicus* Coradetti
A. (*C.*) *daudi* Coluzzi
A. (*C.*) *demeilloni* Evans
A. (*C.*) *distinctus* (Newstead & Carter)
A. (*C.*) *domicolus* Edwards
A. (*C.*) *dthali* Patton
A. (*C.*) *dureni* Edwards
A. (*C.*) *erythraeus* Corradetti
A. (*C.*) *faini* Leleup
A. (*C.*) *flavicosta* Edwards
A. (*C.*) *fontinalis* Gillies & De Meillon
A. (*C.*) *freetownensis* Evans
fmv **A. (C.) funestus** Giles
A. (*C.*) *fuscivenosus* Leeson
fmv **A. (C.) gambiae** Giles

A. (C.) *garnhami* Edwards
A. (C.) *gibbinsi* Evans
A. (C.) *hamoni* Adam
m **A. (C.) hancocki** Edwards
m **A. (C.) hargreavesi** Evans
A. (C.) *harperi* Evans
m **A. (C.) hispaniola** (Theobald)
A. (C.) *jebudensis* Froud
A. (C.) *keniensis* Evans
A. (C.) *kingi* Christophers
A. (C.) *leesoni* Evans
A. (C.) *listeri* De Meillon
A. (C.) *lloreti* Gil Collado
A. (C.) *longipalpis* (Theobald)
A. (C.) *lovettae* Evans
A. (C.) *machardyi* Edwards
f **A. (C.) maculipalpis** Giles
A. (C.) *maliensis* Bailly-Choumara & Adam
A. (C.) *marshallii* (Theobald)
m **A. (C.) melas** (Theobald)
A. (C.) *merus* Dönitz
A. (C.) *mortiauxi* Edwards
m **A. (C.) moucheti** Evans
A. (C.) *mousinhoi* De Meillon & Pereira
A. (C.) *multicolor* Cambouliu
A. (C.) *murphyi* Gillies & De Meillon
A. (C.) *natalensis* (Hill & Haydon)
m **A. (C.) nili** (Theobald)
A. (C.) *njombiensis* Peters
A. (C.) *parensis* Gillies
fmv **A. (C.) pharoensis** Theobald
A. (C.) *pretoriensis* (Theobald)
A. (C.) *rhodesiensis* Theobald
A. (C.) *rivulorum* Leeson
A. (C.) *rodhaini* Leleup & Lips
A. (C.) *ruarinus* Edwards
m **A. (C.) rufipes** (Gough)
A. (C.) *salbaii* Maffi & Coluzzi
A. (C.) *schwetzi* Evans
m **A. (C.) sergentii** (Theobald)
A. (C.) *seydeli* Edwards
A. (C.) *smithii* Theobald
A. (C.) *somalicus* Rivola & Holstein
A. (C.) *squamosus* Theobald
A. (C.) *swahilicus* Gillies
A. (C.) *tchekedii* De Meillon & Leeson
A. (C.) *theileri* Edwards
A. (C.) *turkhudi* Liston
A. (C.) *upemba* Lips
A. (C.) *vanhoofi* Wanson & Lebied
A. (C.) *vernus* Gillies & De Meillon
A. (C.) *vinckei* De Meillon
A. (C.) *walravensi* Edwards
A. (C.) *wellcomei* Theobald
A. (C.) *wilsoni* Evans
Toxorhynchites (*Toxorhynchites*) *aeneus* (Evans)
T. (T.) *barbipes* Edwards
T. (T.) *brevipalpis* Theobald
T. (T.) *erythrurus* (Edwards)
T. (T.) *evansae* (Edwards)
T. (T.) *kaimosi* (Someren)
T. (T.) *lutescens* (Theobald)
T. (T.) *nairobiensis* (Someren)
T. (T.) *phytophagus* Theobald
T. (T.) *ruwenzori* (Someren)
T. (T.) *viridibasis* (Edwards)
Malaya farquharsoni (Edwards)
M. *fraseri* (Edwards)
M. *marceli* (Mattingly)

M. moucheti (Hamon & Adam)
M. taeniarostris (Theobald)
M. trichorostris (Theobald)
Ficalbia (*Ficalbia*) *circumtestacea* (Theobald)
F. (*F.*) *malfeyti* (Newstead)
F. (*F.*) *nigra* (Theobald)
F. (*F.*) *uniformis* (Theobald)
F. (*Mimomyia*) *femorata* Edwards
F. (*M.*) *flavopicta* Edwards
F. (*M.*) *hispida* (Theobald)
F. (*M.*) *lacustris* Edwards
F. (*M.*) *mimomyiaformis* (Newstead)
F. (*M.*) *pallida* (Edwards)
F. (*M.*) *parenti* De Meillon & Lavoipierre
F. (*M.*) *perplexens* Edwards
F. (*M.*) *plumosa* (Theobald)
F. (*M.*) *splendens* (Theobald)
F. (*Etorleptiomyia*) *mediolineata* (Theobald)
F. (*E.*) *xanthozona* Someren
Coquillettidia (*Coquillettidia*) *annettii* (Theobald)
C. (*C.*) *aurea* (Edwards)
v **C. (C.) aurites** (Theobald)
C. (*C.*) *chrysosoma* (Edwards)
C. (*C.*) *cristata* (Theobald)
C. (*C.*) *flavocincta* (Edwards)
C. (*C.*) *fraseri* (Theobald)
v **C. (C.) fuscopennata** (Theobald)
C. (*C.*) *karandalaensis* (Wolfs)
C. (*C.*) *maculipennis* (Theobald)
v **C. (C.) metallica** (Theobald)
C. (*C.*) *microannulata* (Theobald)
C. (*C.*) *nigritarsis* (Wolfs)
C. (*C.*) *nigrithorax* (Theobald)
C. (*C.*) *pseudoconopas* (Theobald)
C. (*C.*) *schoutedeni* (Wolfs)
C. (*C.*) *vanoyei* (Wolfs)
C. (*C.*) *versicolor* (Edwards)
C. (*C.*) *whalbergi* (Edwards)
fv **Mansonia (Mansonioides) africana** (Theobald)
fv **M. (M.) uniformis** (Theobald)
Uranotaenia alba Theobald
U. alboabdominalis Theobald
U. annulata Theobald
U. balfouri Theobald
U. bilineata Theobald
U. browni Mattingly
U. caeruleocephala Theobald
U. caliginosa Philip
U. candidipes Edwards
U. cavernicola Mattingly
U. chorleyi Edwards
U. devemyi Hamon
U. fusca Theobald
U. garnhami Someren
U. henrardi Edwards
U. hopkinsi Edwards
U. lucyae Someren
U. mashonaensis Theobald
U. mayeri Edwards
U. montana Ingram & De Meillon
U. nepenthes (Theobald)
U. nigripes (Theobald)
U. nigromaculata Edwards
U. ornata Theobald
U. pallidocephala Theobald
U. palmeirimi De Meillon & Rebêlo
U. pandani (Theobald)
U. philonuxia Philip
U. pseudohenrardi Peters
U. shillitonis Edwards
U. yovani Someren

Hodgesia cyptopus Theobald
H. nigeriae Edwards
H. psectropus Edwards
H. sanguinae Theobald
Orthopodomyia arboricollis (Charmoy)
Aedeomyia africana Neveu-Lemaire
A. furfurea (Enderlein)
Eretmapodites argyrurus Edwards
E. brottesi Rickenbach
E. caillardi Rickenbach
v **E. chrysogaster** Graham
E. corbeti Hamon
E. dracaenae Edwards
E. forcipulatus Edwards
E. gilletti Someren
E. grahami Edwards
E. grenieri Hamon & Someren
E. haddowi Someren
E. harperi Someren
E. hightoni Someren
E. inornatus Newstead
E. intermedius Edwards
v **E. leucopous** Graham
E. mahaffyi Someren
E. marcelleae Adam & Hamon
E. mattinglyi Hamon & Someren
E. melanopous Graham
E. oedipodeios Graham
E. parvipluma Edwards
E. pauliani Grjebine
E. penicillatus Edwards
E. plioleucus Edwards
E. quinquevittatus Theobald
E. salauni Rickenbach
E. semisimplicipes Edwards
v **E. silvestris** Ingram & De Meillon
E. subsimplicipes Edwards
E. tonsus Edwards
E. vansomereni Hamon
E. wansoni Edwards

Aedes (*Mucidus*) *grahamii* Theobald
A. (*M.*) *mucidus* (Karsch)
A. (*M.*) *nigerrimus* (Theobald)
A. (*M.*) *scatophagoides* (Theobald)
A. (*M.*) *tonkingi* Gebert
A. (*Ochlerotatus*) *breedensis* Muspratt
v **A. (O.) caballus** (Theobald)
v **A. (O.) fryeri** (Theobald)
A. (*O.*) *harrisoni* Muspratt
A. (*Finlaya*) *barnardi* Edwards
A. (*F.*) *embuensis* Edwards
A. (*F.*) *fulgens* (Edwards)
A. (*F.*) *ingrami* Edwards
A. (*F.*) *longipalpis* (Grünberg)
A. (*F.*) *luteostriatus* Robinson
A. (*F.*) *nyasae* Edwards
A. (*F.*) *pulchrithorax* Edwards
A. (*F.*) *wellmanii* (Theobald)
A. (*Skusea*) *lambrechti* Someren
v **A. (S.) pembaensis** Theobald
fv **A. (Stegomyia) aegypti** (Linnaeus)
v **A. (S.) africanus** (Theobald)
v **A. (S.) albopictus** (Skuse)
A. (*S.*) *amaltheus* De Meillon & Lavoipierre
A. (*S.*) *angustus* Edwards
A. (*S.*) *apicoargenteus* (Theobald)
A. (*S.*) *bambusae* Edwards
A. (*S.*) *calceatus* Edwards
A. (*S.*) *chaussieri* Edwards
A. (*S.*) *contiguus* Edwards
A. (*S.*) *deboeri* Edwards
v **A. (S.) demeilloni** Edwards

A. (*S.*) *dendrophilus* Edwards
A. (*S.*) *fraseri* (Edwards)
A. (*S.*) *grantii* (Theobald)
A. (*S.*) *heischi* Someren
A. (*S.*) *keniensis* Someren
A. (*S.*) *kivuensis* Edwards
A. (*S.*) *langata* Someren
v **A. (S.) luteocephalus** (Newstead)
A. (*S.*) *mascarensis* MacGregor
A. (*S.*) *masseyi* Edwards
v **A. (S.) metallicus** (Edwards)
A. (*S.*) *poweri* (Theobald)
A. (*S.*) *pseudoafricanus* Chwatt
A. (*S.*) *pseudonigeria* (Theobald)
A. (*S.*) *ruwenzori* Haddow & Someren
A. (*S.*) *schwetzi* Edwards
v **A. (S.) simpsoni** (Theobald)
A. (*S.*) *soleatus* Edwards
v **A. (S.) strelitziae** Muspratt
A. (*S.*) *subargenteus* Edwards
A. (*S.*) *unilineatus* (Theobald)
A. (*S.*) *usambara* Mattingly
A. (*S.*) *vinsoni* Mattingly
v **A. (S.) vittatus** (Bigot)
A. (*S.*) *woodi* Edwards
v **A. (Aedimorphus) abnormalis** (Theobald)
v **A. (A.) albocephalus** (Theobald)
A. (*A.*) *alboventralis* (Theobald)
A. (*A.*) *apicoannulatus* (Edwards)
A. (*A.*) *arabiensis* (Patton)
v **A. (A.) argenteopunctatus** (Theobald)
A. (*A.*) *bedfordi* Edwards
A. (*A.*) *bevisi* (Edwards)
A. (*A.*) *boneti* Gil Collado
A. (*A.*) *caliginosus* (Graham)
A. (*A.*) *capensis* Edwards
A. (*A.*) *centropunctatus* (Theobald)
A. (*A.*) *chamboni* Cornet
A. (*A.*) *congolensis* Edwards
v **A. (A.) cumminsii** (Theobald)
A. (*A.*) *dalzieli* (Theobald)
A. (*A.*) *dentatus* (Theobald)
A. (*A.*) *domesticus* (Theobald)
A. (*A.*) *durbanensis* (Theobald)
A. (*A.*) *eritreae* Lewis
A. (*A.*) *falabreguesi* Hamon
A. (*A.*) *filicis* Ingram & De Meillon
v **A. (A.) fowleri** (Charmoy)
A. (*A.*) *gibbinsi* Edwards
A. (*A.*) *grjebinei* Hamon, Taufflieb & Maillot
A. (*A.*) *haworthi* Edwards
A. (*A.*) *hirsutus* (Theobald)
A. (*A.*) *holocinctus* Edwards
A. (*A.*) *hopkinsi* Edwards
A. (*A.*) *insolens* Edwards
A. (*A.*) *irritans* (Theobald)
A. (*A.*) *kapretwae* Edwards
A. (*A.*) *kennethi* Muspratt
A. (*A.*) *lamborni* Edwards
A. (*A.*) *leesoni* Edwards
A. (*A.*) *leptolabis* Edwards
A. (*A.*) *leucarthrius* (Speiser)
A. (*A.*) *longiseta* Edwards
v **A. (A.) marshallii** (Theobald)
A. (*A.*) *mattinglyi* Hamon & Rickenbach
A. (*A.*) *microstictus* Edwards
v **A. (A.) minutus** (Theobald)

A. (*A.*) *mixtus* Edwards
v **A. (A.) mutilus** Edwards
A. (*A.*) *natronius* Edwards
A. (*A.*) *neobiannulatus* (Theobald)
A. (*A.*) *ngong* Someren
A. (*A.*) *nigricephalus* (Theobald)
A. (*A.*) *nyounae* Hamon & Adam
A. (*A.*) *ochraceus* (Theobald)
A. (*A.*) *pachyurus* Edwards
A. (*A.*) *phyllolabis* Edwards
A. (*A.*) *pseudotarsalis* Someren
A. (*A.*) *pubescens* Edwards
A. (*A.*) *punctothoracis* (Theobald)
A. (*A.*) *quasiunivittatus* (Theobald)
A. (*A.*) *reali* Hamon & Adam
A. (*A.*) *semlikiensis* Someren
A. (*A.*) *seychellensis* (Theobald)
A. (*A.*) *simulans* (Newstead & Carter)
A. (*A.*) *smithburni* Someren
v **A. (A.) stokesi** Evans
A. (*A.*) *subdentatus* Edwards
v **A. (A.) tarsalis** (Newstead)
A. (*A.*) *teesdalei* Someren
A. (*A.*) *tricholabis* Edwards
v **A. (A.) vexans** (Meigen)
A. (*A.*) *wigglesworthi* Edwards
A. (*A.*) *yangambiensis* De Meillon & Lavoipierre
A. (*A.*) *yvoneae* Edwards
A. (*Neomelaniconion*) *albicosta* (Edwards)
A. (*N.*) *albothorax* (Theobald)
A. (*N.*) *bequaerti* Wolfs
A. (*N.*) *bolensis* Edwards
A. (*N.*) *carteri* Edwards
v **A. (N.) circumluteolus** (Theobald)
A. (*N.*) *crassiforceps* Edwards
A. (*N.*) *ellinorae* Edwards
A. (*N.*) *flavimargo* Edwards
A. (*N.*) *fuscinervis* (Edwards)
A. (*N.*) *jamoti* Hamon & Rickenbach
fv **A. (N.) lineatopennis** (Ludlow)
A. (*N.*) *luteolateralis* (Theobald)
A. (*N.*) *monotrichus* Edwards
A. (*N.*) *palpalis* (Newstead)
A. (*N.*) *pogonurus* Edwards
A. (*N.*) *punctocostalis* (Theobald)
A. (*N.*) *taeniarostris* (Theobald)
A. (*Diceromyia*) *adersi* (Edwards)
A. (*D.*) *bananea* Wolfs
A. (*D.*) *fascipalpis* (Edwards)
A. (*D.*) *flavicollis* Edwards
v **A. (D.) furcifer** (Edwards)
v **A. (D.) taylori** Edwards
A. (*D.*) *zethus* De Meillon & Lavoipierre
A. (*Pseudarmigeres*) *albomarginatus* (Newstead)
A. (*P.*) *argenteoventralis* (Theobald)
A. (*P.*) *kummi* Edwards
A. (*P.*) *michaelikati* Someren
A. (*P.*) *natalensis* Edwards
Culiseta (*allotheobaldia*) *longiareolata* (Macquart)
C. (*Theomyia*) *fraseri* (Edwards)
Culex (*Lutzia*) *tigripes* Grandpré & Charmoy

C. (*Lasiosiphon*) *adairi* Kirkpatrick
C. (*Barraudius*) *richeti* Brunhes & Venhard
C. (*Neoculex*) *acrostichalis* Edwards
C. (*N.*) *adami* (Hamon & Mouchet)
C. (*N.*) *adersianus* Edwards
C. (*N.*) *albertianus* Edwards
C. (*N.*) *albiventris* Edwards
C. (*N.*) *andreanus* Edwards
C. (*N.*) *arbieeni* Salem
C. (*N.*) *avianus* De Meillon
C. (*N.*) *calabarensis* Edwards
C. (*N.*) *galliardi* Edwards
C. (*N.*) *horridus* Edwards
C. (*N.*) *insignis* (Carter)
C. (*N.*) *kanyamwerima* Someren
C. (*N.*) *kilara* Someren
C. (*N.*) *kingianus* Edwards
C. (*N.*) *laplantei* (Hamon, Adam & Mouchet)
C. (*N.*) *peringueyi* Edwards
C. (*N.*) *pulchrithorax* Edwards
C. (*N.*) *rima* Theobald
v **C. (N.) rubinotus** Theobald
C. (*N.*) *salisburiensis* Theobald
C. (*N.*) *stellatus* Someren
C. (*N.*) *subrima* Edwards
C. (*N.*) *sunyaniensis* Edwards
C. (*N.*) *vinckei* Hamon, Holstein & Rivola
C. (*N.*) *wansoni* Wolfs
C. (*N.*) *wigglesworthi* Edwards
C. (*Mochthogenes*) *castor* De Meillon & Lavoipierre
C. (*M.*) *fimbriforceps* Edwards
C. (*M.*) *inconspicuosus* (Theobald)
C. (*M.*) *nyangae* Galliard
C. (*M.*) *simpliciforceps* Edwards
C. (*Culiciomyia*) *cambournaci* Hamon & Gandara
C. (*C.*) *cinerellus* Edwards
C. (*C.*) *cinereus* Theobald
C. (*C.*) *furlongi* Someren
C. (*C.*) *harleyi* Peters
C. (*C.*) *liberiensis* Peters
C. (*C.*) *macfiei* Edwards
C. (*C.*) *mongiro* Someren
C. (*C.*) *nebulosus* Theobald
C. (*C.*) *semibrunneus* Edwards
C. (*C.*) *subaequalis* Edwards
C. (*Culex*) *andersoni* Edwards
v **C. (C.) annulioris** Theobald
v **C. (C.) antennatus** (Becker)
C. (*C.*) *argenteopunctatus* (Ventrillon)
C. (*C.*) *astridianus* De Meillon
C. (*C.*) *ataeniatus* Theobald
C. (*C.*) *aurantapex* Edwards
fv **C. (C.) bitaeniorhynchus** Giles
C. (*C.*) *bukavuensis* Wolfs
C. (*C.*) *calurus* Edwards
C. (*C.*) *castelli* Hamon
C. (*C.*) *chorleyi* Edwards
C. (*C.*) *decens* Theobald
C. (*C.*) *duttoni* Theobald
C. (*C.*) *ethiopicus* Edwards
C. (*C.*) *grahamii* Theobald
C. (*C.*) *guiarti* Blanchard
C. (*C.*) *hancocki* Edwards
C. (*C.*) *hopkinsi* Edwards
C. (*C.*) *ingrami* Edwards
C. (*C.*) *invidiosus* Theobald
C. (*C.*) *mattinglyi* Knight
C. (*C.*) *mirificus* Edwards
C. (*C.*) *moucheti* Evans
C. (*C.*) *musarum* Edwards
C. (*C.*) *nakuruensis* Mattingly

C. (C.) *ninagongoensis* Edwards
C. (C.) *ornatothoracis* Theobald
C. (C.) *perfidiosus* Edwards
C. (C.) *perfuscus* Edwards
C. (C.) *philipi* Edwards
fv **C. (C.) pipiens** Linnaeus
C. (C.) *poicilipes* (Theobald)
C. (C.) *pruina* Theobald
C. (C.) *pseudopruina* Someren
C. (C.) *quasiguiarti* Theobald
C. (C.) *schwetzi* Edwards
C. (C.) *scottii* Theobald
C. (C.) *seldeslachtsi* Wolfs
C. (C.) *shoae* Hamon & Ovazza
C. (C.) *simpsoni* Theobald
C. (C.) *sinaiticus* Kirkpatrick
C. (C.) sitiens Wiedemann
C. (C.) *striatipes* Edwards
C. (C.) *tamsi* Edwards
v **C. (C.) telesilla** De Meillon & Lavoipierre
C. (C.) *tenagius* Someren
C. (C.) *terzii* Edwards
v **C. (C.) thalassius** Theobald
v **C. (C.) theileri** Theobald
C. (C.) *toroensis* Edwards & Gibbins
C. (C.) *trifilatus* Edwards
C. (C.) *trifoliatus* Edwards
fv **C. (C.) tritaeniorhynchus** Giles
C. (C.) *umbripes* Edwards
v **C. (C.) univittatus** Theobald
C. (C.) *vansomereni* Edwards
C. (C.) *watti* Edwards
C. (C.) *weschei* Edwards
C. (C.) *zeltneri* Neveu-Lemaire
C. (C.) *zombaensis* Theobald

APPENDIX 4 MOSQUITOS OF MADAGASCAR

Anopheles (*Anopheles*) *coustani* Laveran
A. (A.) *fuscicolor* Someren
A. (A.) *obscurus* (Grünberg)
A. (*Cellia*) *ardensis* (Theobald)
A. (C.) *argenteolobatus* (Gough)
A. (C.) *arnoulti* Grjebine
A. (C.) *christyi* (Newstead & Carter)
A. (C.) *cordurieri* Grjebine
fmv **A. (C.) funestus** Giles
fmv **A. (C.) gambiae** Giles
m **A. (C.) hancocki** Edwards
A. (C.) *lacani* Grjebine
f **A. (C.) maculipalpis** Giles
A. (C.) *marshallii* (Theobald)
A. (C.) *mascarensis* De Meillon
m **A. (C.) melas** (Theobald)
A. (C.) *milloti* Grjebine & Lacan
A. (C.) *notleyi* Someren
A. (C.) *pauliani* Grjebine
fmv **A. (C.) pharoensis** Theobald
A. (C.) *radama* De Meillon
A. (C.) *ranci* Grjebine
A. (C.) *roubaudi* Grjebine
A. (C.) *squamosus* Theobald
A. (C.) *theileri* Edwards
Toxorhynchites (*Toxorhynchites*) *brevipalpis* Theobald
T. (T.) *pauliani* (Doucet)
Ficalbia (*Ficalbia*) *uniformis* (Theobald)

F. (*Mimomyia*) *hispida* (Theobald)
F. (*M.*) *plumosa* (Theobald)
F. (*Ravenalites*) *aurata* (Doucet)
F. (*R.*) *bernardi* (Doucet)
F. (*R.*) *beytouti* (Doucet)
F. (*R.*) *jeansottei* (Doucet)
F. (*R.*) *martinei* Doucet
F. (*R.*) *roubaudi* (Doucet)
F. (*R.*) *spinosa* (Doucet)
Coquillettidia (*Coquillettidia*) *grandidieri* (Blanchard)
C. (*C.*) *rochei* (Doucet)
Uranotaenia andavakae Doucet
U. balfouri Theobald
U. brumpti Doucet
U. cachani (Doucet)
U. combesi Doucet
U. douceti Grjebine
U. dumonti Doucet
U. fusca Theobald
U. grenieri Doucet
U. kraussi Grjebine
U. lavieri Doucet
U. mashonaensis Theobald
U. neireti Edwards
U. nepenthes (Theobald)
U. ornata Theobald
U. pandani (Theobald)
U. shillitonis Edwards
U. tsaratananae Doucet
Orthopodomyia geberti Grjebine
O. milloti Doucet
O. vernoni Someren
Aedeomyia furfurea (Enderlein)
A. pauliani Grjebine
Eretmapodites oedipodius Graham
E. quinquevittatus Theobald
E. subsimplicipes Edwards
Aedes (*Ochlerotatus*) *dufouri* Hamon
v **A. (O.) fryeri** (Theobald)
A. (*Finlaya*) *madagascarensis* Someren
A. (*F.*) *monetus* Edwards
A. (*F.*) *phillipi* Someren
v **A. (Skusea) pembaensis** Theobald
fv **A. (Stegomyia) aegypti** (Linnaeus)
v **A. (S.) albopictus** (Skuse)
v **A. (Aedimorphus) albocephalus** (Theobald)
A. (*A.*) *tiptoni* Grjebine
A. (*Diceromyia*) *grassei* Doucet
Culex (*Neoculex*) *chauveti* Brunhes & Rambelo
C. (*N.*) *horridus* Edwards
C. (*N.*) *insignis* (Carter)
C. (*N.*) *kingianus* Edwards
C. (*N.*) *salisburiensis* Theobald
C. (*N.*) *seyrigi* Edwards
C. (*Culiciomyia*) *cinerellus* Edwards
C. (*C.*) *cinereus* Theobald
C. (*C.*) *milloti* Doucet
C. (*Culex*) *argenteopunctatus* (Ventrillon)
fv **C. (C.) bitaeniorhynchus** Giles
C. (*C.*) *demeilloni* Doucet
C. (*C.*) *giganteus* Ventrillon
C. (*C.*) *grahamii* Theobald
C. (*C.*) *guiarti* Blanchard
C. (*C.*) *perfidiosus* Edwards
fv **C. (C.) pipiens** Linnaeus
C. (*C.*) *poicilipes* (Theobald)
C. (*C.*) *quasiguiarti* Theobald
C. (*C.*) *robici* Doucet
fv **C. (C.) sitiens** Wiedemann
fv **C. (C.) tritaeniorhynchus** Giles
v **C. (C.) univittatus** Theobald
C. (*C.*) *ventrilloni* Edwards

APPENDIX 5 MOSQUITOS OF THE NEARCTIC REGION (*Canada, USA and the northern highlands of Mexico*)

Anopheles (*Anopheles*) *atropos* Dyar & Knab
A. (*A.*) *barberi* Coquillett
A. (*A.*) *bradleyi* King
v **A. (A.) crucians** Wiedemann
A. (*A.*) *earlei* Vargas
mv **A. (A.) freeborni** Aitken
A. (*A.*) *georgianus* King
A. (*A.*) *judithae* Zavortink
A. (*A.*) *occidentalis* Dyar & Knab
A. (*A.*) *perplexens* Ludlow
mv **A. (A.) pseudopunctipennis** Theobald
A. (*A.*) *punctipennis* (Say)
fm **A. (A.) quadrimaculatus** Say
A. (*A.*) *walkeri* Theobald
Toxorhynchites (*Lynchiella*) *rutilus* (Coquillett)
Wyeomyia (*Wyeomyia*) *haynei* Dodge
W. (*W.*) *smithii* (Coquillett)
v **Coquillettidia (Coquillettidia) perturbans** (Walker)
Mansonia (*Mansonia*) *indubitans* Dyar & Shannon
v **M. (M.) titillans** (Walker)
Uranotaenia anhydor Dyar
U. lowii Theobald
U. sapphirina (Osten Sacken)
Orthopodomyia alba Baker
v **O. signifera** (Coquillett)
O. waverleyi (Grabham)
v **Psorophora (Psorophora) ciliata** (Fabricius)
P. (*P.*) *howardii* Coquillett
P. (*Janthinosoma*) *cyanescens* (Coquillett)
v **P. (J.) ferox** (Humboldt)
P. (*J.*) *horrida* (Dyar & Knab)
P. (*J.*) *longipalpis* Roth
P. (*J.*) *mexicana* (Bellardi)
P. (*J.*) *varipes* (Coquillett)
fv **P. (Grabhamia) confinis** (Lynch Arribálzaga)
fv **P. (G.) discolor** (Coquillett)
P. (*G.*) *signipennis* (Coquillett)
Aedes (*Ochlerotatus*) *aboriginis* Dyar
A. (*O.*) *abserratus* (Felt & Young)
A. (*O.*) *aloponotum* Dyar
v **A. (O.) atlanticus** Dyar & Knab
A. (*O.*) *aurifer* (Coquillett)
A. (*O.*) *barri* Rueger
A. (*O.*) *bicristatus* Thurman & Winkler
A. (*O.*) *bimaculatus* (Coquillett)
A. (*O.*) *campestris* Dyar & Knab
A. (*O.*) *canadensis* (Theobald)
A. (*O.*) *cantator* (Coquillett)
A. (*O.*) *cataphylla* Dyar
A. (*O.*) *communis* (De Geer)
A. (*O.*) *decticus* Howard, Dyar & Knab
A. (*O.*) *deserticola* Zavortink
A. (*O.*) *diantaeus* Howard, Dyar & Knab
v **A. (O.) dorsalis** (Meigen)
A. (*O.*) *dupreei* (Coquillett)
A. (*O.*) *excrucians* (Walker)
A. (*O.*) *fitchii* (Felt & Young)
A. (*O.*) *flavescens* (Müller)
A. (O.) fulvus (Wiedemann)

A. (*O.*) *grossbecki* Dyar & Knab
A. (*O.*) *hexodontus* Dyar
A. (*O.*) *idahoensis* (Theobald)
A. (*O.*) *impiger* (Walker)
A. (*O.*) *implicatus* Vockeroth
A. (*O.*) *increpitus* Dyar
v **A. (O.) infirmatus** Dyar & Knab
A. (*O.*) *intrudens* Dyar
v **A. (O.) melanimon** Dyar
v **A. (O.) mitchellae** (Dyar)
A. (*O.*) *monticola* Belkin & McDonald
A. (*O.*) *muelleri* Dyar
A. (*O.*) *nigripes* (Zetterstedt)
v **A. (O.) nigromaculis** (Ludlow)
A. (*O.*) *niphadopsis* Dyar & Knab
A. (*O.*) *pionips* Dyar
A. (*O.*) *pullatus* (Coquillett)
A. (*O.*) *punctodes* Dyar
A. (*O.*) *punctor* (Kirby)
A. (*O.*) *purpureipes* Aitken
A. (*O.*) *rempeli* Vockeroth
A. (*O.*) *riparius* Dyar & Knab
v **A. (O.) scapularis** (Rondani)
A. (*O.*) *schizopinax* Dyar
A. (*O.*) *sierrensis* (Ludlow)
v **A. (O.) sollicitans** (Walker)
A. (*O.*) *spencerii* (Theobald)
A. (*O.*) *squamiger* (Coquillett)
v **A. (O.) sticticus** (Meigen)
v **A. (O.) stimulans** (Walker)
v **A. (O.) taeniorhynchus** (Wiedemann)
A. (*O.*) *thelcter* Dyar
A. (*O.*) *thibaulti* Dyar & Knab
A. (*O.*) *tormentor* Dyar & Knab
A. (*O.*) *trichurus* (Dyar)
v **A. (O.) trivattatus** (Coquillett)
A. (*O.*) *varipalpus* (Coquillett)
A. (*O.*) *ventrovittis* Dyar
A. (*Finlaya*) *atropalpus* (Coquillett)
fv **A. (F.) triseriatus** (Say)
A. (*F.*) *zoosophus* Dyar & Knab
fv **A. (Stegomyia) aegypti** (Linnaeus)
v **A. (Aedimorphus) vexans** (Meigen)
A. (*Aedes*) *cinereus* Meigen
Culiseta (*Culiseta*) *alaskaensis* (Ludlow)
C. (*C.*) *impatiens* (Walker)
v **C. (C.) incidens** (Thomson)
v **C. (C.) inornata** (Williston)
C. (*C.*) *particeps* (Adams)
C. (*Culicella*) *minnesotae* Barr
C. (*C.*) *morsitans* (Theobald)
C. (*C.*) *parodites* (Dyar)
v **C. (Climacura) melanura** (Coquillett)
Culex (*Neoculex*) *apicalis* Adams
C. (*N.*) *arizonensis* Bohart
C. (*N.*) *boharti* Brookman & Reeves
C. (*N.*) *reevesi* Wirth
C. (*N.*) *territans* Walker
C. (*Culex*) *erythrothorax* Dyar
fv **C. (C.) nigripalpus** Theobald
v **C. (C.) peus** Speiser
fv **C. (C.) pipiens** Linnaeus
fv **C. (C.) restuans** Theobald
fv **C. (C.) salinarius** Coquillett
fv **C. (C.) tarsalis** Coquillett
C. (*C.*) *thriambus* Dyar

C. (*Melanoconion*) *abominator* Dyar & Knab
C. (*M.*) *anips* Dyar
C. (*M.*) *erraticus* (Dyar & Knab)
C. (*M.*) *peccator* Dyar & *Knab*
C. (*Mochlostyrax*) *pilosus* (Dyar & Knab)
C. (*Tinolestes*) *latisquama* (Coquillett)
Deinocerites mathesoni Belkin & Hogue

APPENDIX 6 MOSQUITOS OF THE NEOTROPICAL REGION

(*Central and South America and neighbouring islands*)

Chagasia bathana (Dyar)
C. bonneae Root
C. fajardoi (Lutz)
C. rozeboomi Causey, Deane & Deane
Anopheles (*Stethomyia*) *acanthotorynus* Komp
A. (*S.*) *canorii* Flock & Abonnenc
A. (*S.*) *kompi* Edwards
v **A. (S.) nimbus** (Theobald)
A. (*S.*) *thomasi* Shannon
A. (*Anopheles*) *anchietai* Correa & Ramalho
A. (*A.*) *annulipalpis* Lynch Arribálzaga
A. (*A.*) *apicimacula* Dyar & Knab
A. (*A.*) *arboricola* Zavortink
m **A. (A.) aztecus** Hoffmann
A. (*A.*) *bonnei* Fonseca & Ramos
A. (*A.*) *bustamentei* Galvão
v **A. (A.) crucians** Wiedemann
A. (*A.*) *eiseni* Coquillett
A. (*A.*) *evandroi* Lima
A. (*A.*) *fausti* Vargas
A. (*A.*) *fluminensis* Root
A. (*A.*) *gabaldoni* Vargas
v **A. (A.) grabhamii** Theobald
A. (*A.*) *guarao* Anduze & Capdevielle
A. (*A.*) *hectoris* Giaquinto-Mira
A. (*A.*) *intermedius* (Peryassú)
A. (*A.*) *maculipes* (Theobald)
A. (*A.*) *mattogrossensis* Lutz & Neiva
A. (*A.*) *mediopunctatus* (Theobald)
A. (*A.*) *minor* Lima
A. (*A.*) *neomaculipalpus* Curry
A. (*A.*) *parapunctipennis* Martini
A. (*A.*) *peryassui* Dyar & Knab
A. (*A.*) *powderi* Zavortink
A. (*A.*) *pseudomaculipes* (Peryassú)
mv **A. (A.) pseudopunctipennis** Theobald
m **A. (A.) punctimacula** Dyar & Knab
A. (*A.*) *rachoui* Galvão
A. (*A.*) *shannoni* Davis
A. (*A.*) *tibiamaculatus* (Neiva)
A. (*A.*) *vestitipennis* Dyar & Knab
A. (*A.*) *xelajuensis* de Leon
fmv **A. (Nyssorhynchus) albimanus** Wiedemann
fm **A. (N.) albitarsis** Lynch Arribálzaga

A. (*N.*) *anomalophyllus* Komp
A. (*N.*) *antunesi* Galvão & Amaral
fm **A. (N.) aquasalis Curry**
A. (*N.*) *argyritarsis* Robineau-Desvoidy
A. (*N.*) *benarrochi* Gabaldon, Cova-Garcia & Lopez
A. (*N.*) *braziliensis* (Chagas)
fm **A. (N.) darlingi Root**
A. (*N.*) *evansae* (Brèthes)
A. (*N.*) *galvaoi* Causey, Deane & Deane
A. (*N.*) *ininii* Senevet & Abonnenc
A. (*N.*) *lanei* Galvão & Amaral
A. (*N.*) *lutzli* Cruz
A. (*N.*) *nigritarsis* (Chagas)
A. (*N.*) *noroestensis* Galvão & Lane
A. (*N.*) *nuneztovari* Gabaldon
A. (*N.*) *oswaldoi* (Peryassú)
A. (*N.*) *parvus* (Chagas)
A. (*N.*) *pictipennis* (Philippi)
A. (*N.*) *rangeli* Gabaldon, Cova-Garcia & Lopez
A. (*N.*) *rondoni* (Neiva & Pinto)
A. (*N.*) *sanctielii* Senevet & Abonnenc
A. (*N.*) *triannulatus* (Neiva & Pinto)
A. (*Kerteszia*) *bambusicola* Komp
m **A. (K.) bellator Dyar & Knab**
v **A. (K.) boliviensis (Theobald)**
mv **A. (K.) cruzii Dyar & Knab**
A. (*K.*) *homunculus* Komp
v **A. (K.) neivai Howard, Dyar & Knab**
A. (*Lophopodomyia*) *gilesi* (Peryassú)
A. (*L.*) *gomezdelatorrei* Levi-Castillo
A. (*L.*) *oiketorakras* Osorno-Mesa
A. (*L.*) *pseudotibiamaculatus* Galvão & Barretto
A. (*L.*) *squamifemur* Antunes
A. (*L.*) *vargasi* Gabaldon, Cova-Garcia & Lopez
Toxorhynchites (*Ankylorhynchus*) *catherinensis* (Lima, Guitton & Ferreira)
T. (*A.*) *hexacis* (Martini)
T. (*A.*) *purpureus* (Theobald)
T. (*A.*) *trichopygus* (Wiedemann)
T. (*Lynchiella*) *bambusicola* (Lutz & Neiva)
T. (*L.*) *cavalierii* García & Casal
T. (*L.*) *grandiosus* (Williston)
T. (*L.*) *guadeloupensis* (Dyar & Knab)
T. (*L.*) *haemorrhoidalis* (Fabricius)
T. (*L.*) *mariae* (Bourroul)
T. (*L.*) *portoricensis* (Röder)
T. (*L.*) *pusillus* (Lima)
T. (*L.*) *solstitialis* (Lutz)
T. (*L.*) *theobaldi* (Dyar & Knab)
T. (*L.*) *violaceus* (Wiedemann)
Trichoprosopon (*Trichoprosopon*) *andinum* Levi-Castillo
T. (*T.*) *compressum* Lutz
v **T. (T.) digitatum (Rondani)**
T. (*T.*) *obscurum* Lane & Cerqueira

T. (*T.*) *soaresi* Lane & Cerqueira
T. (*Limamyia*) *brevipes* (Lima)
T. (*Vonplessenia*) *vonplesseni* (Dyar & Knab)
T. (*Shannoniana*) *fluviatilis* (Theobald)
T. (*S.*) *moralesi* (Dyar & Knab)
T. (*S.*) *schedocyclium* (Dyar & Knab)
T. (*Isostomyia*) *espini* (Martini)
T. (*I.*) *luederwaldti* (Lane)
T. (*I.*) *perturbans* (Williston)
T. (*Ctenogoeldia*) *magnum* (Theobald)
T. (*C.*) *walcotti* Lane & Cerqueira
T. (*Runchomyia*) *castroi* Lane & Cerqueira
T. (*R.*) *cerqueirai* Stone
T. (*R.*) *cotopaxensis* Levi-Castillo
T. (*R.*) *edwardsianum* Lane & Cerqueira
T. (*R.*) *evansae* Lane & Cerqueira
v **T. (R.) frontosum** (Theobald)
T. (*R.*) *humboldti* Lane & Cerqueira
T. (*R.*) *hyperleucum* (Martini)
T. (*R.*) *lampropus* (Howard, Dyar & Knab)
T. (*R.*) *lanei* (Antunes)
T. (*R.*) *leucopus* (Dyar & Knab)
v **T. (R.) longipes** (Fabricius)
T. (*R.*) *lunatum* (Theobald)
T. (*R.*) *pallidiventer* (Lutz)
T. (*R.*) *reversum* Lane & Cerqueira
T. (*R.*) *simile* Lane & Cerqueira
v **T. (R.) theobaldi** Lane & Cerqueira
Wyeomyia (*Wyeomyia*) *abebela* Dyar & Knab
W. (*W.*) *adelpha* Dyar & Knab
W. (*W.*) *aphobema* Dyar
W. (*W.*) *arthrostigma* (Lutz)
W. (*W.*) *bahama* Dyar & Knab
W. (*W.*) *caracula* Dyar & Nuñez Tovar
W. (*W.*) *celaenocephala* Dyar & Knab
W. (*W.*) *charmion* Dyar
W. (*W.*) *cadiocampa* Dyar & Knab
W. (*W.*) *downsi* Lane
W. (*W.*) *florestan* Dyar
W. (*W.*) *gaudians* Dyar & Nuñez Tovar
W. (*W.*) *gausapata* Dyar & Nuñez Tovar
W. (*W.*) *hemisagnosta* Dyar & Knab
W. (*W.*) *hirsuta* (Hill & Hill)
W. (*W.*) *hosautos* Dyar & Knab
W. (*W.*) *limai* Lane & Cerqueira
W. (*W.*) *lutzi* (Lima)
v **W. (W.) medioalbipes** Lutz
W. (*W.*) *melanopus* Dyar
W. (*W.*) *mitchellii* (Theobald)
W. (*W.*) *nigritubus* Galindo, Carpenter & Trapido
W. (*W.*) *oblita* (Lutz)
W. (*W.*) *pertinans* (Williston)
W. (*W.*) *quasiluteoventralis* (Theobald)
W. (*W.*) *robusta* Senevet & Abonnenc

W. (*W.*) *sabethea* Lane & Cerqueira
W. (*W.*) *scotinomus* (Dyar & Knab)
W. (*W.*) *serrata* (Lutz)
W. (*W.*) *simmsi* (Dyar & Knab)
W. (*W.*) *stonei* Vargas & Martinez Palacios
W. (*W.*) *vanduzeei* Dyar & Knab
W. (*Nunezia*) *bicornis* (Root)
W. (*N.*) *lateralis* Petrocchi
W. (*Cruzmyia*) *dyari* Lane & Cerqueira
W. (*C.*) *kummi* Lane & Cerqueira
W. (*C.*) *mattinglyi* Lane
W. (*Davismyia*) *arborea* Galindo, Carpenter & Trapido
W. (*D.*) *petrocchiae* (Shannon & Del Ponte)
W. (*D.*) *schnusei* (Martini)
W. (*Menolepis*) *leucostigma* Lutz
W. (*Antunesmyia*) *alani* Lane & Cerqueira
W. (*A.*) *colombiana* Lane
W. (*A.*) *flavifacies* Edwards
W. (*Dendromyia*) *airosai* Lane & Cerqueira
W. (*D.*) *albosquamata* Bonne-Wepster & Bonne
v **W. (D.) aporonoma** Dyar & Knab
W. (*D.*) *argenteorostris* (Bonne-Wepster & Bonne)
W. (*D.*) *autocratica* Dyar & Knab
W. (*D.*) *bourrouli* (Lutz)
W. (*D.*) *cesari* Del Ponte & Cerqueira
W. (*D.*) *chalcocephala* Dyar & Knab
W. (*D.*) *circumcincta* Dyar & Knab
W. (*D.*) *clasoleuca* Dyar & Knab
W. (*D.*) *coenonus* Howard, Dyar & Knab
W. (*D.*) *complosa* (Dyar)
W. (*D.*) *compta* Senevet & Abonnenc
W. (*D.*) *confusa* (Lutz)
W. (*D.*) *felicia* (Dyar & Nuñez Tovar)
W. (*D.*) *finlayi* Lane & Cerqueira
W. (*D.*) *howardi* Lane & Cerqueira
W. (*D.*) *jocosa* (Dyar & Knab)
W. (*D.*) *kerri* Del Ponte & Cerqueira
W. (*D.*) *knabi* Lane & Cerqueira
W. (*D.*) *lamellata* (Bonne-Wepster & Bonne)
W. (*D.*) *luteoventralis* Theobald
v **W. (D.) melanocephala** Dyar & Knab
W. (*D.*) *moerbista* (Dyar & Knab)
W. (*D.*) *mystes* Dyar
W. (*D.*) *negrensis* Gordon & Evans
W. (*D.*) *occulta* Bonne-Wepster & Bonne
W. (*D.*) *pampithes* (Dyar & Nuñez Tovar)
W. (*D.*) *personata* (Lutz)
W. (*D.*) *phroso* Howard, Dyar & Knab
W. (*D.*) *pseudopecten* Dyar & Knab
W. (*D.*) *rooti* (Del Ponte)
W. (*D.*) *roucouyana* (Bonne-Wepster & Bonne)

W. (*D.*) *serratoria* (Dyar & Nuñez Tovar)
W. (*D.*) *shannoni* Lane & Cerqueira
W. (*D.*) *subcomplosa* (Del Ponte)
W. (*D.*) *tarsata* Lane & Cerqueira
W. (*D.*) *taurepana* Anduze
W. (*D.*) *testei* Senevet & Abonnenc
W. (*D.*) *ulocoma* (Theobald)
W. (*D.*) *undulata* Del Ponte & Cerqueira
v **W. (D.) ypsipola** Dyar
Phoniomyia antunesi (Lane & Guimarães)
P. bonnei Lane & Cerqueira
P. davisi Lane & Cerqueira
P. diabolica Lane & Forattini
P. edwardsi Lane & Cerqueira
P. esmeraldasi Levi-Castillo
P. flabellata Lane & Cerqueira
P. fuscipes (Edwards)
P. galvaoi Corrêa & Ramalho
P. incaudata (Root)
P. lassalli (Bonne-Wepster & Bonne)
P. longirostris (Theobald)
P. lopesi Corrêa & Ramalho
P. muehlensi (Petrocchi)
P. pallidoventer Theobald
P. palmata Lane & Cerqueira
P. pilicauda (Root)
P. quasilongirostris Theobald
P. splendida (Bonne-Wepster & Bonne)
P. theobaldi Lane & Cerqueira
v **P. trinidadensis** (Theobald)
P. tripartita (Bonne-Wepster & Bonne)
Limatus andinus Levi-Castillo
L. asulleptus (Theobald)
v **L. durhamii** Theobald
L. flavisetosus Castro
L. guayasi Levi-Castillo
L. hoffmani Root
L. martiali Senevet & Abonnenc
L. pseudomethysticus (Bonne-Wepster & Bonne)
Sabethes (*Sabethes*) *albiprivus* Theobald
S. (*S.*) *amazonicus* Gordon & Evans
S. (*S.*) *batesi* Lane & Cerqueira
v **S. (S.) belisarioi** Neiva
S. (*S.*) *bipartipes* Dyar & Knab
S. (*S.*) *cyaneus* (Fabricius)
S. (*S.*) *forattinii* Cerqueira
S. (*S.*) *lutzii* Theobald
S. (*S.*) *paraitepuyensis* Anduze
S. (*S.*) *quasicyaneus* Peryassú
S. (*S.*) *tarsopus* Dyar & Knab
v **S. (Sabethoides) chloropterus** (Humboldt)
S. (*S.*) *glaucodaemon* (Dyar & Shannon)
S. (*Sabethinus*) *aurescens* (Lutz)
S. (*S.*) *fabricii* Lane & Cerqueira
S. (*S.*) *identicus* Dyar & Knab
S. (*S.*) *intermedius* (Lutz)
S. (*S.*) *melanonymphe* Dyar
S. (*S.*) *soperi* Lane & Cerqueira
S. (*S.*) *undosus* (Coquillett)
S. (*S.*) *whitmani* Lane & Cerqueira
Coquillettidia (*Rhynchotaenia*) *albicosta* (Peryassú)
C. (*R.*) *albifera* (Prado)
v **C. (R.) arribalzagai** (Theobald)
C. (*R.*) *chrysonotum* (Peryassú)

C. (*R.*) *fasciolata* (Lynch Arribálzaga)
C. (*R.*) *hermanoi* (Lane & Coutinho)
C. (*R.*) *juxtamansonia* (Chagas)
C. (*R.*) *lynchi* (Shannon)
C. (*R.*) *nigricans* (Coquillett)
C. (*R.*) *nitens* (Cerqueira)
C. (*R.*) *shannoni* (Lane & Antunes)
v **C. (R.) venezuelensis** (Theobald)
Mansonia (*Mansonia*) *amazonensis* (Theobald)
M. (*M.*) *cerqueirai* (Barreto & Coutinho)
M. (*M.*) *chagasi* (Lima)
M. (*M.*) *flaveola* (Coquillett)
M. (*M.*) *humeralis* Dyar & Knab
M. (*M.*) *indubitans* Dyar & Shannon
M. (*M.*) *pessoai* (Barreto & Coutinho)
M. (*M.*) *pseudotitillans* (Theobald)
v **M. (M.) titillans** (Walker)
M. (*M.*) *wilsoni* (Barreto & Coutinho)
Uranotaenia aequatorianna Levi-Castillo
U. apicalis Theobald
U. briseis Dyar
U. calosomata Dyar & Knab
U. coatzacoalcos Dyar & Knab
U. cooki Root
U. davisi Lane
U. ditaenionota Prado
U. geometrica Theobald
U. hystera Dyar & Knab
U. incognita Galindo, Blanton & Peyton
U. lanei Martínez & Prosen
U. leucoptera (Theobald)
U. lowii Theobald
U. mathesoni Lane
U. micromelas Edwards
U. nataliae Lynch Arribálzaga
U. orthodoxa Dyar
U. pallidoventer Theobald
U. paludosa Galindo, Blanton & Peyton
U. pulcherrima Lynch Arribálzaga
U. socialis Theobald
U. telmatophila Galindo, Blanton & Peyton
U. trapidoi Galindo, Blanton & Peyton
U. typhlosomata Dyar & Knab
Orthopodomyia albicosta (Lutz)
O. fascipes (Coquillett)
O. kummi Edwards
O. phyllozoa (Dyar & Knab)
O. sampaioi Lima
v **Aedeomyia squamipennis** (Lynch Arribálzaga)
v **Psorophora (Psorophora) ciliata** (Fabricius)
P. (*P.*) *cilipes* (Fabricius)
P. (*P.*) *holmbergii* Lynch Arribálzaga
P. (*P.*) *howardii* Coquillett
P. (*P.*) *lineata* (Humboldt)
P. (*P.*) *pallescens* Edwards
P. (*P.*) *stonei* Vargas
v **P. (Janthinosoma) albipes** (Theobald)
P. (*J.*) *champerico* (Dyar & Knab)
P. (*J.*) *circumflava* Cerqueira
P. (*J.*) *cyanescens* (Coquillett)
P. (*J.*) *discrucians* (Walker)
v **P. (J.) ferox** (Humboldt)
P. (*J.*) *forceps* Cerqueira

P. (*J.*) *johnstonii* (Grabham)
P. (*J.*) *lanei* Shannon & Cerqueira
P. (*J.*) *lutzii* (Theobald)
P. (*J.*) *melanota* Cerqueira
P. (*J.*) *totonaci* Lassmann
P. (*J.*) *varipes* (Coquillett)
P. (*Grabhamia*) *cingulata* (Fabricius)
fv **P. (G.) confinis** (Lynch Arribálzaga)
P. (*G.*) *dimidiata* Cerqueira
P. (*G.*) *insularia* (Dyar & Knab)
P. (*G.*) *leucocnemis* Martini
P. (*G.*) *paulli* Paterson & Shannon
P. (*G.*) *pruinosa* Martini
P. (*G.*) *pygmaea* (Theobald)
P. (*G.*) *varinervis* Edwards
Aedes (*Ochlerotatus*) *aenigmaticus* Cerqueira & Costa
A. (*O.*) *albifasciatus* (Macquart)
A. (*O.*) *angustivittatus* Dyar & Knab
A. (*O.*) *annuliferus* (Blanchard)
A. (*O.*) *bimaculatus* (Coquillett)
A. (*O.*) *camposanus* Dyar
A. (*O.*) *condolescens* Dyar & Knab
A. (*O.*) *crinifer* (Theobald)
A. (*O.*) *eucephalaeus* Dyar
A. (*O.*) *euiris* Dyar
A. (*O.*) *euplocamus* Dyar & Knab
v **A. (O.) fulvus** (Wiedemann)
A. (*O.*) *hastatus* Dyar
A. (*O.*) *hortator* Dyar & Knab
A. (*O.*) *jacobinae* Serafim & Davis
A. (*O.*) *lepidus* Cerqueira & Paraense
A. (*O.*) *martineti* Senevet
A. (*O.*) *meprai* Martínez & Prosen
A. (*O.*) *milleri* Dyar
A. (*O.*) *muelleri* Dyar
A. (*O.*) *obturbator* Dyar & Knab
A. (*O.*) *oligopistus* Dyar
A. (*O.*) *pennai* Antunes & Lane
A. (*O.*) *perventor* Cerqueira & Costa
A. (*O.*) *pix* Martini
A. (*O.*) *raymondi* Del Ponte, Castro & Garcia
A. (*O.*) *rhyacophilus* Lima
v **A. (O.) scapularis** (Rondani)
v **A. (O.) serratus** (Theobald)
A. (*O.*) *shannoni* Vargas & Downs
A. (*O.*) *stigmaticus* Edwards
v **A. (O.) taeniorhynchus** (Wiedemann)
A. (*O.*) *tormentor* Dyar & Knab
A. (*O.*) *tortilis* (Theobald)
A. (*Finlaya*) *argyrothorax* Bonne-Wepster & Bonne
v **A. (F.) atropalpus** (Coquillett)
v **A. (F.) fluviatilis** (Lutz)
A. (*F.*) *knabi* (Coquillett)
A. (*F.*) *kompi* Vargas & Downs
v **A. (F.) leucocelaenus** Dyar & Shannon
A. (*F.*) *leucophoebus* Galindo, Carpenter & Trapido
A. (*F.*) *leucotaeniatus* Komp
A. (*F.*) *lithoecetor* Dyar & Knab
A. (*F.*) *mediovittatus* (Coquillett)
A. (*F.*) *ramirezi* Vargas & Downs

A. (*F.*) *scutellalbum* Boshell-Manrique
A. (*F.*) *terrens* (Walker)
A. (*F.*) *upatensis* Anduze & Hecht
A. (*Howardina*) *albonotatus* (Coquillett)
A. (*H.*) *allotecnon* Kumm, Komp & Ruiz
v **A. (H.) arborealis** Bonne-Wepster & Bonne
A. (*H.*) *argyrites* Dyar & Nuñez Tovar
A. (*H.*) *aureolineatus* Berlin
A. (*H.*) *aurites* (Theobald)
A. (*H.*) *aurivittatus* Cerqueira
A. (*H.*) *bahamensis* Berlin
A. (*H.*) *brevis* Berlin
A. (*H.*) *brevitatus* Berlin
A. (*H.*) *busckii* (Coquillett)
A. (*H.*) *cozumelensis* Diaz Nájera
A. (*H.*) *ecuadoriensis* Berlin
A. (*H.*) *eleanorae* Berlin
A. (*H.*) *fulvithorax* (Lutz)
A. (*H.*) *grabhami* Berlin
A. (*H.*) *guatemala* Berlin
A. (*H.*) *guerrero* Berlin
A. (*H.*) *inaequalis* (Grabham)
A. (*H.*) *ioliota* Dyar & Knab
A. (*H.*) *leei* Berlin
A. (*H.*) *lorraineae* Berlin
A. (*H.*) *marinkellei* Berlin
A. (*H.*) *martinezi* Berlin
A. (*H.*) *osornoi* Berlin
A. (*H.*) *pseudodominicii* Komp
A. (*H.*) *quadrivittatus* (Coquillett)
A. (*H.*) *septemstriatus* Dyar & Knab
v **A. (H.) sexlineatus** (Theobald)
A. (*H.*) *spinosus* Berlin
A. (*H.*) *stenei* Thompson
A. (*H.*) *vanemdeni* Martini
A. (*H.*) *walkeri* (Theobald)
A. (*H.*) *whitmorei* Dunn
fv **A. (Stegomyia) aegypti** (Linnaeus)
v **Haemagogus (Longipalpifer) equinus** Theobald
H. (*L.*) *panarchys* Dyar
H. (*L.*) *soperi* Levi-Castillo
H. (*L.*) *tropicalis* Cerqueira & Antunes
v **H. (Stegoconops) albomaculatus** Theobald
H. (*S.*) *anastasionis* Dyar
H. (*S.*) *andinus* Osorno-Mesa
v **H. (S.) capricornii** Lutz
v **H. (S.) mesodentatus** Komp & Kumm
v **H. (S.) spegazzinii** Brèthes
H. (*S.*) *uriartei* Shannon & Del Ponte
H. (*Haemagogus*) *aeritinctus* Galindo & Trapido
H. (*H.*) *argyromeris* Dyar & Ludlow
H. (*H.*) *boshelli* Osorno-Mesa
H. (*H.*) *chalcospilans* Dyar
H. (*H.*) *garciai* Levi-Castillo
H. (*H.*) *iridicolor* Dyar
H. (*H.*) *lucifer* (Howard, Dyar & Knab)
H. (*H.*) *regalis* Dyar & Knab
v **H. (H.) splendens** Williston
Culiseta (*Culiseta*) *particeps* (Adams)
Culex (*Lutzia*) *allostigma* (Howard, Dyar & Knab)
C. (*L.*) *bigoti* Bellardi
C. (*Neoculex*) *apicalis* Adams
C. (*N.*) *derivator* Dyar & Knab
C. (*Allimanta*) *tramazayguesi* Duret

C. (*Culex*) *abnormalis* Lane
C. (*C.*) *acharistus* Root
C. (*C.*) *airozai* Lane
C. (*C.*) *alticola* Martini
C. (*C.*) *ameliae* Casal
C. (*C.*) *apicinus* Philippi
C. (*C.*) *articularis* Philippi
C. (*C.*) *azuayus* Levi-Castillo
C. (*C.*) *bahamensis* Dyar & Knab
C. (*C.*) *beauperthuyi* Anduze
C. (*C.*) *bonneae* Dyar & Knab
C. (*C.*) *brami* Forattini, Rabello & Lopes
C. (*C.*) *brethesi* Dyar
C. (*C.*) *brevispinosus* Bonne-Wepster & Bonne
C. (*C.*) *carcinoxenus* Castro
C. (*C.*) *castroi* Casal & García
C. (*C.*) *chidesteri* Dyar
C. (*C.*) *chitae* Duret
C. (*C.*) *corniger* Theobald
v **C. (C.) coronator** Dyar & Knab
C. (*C.*) *cuyanus* Duret
C. (*C.*) *declarator* Dyar & Knab
C. (*C.*) *delys* Howard, Dyar & Knab
C. (*C.*) *diplophyllum* Dyar
C. (*C.*) *dolosus* (Lynch Arribálzaga)
C. (*C.*) *duplicator* Dyar & Knab
C. (*C.*) *eduardoi* Casal & García
C. (*C.*) *foliaceus* Lane
C. (*C.*) *guayasi* Levi-Castillo
C. (*C.*) *habilitator* Dyar & Knab
C. (*C.*) *hepperi* Casal & García
C. (*C.*) *inflictus* Theobald
C. (*C.*) *interfor* Dyar
C. (*C.*) *interrogator* Dyar & Knab
C. (*C.*) *janitor* Theobald
C. (*C.*) *laticlasper* Galindo & Blanton
C. (*C.*) *levicastilloi* Lane
C. (*C.*) *lygrus* Root
C. (*C.*) *maracayensis* Evans
C. (*C.*) *mauesensis* Lane
C. (*C.*) *maxi* Dyar
C. (*C.*) *mollis* Dyar & Knab
fv **C. (C.) nigripalpus** Theobald
C. (*C.*) *paramaxi* Duret
v **C. (C.) peus** Speiser
C. (*C.*) *pinarocampa* Dyar & Knab
fv **C. (C.) pipiens** Linnaeus
C. (*C.*) *pseudojanthinosoma* Senevet & Abonnenc
C. (*C.*) *quitensis* Levi-Castillo
C. (*C.*) *riojanus* Duret
C. (*C.*) *saltanensis* Dyar
C. (*C.*) *scutatus* Rozeboom & Komp
C. (*C.*) *secutor* Theobald
C. (*C.*) *sphinx* Howard, Dyar & Knab
C. (*C.*) *spinosus* Lutz
C. (*C.*) *stenolepis* Dyar & Knab
C. (*C.*) *surinamensis* Dyar
C. (*C.*) *tisseuilli* Senevet
v **C. (C.) virgultus** Theobald
C. (*Melanoconion*) *aikenii* (Aiken & Rowland)
C. (*M.*) *albinensis* Bonne-Wepster & Bonne
C. (*M.*) *alcocci* Bonne-Wepster & Bonne
C. (*M.*) *aliciae* Duret
C. (*M.*) *amitis* Komp
C. (*M.*) *andricus* Root
C. (*M.*) *atratus* Theobald
C. (*M.*) *aureonotatus* Duret & Barreto

C. (*M.*) *bahiensis* Duret
C. (*M.*) *bastagarius* Dyar & Knab
C. (*M.*) *batesi* Rozeboom & Komp
C. (*M.*) *bejaranoi* Duret
C. (*M.*) *bequaerti* Dyar & Shannon
C. (*M.*) *bifoliolatus* Duret & Barreto
C. (*M.*) *breviculus* Senevet & Abonnenc
C. (*M.*) *carcinophilus* Dyar & Knab
C. (*M.*) *caribeanus* Galindo & Blanton
C. (*M.*) *cauchensis* Floch & Abonnenc
C. (*M.*) *changuinolae* Galindo & Blanton
C. (*M.*) *chrysonotum* Dyar & Knab
C. (*M.*) *chrysothorax* (Peryassú)
C. (*M.*) *comatus* Senevet & Abonnenc
C. (*M.*) *commevynensis* Bonne-Wepster & Bonne
C. (*M.*) *comminutor* Dyar
C. (*M.*) *confundior* Komp & Rozeboom
C. (*M.*) *conspirator* Dyar & Knab
C. (*M.*) *contei* Duret
C. (*M.*) *coppenamensis* Bonne-Wepster & Bonne
C. (*M.*) *corentynensis* Dyar
C. (*M.*) *creole* Anduze
C. (*M.*) *cristovai* Duret
C. (*M.*) *decorator* Dyar & Knab
C. (*M.*) *delpontei* Duret
C. (*M.*) *distinguendus* Dyar
C. (*M.*) *dunni* Dyar
C. (*M.*) *dureti* Casal & García
C. (*M.*) *dyius* Root
C. (*M.*) *eastor* Dyar
C. (*M.*) *educator* Dyar & Knab
C. (*M.*) *egcymon* Dyar
C. (*M.*) *elephas* Komp
C. (*M.*) *elevator* Dyar & Knab
C. (*M.*) *epanastasis* Dyar
C. (*M.*) *epirus* Aiken
C. (*M.*) *equinoxialis* Floch & Abonnenc
C. (*M.*) *ernanii* Duret
C. (*M.*) *ernsti* Anduze
C. (*M.*) *erraticus* (Dyar & Knab)
C. (*M.*) *evansae* Root
C. (*M.*) *fairchildi* Galindo & Blanton
C. (*M.*) *fasciolatus* (Lutz)
C. (*M.*) *faurani* Duret
C. (*M.*) *ferreri* Duret
C. (*M.*) *flabellifer* Komp
C. (*M.*) *flochi* Duret
C. (*M.*) *galindoi* Komp & Rozeboom
C. (*M.*) *gravitator* Dyar & Knab
C. (*M.*) *humilis* Theobald
C. (*M.*) *idottus* Dyar
C. (*M.*) *implicatus* Senevet & Abonnenc
C. (*M.*) *inadmirabilis* Dyar
C. (*M.*) *indecorabilis* (Theobald)
C. (*M.*) *inhibitator* Dyar & Knab
C. (*M.*) *intonsus* Galindo & Blanton
C. (*M.*) *intrincatus* Brèthes
C. (*M.*) *iolambdis* Dyar
C. (*M.*) *isabelae* Duret
C. (*M.*) *jocasta* Komp & Rozeboom
C. (*M.*) *johnnyi* Duret

C. (*M.*) *jubifer* Komp & Brown
C. (*M.*) *kerri* Duret
C. (*M.*) *kummi* Komp & Rozeboom
C. (*M.*) *limacifer* Komp
C. (*M.*) *lucifugus* Komp
C. (*M.*) *lugens* Lutz
C. (*M.*) *madininensis* Senevet
C. (*M.*) *martinezi* Casal & García
C. (*M.*) *maxinocca* Dyar
C. (*M.*) *misionensis* Duret
C. (*M.*) *mistura* Komp & Rozeboom
v **C. (M.) mojuensis** Duret & Damasceno
C. (*M.*) *mulrennani* Basham
C. (*M.*) *mutator* Dyar & Knab
C. (*M.*) *nicaroensis* Duret
C. (*M.*) *nicceriensis* Bonne-Wepster & Bonne
C. (*M.*) *nigrescens* (Theobald)
C. (*M.*) *nigricorpus* (Theobald)
C. (*M.*) *nigrimacula* Lane & Whitman
C. (*M.*) *ocellatus* Theobald
C. (*M.*) *oedipus* Root
C. (*M.*) *opisthopus* Komp
C. (*M.*) *orfilai* Duret
C. (*M.*) *paracrybda* Komp
C. (*M.*) *patientiae* Floch & Fauran
C. (*M.*) *pavlovskyi* Casal & García
C. (*M.*) *peccator* Dyar & Knab
C. (*M.*) *pereyrai* Duret
C. (*M.*) *phlabistus* Dyar
C. (*M.*) *phlogistus* Dyar
C. (*M.*) *pifanoi* Anduze
C. (*M.*) *plectoporpe* Root
C. (*M.*) *productus* Senevet & Abonnenc
C. (*M.*) *psatharus* Dyar
C. (*M.*) *pseudotaeniopus* Galindo & Blanton
C. (*M.*) *punctiscapularis* Floch & Abonnenc
C. (*M.*) *putumayensis* Matheson
C. (*M.*) *quadrifoliatus* Komp
C. (*M.*) *quasihibridus* Galindo & Blanton
C. (*M.*) *rabanicolus* Floch & Abonnenc
C. (*M.*) *rachoui* Duret
C. (*M.*) *ronderosi* Quiroga de Linero
C. (*M.*) *rorotaensis* Floch & Abonnenc
C. (*M.*) *saramaccensis* Bonne-Wepster & Bonne
C. (*M.*) *sardinerae* Fox
C. (*M.*) *serratimarge* Root
C. (*M.*) *silvai* Duret
C. (*M.*) *simulator* Dyar & Knab
v **C. (M.) spissipes** (Theobald)
C. (*M.*) *sursumptor* Dyar
v **C. (M.) taeniopus** Dyar & Knab
C. (*M.*) *tecmarsis* Dyar
C. (*M.*) *terebor* Dyar
C. (*M.*) *theobaldi* (Lutz)
C. (*M.*) *thomasi* Evans
C. (*M.*) *tournieri* Senevet & Abonnenc
C. (*M.*) *trifidus* Dyar
C. (*M.*) *trilobulatus* Duret & Barreto
C. (*M.*) *venezuelensis* Anduze
C. (*M.*) *vidali* Floch & Fauran
v **C. (M.) vomerifer** Komp
C. (*M.*) *wepsterae* Komp & Rozeboom

C. (*M.*) *ybarmis* Dyar
C. (*M.*) *zeteki* Dyar
C. (*Mochlostyrax*) *alogistus* Dyar
v **C. (M.) caudelli** (Dyar & Knab)
C. (*M.*) *foliafer* Komp & Rozeboom
C. (*M.*) *galvaoi* Duret
C. (*M.*) *garcesi* Duret
C. (*M.*) *innovator* Evans
C. (*M.*) *lacertosus* Komp & Rozeboom
C. (*M.*) *palaciosi* Duret
C. (*M.*) *pilosus* (Dyar & Knab)
C. (*M.*) *rooti* Rozeboom
C. (*M.*) *unicornis* Root
C. (*M.*) *vexillifer* Komp
C. (*Microculex*) *albipes* Lutz
C. (*M.*) *aphylactus* Root
C. (*M.*) *aureus* Lane & Whitman
C. (*M.*) *azymus* Dyar & Knab
C. (*M.*) *carioca* Lane & Whitman
C. (*M.*) *chryselatus* Dyar & Knab
C. (*M.*) *consolator* Dyar & Knab
C. (*M.*) *đaumastocampa* Dyar & Knab
C. (*M.*) *davisi* Kumm
C. (*M.*) *dubitans* Lane & Whitman
C. (*M.*) *elongatus* Rozeboom & Komp
C. (*M.*) *gairus* Root
C. (*M.*) *gaudeator* Dyar & Knab
C. (*M.*) *hedys* Root
C. (*M.*) *imitator* Theobald
C. (*M.*) *inimitabilis* Dyar & Knab
C. (*M.*) *intermedius* Lane & Whitman
C. (*M.*) *jenningsi* Dyar & Knab
C. (*M.*) *kukenan* Anduze
C. (*M.*) *microphyllus* Root
C. (*M.*) *neglectus* Lutz
C. (*M.*) *pleuristriatus* Theobald
C. (*M.*) *reducens* Lane & Whitman
C. (*M.*) *reginae* Floch & Fauran
C. (*M.*) *rejector* Dyar & Knab
C. (*M.*) *restrictor* Dyar & Knab
C. (*M.*) *stonei* Lane & Whitman
C. (*M.*) *worontzowi* Pessôa & Galvão
v **C. (Aedinus) accelerans** Root
v **C. (A.) amazonensis** (Lutz)
C. (*A.*) *americanus* (Neveu-Lemaire)
C. (*A.*) *clastrieri* Casal & García
C. (*Anoedioporpa*) *bamborum* Rozeboom & Komp
C. (*A.*) *belemensis* Duret & Damasceno
C. (*A.*) *browni* Komp
C. (*A.*) *canaanensis* Lane & Whitman
C. (*A.*) *chaguanco* Casal, García & Fernandez
C. (*A.*) *conservator* Dyar & Knab
C. (*A.*) *corrigani* Dyar & Knab
C. (*A.*) *damascenoi* Duret
C. (*A.*) *luteopleurus* (Theobald)
C. (*A.*) *originator* Gordon & Evans
C. (*A.*) *paganus* Evans

C. (*Micraedes*) *antillummagnorum* Dyar
C. (*M.*) *arawak* Berlin
C. (*M.*) *bisulcatus* (Coquillett)
C. (*M.*) *erethyzonfer* Galindo & Blanton
C. (*M.*) *sandrae* Berlin
C. (*M.*) *schicki* Berlin
v **C. (Tinolestes) latisquama** (Coquillett)
C. (*Carrollia*) *anduzei* Cerqueira & Lane
C. (*C.*) *antunesi* Lane & Whitman
C. (*C.*) *babahoyensis* Levi-Castillo
C. (*C.*) *bihaicolus* Dyar & Nuñez Tovar
C. (*C.*) *bonnei* Dyar
C. (*C.*) *infoliatus* Bonne-Wepster & Bonne
C. (*C.*) *iridescens* (Lutz)
C. (*C.*) *manaensis* Floch & Fauran
C. (*C.*) *mathesoni* Anduze
C. (*C.*) *metempsytus* Dyar
C. (*C.*) *secundus* Bonne-Wepster & Bonne
C. (*C.*) *soperi* Antunes & Lane
C. (*C.*) *urichii* (Coquillett)
C. (*C.*) *wilsoni* Lane & Whitman
C. (*Galindomyia*) *leei* Stone & Barreto
Deinocerites cancer Theobald
D. costaricensis Adames & Hogue
D. dyari Belkin & Hogue
D. epitedeus (Knab)
D. howardi Belkin & Hogue
D. magnus (Theobald)
D. mcdonaldi Belkin & Hogue
D. melanophylum Dyar & Knab
D. nicoyae Adames & Hogue
D. pseudes Dyar & Knab
D. spanius (Dyar & Knab)

APPENDIX 7 MOSQUITOS OF THE ORIENTAL REGION

(Tropical Asia east of Afghanistan, including most of India and Pakistan, southern China, southern Japan, Malaysia, Thailand, Indonesia and Philippines)

Anopheles (*Anopheles*) *acaci* Baisas
A. (*A.*) *aitkenii* James
A. (*A.*) *albotaeniatus* (Theobald)
A. (*A.*) *alongensis* Venhuis
A. (*A.*) *annandalei* Prashad
A. (*A.*) *argyropus* (Swellengrebel)
A. (*A.*) *asiaticus* Leicester
A. (*A.*) *baezai* Gater
A. (*A.*) *balerensis* Mendoza
fm **A. (A.) barbirostris** van der Wulp
A. (*A.*) *barbumbrosus* Strickland & Choudhury
A. (*A.*) *barianensis* James
A. (*A.*) *bengalensis* Puri
A. (*A.*) *borneensis* McArthur
A. (*A.*) *brevipalpis* Roper
A. (*A.*) *brevirostris* Reid
A. (*A.*) *bulkleyi* Causey
fm **A. (A.) campestris** Reid
A. (*A.*) *collessi* Reid

A. (*A.*) *crawfordi* Reid
A. (*A.*) *culiciformis* Cogill
f **A. (A.) donaldi** Reid
A. (*A.*) *ejercitoi* Mendoza
A. (*A.*) *fragilis* Theobald
A. (*A.*) *gigas* Giles
A. (*A.*) *habibi* Mulligan & Puri
A. (*A.*) *hunteri* (Strickland)
fm **A. (A.) hyrcanus** (Pallas)
A. (*A.*) *indiensis* Theobald
A. (*A.*) *insulaeflorum* (Swellengrebel & Swellengrebel de Graaf)
A. (*A.*) *interruptus* Puri
A. (*A.*) *koreicus* Yamada & Watanabe
A. (*A.*) *kweiyangensis* Yao & Wu
A. (*A.*) *kyondawensis* Abraham
A. (*A.*) *lesteri* Baisas & Hu
m **A. (A.) letifer** Sandosham
A. (*A.*) *lindesayi* Giles
A. (*A.*) *montanus* Stanton & Hacker
m **A. (A.) nigerrimus** Giles
A. (*A.*) *noniae* Reid
A. (*A.*) *ohamai* Ohama
A. (*A.*) *palmatus* (Rodenwaldt)
A. (*A.*) *peditaeniatus* (Leicester)
A. (*A.*) *pinjaurensis* Barraud
A. (*A.*) *pseudobarbirostris* Ludlow
A. (*A.*) *pseudosinensis* Baisas
A. (*A.*) *pullus* Yamada
A. (*A.*) *pursati* Laveran
A. (*A.*) *roperi* Reid
A. (*A.*) *samarensis* Rozeboom
A. (*A.*) *saperoi* Bohart & Ingram
A. (*A.*) *separatus* (Leicester)
A. (*A.*) *similissimus* Strickland & Choudhury
m **A. (A.) sinensis** Wiedemann
A. (*A.*) *sineroides* Yamada
A. (*A.*) *sintoni* Puri
A. (*A.*) *sintonoides* Ho
A. (*A.*) *tigertti* Scanlon & Peyton
m **A. (A.) umbrosus** (Theobald)
A. (*A.*) *vanus* Walker
f **A. (A.) whartoni** Reid
A. (*A.*) *wellingtonianus* Alcock
A. (*A.*) *yatsushiroensis* Miyazeki
fm **A. (Cellia) aconitus** Dönitz
fm **A. (C.) annularis** van der Wulp
A. (*C.*) *aurirostris* (Watson)
m **A. (C.) balabacensis** Baisas
A. (*C.*) *cristatus* King & Baisas
m **A. (C.) culicifacies** Giles
A. (*C.*) *elegans* (James)
A. (*C.*) *errabundus* (Swellengrebel)
A. (*C.*) *filipinae* Manalang
m **A. (C.) fluviatilis** James
m **A. (C.) hackeri** Edwards
A. (*C.*) *indefinitus* (Ludlow)
A. (*C.*) *jamesii* Theobald
fm **A. (C.) jeyporiensis** James
A. (*C.*) *karwari* (James)
m **A. (C.) kochi** Dönitz
A. (*C.*) *kolambuganensis* Baisas
m **A. (C.) leucosphyrus** Dönitz
m **A. (C.) litoralis** King
m **A. (C.) ludlowae** (Theobald)
fm **A. (C.) maculatus** Theobald
A. (*C.*) *majidi* Young & Majid
m **A. (C.) mangyanus** (Banks)
fm **A. (C.) minimus** Theobald

A. (*C.*) *moghulensis* Christophers
A. (*C.*) *nivipes* Theobald
A. (*C.*) *pallidus* Theobald
A. (*C.*) *pampanai* Büttiker & Beales
A. (*C.*) *parangensis* (Ludlow)
m **A. (C.) pattoni** Christophers
mv **A. (C.) philippinensis** Ludlow
A. (*C.*) *pujutensis* Colless
m **A. (C.) pulcherrimus** Theobald
f **A. (C.) ramsayi** Covell
A. (*C.*) *riparis* King & Baisas
A. (*C.*) *saungi* Colless
A. (*C.*) *schueffneri* Stanton
A. (*C.*) *splendidus* Koizumi
m **A. (C.) stephensi** Liston
A. (*C.*) *stookesi* Colless
fm **A. (C.) subpictus** Grassi
m **A. (C.) sundaicus** (Rodenwaldt)
m **A. (C.) tessellatus** Theobald
A. (*C.*) *theobaldi* Giles
A. (*C.*) *turkhudi* Liston
A. (*C.*) *vagus* Dönitz
m **A. (C.) varuna** Iyengar
A. (*C.*) *watsonii* (Leicester)
Toxorhynchites (*Toxorhynchites*) *acaudatus* (Leicester)
T. (*T.*) *albipes* (Edwards)
T. (*T.*) *amboinensis* (Doleschall)
T. (*T.*) *ater* (Daniels)
T. (*T.*) *aurifluus* (Edwards)
T. (*T.*) *auripes* (Edwards)
T. (*T.*) *coeruleus* (Brug)
T. (*T.*) *edwardsi* (Barraud)
T. (*T.*) *funestus* (Leicester)
T. (*T.*) *gigantulus* (Dyar & Shannon)
T. (*T.*) *gravelyi* (Edwards)
T. (*T.*) *kempi* (Edwards)
T. (*T.*) *klossi* (Edwards)
T. (*T.*) *leicesteri* Theobald
T. (*T.*) *magnificus* (Leicester)
T. (*T.*) *manicatus* (Edwards)
T. (*T.*) *metallicus* Leicester
T. (*T.*) *minimus* (Theobald)
T. (*T.*) *nepenthis* (Dyar & Shannon)
T. (*T.*) *nigripes* (Edwards)
T. (*T.*) *pendleburyi* (Edwards)
T. (*T.*) *quasiferox* (Leicester)
T. (*T.*) *raris* (Leicester)
T. (*T.*) *speciosus* (Skuse)
T. (*T.*) *splendens* (Wiedemann)
T. (*T.*) *sumatranus* (Brug)
T. (*T.*) *towadensis* (Matsumura)
T. (*T.*) *yaeyamae* Bohart
T. (*T.*) *yamadai* (Ôuchi)
Tripteroides (*Tripteroides*) *aeneus* (Edwards)
T. (*T.*) *affinis* (Edwards)
T. (*T.*) *antennalis* Bohart & Farner
T. (*T.*) *apoensis* Baisas & Ubaldo-Pagayon
T. (*T.*) *aranoides* (Theobald)
T. (*T.*) *bambusa* (Yamada)
T. (*T.*) *barraudi* Baisas & Ubaldo-Pagayon
T. (*T.*) *belkini* Baisas & Ubaldo-Pagayon
T. (*T.*) *caeruleocephalus* (Leicester)
T. (*T.*) *cheni* Lien
T. (*T.*) *christophersi* Baisas & Ubaldo-Pagayon
T. (*T.*) *claggi* Bohart & Farner
T. (*T.*) *delpilari* Baisas & Ubaldo-Pagayon
T. (*T.*) *denticulatus* Delfinado & Hodges

T. (T.) *dofleini* (Günther)
T. (T.) *dyari* Bohart & Farner
T. (T.) *dyi* Baisas & Ubaldo-Pagayon
T. (T.) *edwardsi* (Barraud)
T. (T.) *erlindae* Baisas & Ubaldo-Pagayon
T. (T.) *hoogstraali* Baisas
T. (T.) *hybridus* (Leicester)
T. (T.) *indeterminatus* Baisas & Ubaldo-Pagayon
T. (T.) *intermediatus* Baisas & Ubaldo-Pagayon
T. (T.) *knighti* Baisas & Ubaldo-Pagayon
T. (T.) *malayi* Delfinado & Hodges
T. (T.) *malvari* Baisas & Ubaldo-Pagayon
T. (T.) *mendacis* (Daniels)
T. (T.) *microcala* (Dyar)
T. (T.) *monetifer* (Dyar)
T. (T.) *nepenthicola* (Banks)
T. (T.) *nepenthis* (Edwards)
T. (T.) *nitidoventer* (Giles)
T. (T.) *plumosus* (Brug)
T. (T.) *powelli* (Ludlow)
T. (T.) *proximus* (Edwards)
T. (T.) *roxasi* Baisas & Ubaldo-Pagayon
T. (T.) *rozeboomi* Baisas & Ubaldo-Pagayon
T. (T.) *similis* (Leicester)
T. (T.) *simulatus* Baisas & Ubaldo-Pagayon
T. (T.) *sullivanae* Baisas & Ubaldo-Pagayon
T. (T.) *tarsalis* Delfinado & Hodges
T. (T.) *toffaletii* Baisas & Ubaldo-Pagayon
T. (T.) *uichancoi* Baisas & Ubaldo-Pagayon
T. (T.) *vicinus* (Edwards)
T. (T.) *werneri* Baisas & Ubaldo-Pagayon
T. (*Rachisoura*) *mabinii* Baisas & Ubaldo-Pagayon
Malaya genurostris Leicester
M. *jacobsoni* (Edwards)
M. *splendens* (Meijere)
Topomyia argenteoventralis Leicester
T. *argyropalpis* Leicester
T. *aureoventer* (Theobald)
T. *auriceps* Brug
T. *barbus* Baisas
T. *decorabilis* Leicester
T. *dejesusi* Baisas & Feliciano
T. *dubitans* Leicester
T. *gracilis* Leicester
T. *hernandoi* Baisas & Feliciano
T. *houghtoni* Feng
T. *imitatus* Baisas
T. *minor* Leicester
T. *nigra* Leicester
T. *pilosa* Brug
T. *pseudobarbus* Baisas
T. *rubithoracis* Leicester
T. *spathulirostris* Edwards
T. *tenuis* Edwards
T. *tipuliformis* Leicester
T. *trifida* Edwards
Ficalbia (*Ficalbia*) *jacksoni* Mattingly
F. (F.) *ludlowae* Brunetti
F. (F.) *minima* (Theobald)
F. (*Mimomyia*) *aurea* (Leicester)
F. (M.) *chamberlaini* (Ludlow)
F. (M.) *hybrida* (Leicester)
F. (M.) *intermedia* (Barraud)
F. (*Etorleptiomyia*) *elegans* (Taylor)
F. (E.) *luzonensis* (Ludlow)
F. (*Ravenalites*) *deguzmanae* Mattingly
F. (R.) *fusca* (Leicester)

F. (*R.*) *kiriromi* Klein
Coquillettidia (*Coquillettidia*) *aureosquammata* (Ludlow)
C. (*C.*) *crassipes* (van der Wulp)
C. (*C.*) *giblini* (Taylor)
C. (*C.*) *hodgkini* (Wharton)
C. (*C.*) *iracunda* (Walker)
C. (*C.*) *nigrochracea* (Bonne-Wepster)
C. (*C.*) *nigrosignata* (Edwards)
C. (*C.*) *novochracea* (Barraud)
C. (*C.*) *ochracea* (Theobald)
f **Mansonia (Mansonioides) annulata** Leicester
f **M. (M.) annulifera** (Theobald)
f **M. (M.) bonneae** Edwards
f **M. (M.) dives** (Schiner)
f **M. (M.) indiana** Edwards
fv **M. (M.) uniformis** (Theobald)
Uranotaenia alboannulata (Theobald)
U. annandalei Barraud
U. arguellesi Baisas
U. argyrotarsis Leicester
U. ascidiicola Meijere
U. bicolor Leicester
U. bimaculata Leicester
U. bimaculiala Leicester
U. brevirostris Edwards
U. campestris Leicester
U. christophersi Barraud
U. edwardsi Barraud
U. gigantea Brug
U. hebes Barraud
U. heiseri Baisas
U. hongayi Galliard & Ngu
U. husaini Qutubuddin
U. jacksoni Edwards
U. kalabahensis Haga
U. lagunensis Baisas
U. lateralis Ludlow
U. longirostris Leicester
U. ludlowae Dyar & Shannon
U. lui Lien
U. luteola Edwards
U. lutescens Leicester
U. macfarlanei Edwards
U. maculipleura Leicester
U. mattinglyi Qutubuddin
U. maxima Leicester
U. mendiolai Baisas
U. metatarsata Edwards
U. micans Leicester
U. modesta Leicester
U. moultoni Edwards
U. nii Lien
U. nivea Leicester
U. nivipleura Leicester
U. novobscura Barraud
U. obscura Edwards
U. orientalis Barraud
U. pylei Baisas
U. quinquemaculata Bonne-Wepster
U. recondita Edwards
U. reyi Baisas
U. rutherfordi Edwards
U. stonei Bohart & Ingram
U. stricklandi Barraud
U. subnormalis Martini
U. testacea Theobald
U. trilineata Leicester
U. tubanguii Baisas
U. unimaculiala Leicester
U. xanthomelaena Edwards
Hodgesia bailyi Barraud
H. malayi Leicester
H. quasisanguinae Leicester
Zeugnomyia aguilari Baisas & Feliciano
Z. fajardoi Baisas & Feliciano
Z. gracilis Leicester
Z. lawtoni Baisas
Orthopodomyia albipes Leicester
O. andamanensis Barraud
O. anopheloides (Giles)

O. flavicosta Barraud
O. flavithorax Barraud
O. lanyuensis Lien
O. madrensis Baisas
O. mcgregori (Banks)
O. papuensis Zavortink
O. siamensis Zavortink
O. wilsoni Macdonald
v **Aedeomyia catasticta** Knab
Heizmannia achaetae (Leicester)
H. aureochaeta (Leicester)
H. chandi Edwards
H. chengi Lien
H. cheni Lien
H. communis (Leicester)
H. complex (Theobald)
H. covelli Barraud
H. greenii (Theobald)
H. himalayensis Edwards
H. indica (Theobald)
H. macdonaldi Mattingly
H. metallica (Leicester)
H. nivirostris Lien
H. reidi Mattingly
H. scintillans Ludlow
H. stonei Mattingly
H. taiwanensis Lien
H. viridis Barraud
Mattinglyia catesi Lien
Udaya argyrurus (Edwards)
Aedes (*Mucidus*) *aurantius* (Theobald)
A. (*M.*) *ferinus* Knight
A. (*M.*) *laniger* (Wiedemann)
A. (*M.*) *scatophagoides* (Theobald)
A. (*Ochlerotatus*) *pulchritarsis* (Rondani)
f **A. (O.) vigilax** (Skuse)
A. (*Finlaya*) *albocinctus* (Barraud)
A. (*F.*) *albolateralis* (Theobald)
A. (*F.*) *alboniveus* Barraud
A. (*F.*) *albotaeniatus* (Leicester)
A. (*F.*) *alongi* Galliard & Ngu
A. (*F.*) *ananae* Knight & Laffoon
A. (*F.*) *aranetana* Banks
A. (*F.*) *assamensis* (Theobald)
A. (*F.*) *aureostriatus* (Doleschall)
A. (*F.*) *auronitens* Edwards
A. (*F.*) *avistylus* Brug
A. (*F.*) *banksi* Edwards
A. (*F.*) *cacharanus* (Barraud)
A. (*F.*) *christophersi* Edwards
A. (*F.*) *chrysolineatus* (Theobald)
A. (*F.*) *chungi* Lien
A. (*F.*) *cogilli* Edwards
A. (*F.*) *croceus* Knight & Laffoon
A. (*F.*) *crossi* Lien
A. (*F.*) *deccanus* (Barraud)
A. (*F.*) *dissimilis* (Leicester)
A. (*F.*) *elsiae* (Barraud)
A. (*F.*) *feegradei* Barraud
A. (*F.*) *flavipennis* (Giles)
A. (*F.*) *formosensis* Yamada
A. (*F.*) *ganapathi* Colless
A. (*F.*) *gilli* (Barraud)
A. (*F.*) *gubernatoris* (Giles)
A. (*F.*) *harperi* Knight
A. (*F.*) *harveyi* (Barraud)
A. (*F.*) *hegneri* Causey
A. (*F.*) *hurlbuti* Lien
A. (*F.*) *idjenensis* Brug
A. (*F.*) *inermis* Colless
A. (*F.*) *inquinatus* Edwards
A. (*F.*) *jugraensis* (Leicester)
A. (*F.*) *khazani* Edwards
A. (*F.*) *lacteus* Knight
A. (*F.*) *laoagensis* Knight
A. (*F.*) *lepchanus* (Barraud)
A. (*F.*) *leucomeres* (Giles)
A. (*F.*) *leucopleurus* Rozeboom

A. (*F.*) *litoreus* Colless
A. (*F.*) *loi* Lien
A. (*F.*) *lophoventralis* (Theobald)
A. (*F.*) *luteus* (Ludlow)
A. (*F.*) *luzonensis* Rozeboom
A. (*F.*) *macdougalli* Edwards
A. (*F.*) *macfarlanei* (Edwards)
A. (*F.*) *medleri* Knight & Laffoon
A. (*F.*) *melanopterus* (Giles)
A. (*F.*) *mjoebergi* (Edwards)
A. (*F.*) *mohani* Knight
A. (*F.*) *nigrorhynchus* Brug
A. (*F.*) *niveoides* Barraud
A. (F.) niveus (Ludlow)
A. (*F.*) *notoscriptus* (Skuse)
A. (*F.*) *novoniveus* Barraud
A. (*F.*) *omorii* Lien
A. (*F.*) *oreophilus* (Edwards)
A. (*F.*) *paradissimilis* Rozeboom
A. (*F.*) *pexus* Colless
f **A. (F.) poecilus** (Theobald)
A. (*F.*) *prominens* (Barraud)
A. (*F.*) *pseudoniveus* (Theobald)
A. (*F.*) *pseudotaeniatus* (Giles)
A. (*F.*) *pulchriventer* (Giles)
A. (*F.*) *rizali* (Banks)
A. (*F.*) *saperoi* Knight
A. (*F.*) *saxicola* Edwards
A. (*F.*) *sherki* Knight
A. (*F.*) *shintienensis* Tsai & Lien
A. (*F.*) *shortti* (Barraud)
A. (*F.*) *simlensis* Edwards
A. (*F.*) *sinensis* Chow
A. (*F.*) *sintoni* (Barraud)
A. (*F.*) *stevensoni* (Barraud)
A. (*F.*) *stonei* Knight & Laffoon
A. (*F.*) *subniveus* Edwards
A. (*F.*) *subsimilis* (Barraud)
A. (*F.*) *suffusus* Edwards
fv **A. (F.) togoi** (Theobald)
A. (*F.*) *tonkinensis* Galliard & Ngu
A. (*F.*) *unicinctus* Edwards
A. (*F.*) *vanus* Colless
A. (*F.*) *watteni* Lien
A. (*Skusea*) *amesii* (Ludlow)
A. (*S.*) *fumidus* Edwards
A. (*Christophersiomyia*) *annulirostris* (Theobald)
A. (*C.*) *brayi* Knight
A. (*C.*) *ibis* Barraud
A. (*C.*) *thomsoni* (Theobald)
A. (*Geoskusea*) *baisasi* Knight & Hull
A. (*G.*) *kabaenensis* Brug
A. (*Rhinoskusea*) *longirostris* (Leicester)
A. (*R.*) *pillaii* Mattingly
fv **A. (Stegomyia) aegypti** (Linnaeus)
A. (*S.*) *albolineatus* (Theobald)
v **A. (S.) albopictus** (Skuse)
A. (*S.*) *annandalei* (Theobald)
A. (*S.*) *arboricolus* Knight & Rozeboom
A. (*S.*) *aurotaeniatus* Edwards
A. (*S.*) *bambusicolus* Knight & Rozeboom
A. (*S.*) *boharti* Knight & Rozeboom
A. (*S.*) *brevitibia* (Edwards)
A. (*S.*) *chemulpoensis* Yamada
A. (*S.*) *craggi* (Barraud)
A. (*S.*) *desmotes* (Giles)
A. (*S.*) *edwardsi* (Barraud)
A. (*S.*) *flavopictus* Yamada
A. (*S.*) *gardnerii* (Ludlow)
A. (*S.*) *hoogstraali* Knight & Rozeboom
A. (*S.*) *horishensis* Yamada

A. (*S.*) *impatibilis* (Walker)
A. (*S.*) *indosinensis* (Borel)
A. (*S.*) *laffooni* Knight & Rozeboom
A. (*S.*) *mediopunctatus* (Theobald)
A. (*S.*) *meronephada* (Dyar & Shannon)
A. (*S.*) *novalbopictus* Barraud
A. (*S.*) *patriciae* Mattingly
A. (*S.*) *paullusi* Stone & Farner
A. (*S.*) *pseudalbolineatus* Brug
A. (*S.*) *pseudalbopictus* (Borel)
A. (*S.*) *riversi* Bohart & Ingram
A. (*S.*) *seato* Huang
A. (*S.*) *subalbopictus* Barraud
A. (*S.*) *treubi* (Meijere)
A. (*S.*) *unilineatus* (Theobald)
v **A. (S.) vittatus** (Bigot)
A. (*S.*) *w-albus* (Theobald)
A. (*Aedimorphus*) *alboscutellatus* (Theobald)
A. (*A.*) *argenteoscutellatus* Carter & Wijesundara
A. (*A.*) *caecus* (Theobald)
A. (*A.*) *culicinus* Edwards
A. (*A.*) *jamesi* (Edwards)
A. (*A.*) *lowisii* (Theobald)
A. (*A.*) *mediolineatus* (Theobald)
A. (*A.*) *mindoroensis* Knight & Hull
A. (*A.*) *nigrostriatus* (Barraud)
A. (*A.*) *niveoscutellum* (Theobald)
A. (*A.*) *nummatus* Edwards
A. (*A.*) *orbitae* Edwards
A. (*A.*) *pallidostriatus* (Theobald)
A. (*A.*) *pampangensis* (Ludlow)
A. (*A.*) *pipersalatus* (Giles)
A. (*A.*) *punctifemoris* (Ludlow)
A. (*A.*) *stenoetrus* (Theobald)
A. (*A.*) *syntheticus* Barraud
f **A. (A.) taeniorhynchoides** (Christophers)
A. (*A.*) *trimaculatus* (Theobald)
v **A. (A.) vexans** (Meigen)
A. (*A.*) *wainwrighti* Baisas
A. (*Paraedes*) *barraudi* (Edwards)
A. (*P.*) *bonneae* Mattingly
A. (*P.*) *collessi* Mattingly
A. (*P.*) *menoni* Mattingly
A. (*P.*) *ostentatio* (Leicester)
A. (*P.*) *pagei* (Ludlow)
A. (*Neomelaniconion*) *imprimens* (Walker)
fv **A. (N.) lineatopennis** (Ludlow)
A. (*Diceromyia*) *iyengari* Edwards
A. (*D.*) *micropterus* (Giles)
A. (*D.*) *periskalatus* (Giles)
A. (*D.*) *punctipes* Edwards
A. (*D.*) *ramachandrai* Reuben
A. (*D.*) *reginae* Edwards
A. (*Aedes*) *abditus* Barraud
A. (*A.*) *agrestis* Barraud
v **A. (A.) butleri** Theobald
A. (*A.*) *comatus* Barraud
A. (*A.*) *lankaensis* Stone & Knight
A. (*A.*) *malayi* (Leicester)
A. (*A.*) *perditus* (Leicester)
A. (*A.*) *petroelephantus* Wijesundara

f **A. (A.) pseudomediofasciatus** (Theobald)
A. (*A.*) *rami* Barraud
A. (*A.*) *seculatus* Menon
A. (*A.*) *spermathecus* Wijesundara
A. (*A.*) *uniformis* (Theobald)
A. (*A.*) *yerburyi* Edwards
A. (*Verrallina*) *panayensis* Ludlow
A. (*Neomacleaya*) *andamanensis* Edwards
A. (*N.*) *atrius* Barraud
A. (*N.*) *campylostylus* Laffoon
A. (*N.*) *cautus* Barraud
A. (*N.*) *clavatus* Barraud
A. (*N.*) *cretatus* Delfinado
A. (*N.*) *cyrtolabis* Edwards
A. (*N.*) *dermajoensis* Brug
A. (*N.*) *dux* Dyar & Shannon
A. (*N.*) *fragilis* (Leicester)
A. (*N.*) *gibbosus* Delfinado
A. (*N.*) *hamistylus* Laffoon
A. (*N.*) *hirsutipleura* Barraud
A. (*N.*) *hispidus* Delfinado
A. (*N.*) *incertus* Edwards
A. (*N.*) *indecorabilis* (Leicester)
A. (*N.*) *indicus* (Theobald)
A. (*N.*) *ishigakiensis* Bohart
A. (*N.*) *johnsoni* Laffoon
A. (*N.*) *latipennis* Delfinado
A. (*N.*) *leicesteri* Edwards
A. (*N.*) *lugubris* Barraud
A. (*N.*) *macrodixoa* Dyar & Shannon
A. (*N.*) *margarsen* Dyar & Shannon
A. (*N.*) *nigrotarsis* (Ludlow)
A. (*N.*) *notabilis* Delfinado
A. (*N.*) *nubicolus* Laffoon
A. (*N.*) *pahangi* Delfinado
A. (*N.*) *philippinensis* Delfinado
A. (*N.*) *prioekanensis* Brug
A. (*N.*) *protuberans* Delfinado
A. (*N.*) *pseudodiurnus* (Theobald)
A. (*N.*) *rarus* Delfinado
A. (*N.*) *robertsi* Laffoon
A. (*N.*) *siamensis* Delfinado
A. (*N.*) *singularis* (Leicester)
A. (*N.*) *torosus* Delfinado
A. (*N.*) *uncus* (Theobald)
A. (*N.*) *vallistris* Barraud
A. (*N.*) *varietas* (Leicester)
A. (*N.*) *virilis* (Leicester)
A. (*N.*) *yusafi* Barraud
A. (*Cancraedes*) *cancricomes* Edwards
A. (*C.*) *curtipes* Edwards
A. (*C.*) *indonesiae* Mattingly
A. (*C.*) *kohkutensis* Mattingly
A. (*C.*) *mamoedjoensis* Mattingly
A. (*C.*) *masculinus* Mattingly
A. (*C.*) *palawanicus* Mattingly
A. (*C.*) *penghuensis* Lien
A. (*C.*) *simplex* (Theobald)
A. (*C.*) *thurmanae* Mattingly
Ayurakitia griffithi Thurman
Armigeres (*Armigeres*) *apoensis* Bohart & Farner
A. (*A.*) *aureolineatus* (Leicester)
A. (*A.*) *baisasi* Stone & Thurman
A. (*A.*) *bhayungi* Thurman & Thurman
A. (*A.*) *candelabrifer* Brug
A. (*A.*) *confusus* Edwards
A. (*A.*) *conjungens* Edwards
A. (*A.*) *durhami* (Edwards)
A. (*A.*) *ejercitoi* Baisas

A. (*A.*) *fimbriatus* Edwards
A. (*A.*) *foliatus* Brug
A. (*A.*) *giveni* Edwards
A. (*A.*) *hybridus* Edwards
A. (*A.*) *joloensis* (Ludlow)
A. (*A.*) *jugraensis* (Leicester)
A. (*A.*) *kuchingensis* Edwards
A. (*A.*) *maiae* (Edwards)
A. (*A.*) *malayi* (Theobald)
A. (*A.*) *manalangi* Baisas
A. (*A.*) *maximus* Edwards
A. (*A.*) *moultoni* Edwards
A. (*A.*) *subalbatus* (Coquillett)
A. (*A.*) *theobaldi* Barraud
A. (*Leicesteria*) *annulipalpis* (Theobald)
A. (*L.*) *annulitarsis* (Leicester)
A. (*L.*) *cingulatus* (Leicester)
A. (*L.*) *dentatus* Barraud
A. (*L.*) *digitatus* (Edwards)
A. (*L.*) *dolichocephalus* (Leicester)
A. (*L.*) *inchoatus* Barraud
A. (*L.*) *longipalpis* (Leicester)
A. (*L.*) *magnus* (Theobald)
A. (*L.*) *omissus* (Edwards)
A. (*L.*) *pectinatus* (Edwards)
A. (*L.*) *pendulus* (Edwards)
A. (*L.*) *traubi* Macdonald
A. (*L.*) *vimoli* Thurman & Thurman
A. (*Leicesteriomyia*) *flavus* (Leicester)
Culiseta (*Culiseta*) *lishanensis* Lien
C. (*Climacura*) *marchettei* Garcia, Jeffery & Rudnick
C. (*Allotheobaldia*) *indica* (Edwards)
C. (*A.*) *longiareolata* (Macquart)
f **Culex (Lutzia) fuscanus** Wiedemann
C. (*L.*) *halifaxii* Theobald
C. (*Neoculex*) *brevipalpis* (Giles)
C. (*N.*) *hayashii* Yamada
C. (*N.*) *lini* Lien
C. (*N.*) *nematoides* Dyar & Shannon
C. (*N.*) *okinawae* Bohart
C. (*N.*) *quettensis* Mattingly
C. (*N.*) *simplicicornis* Edwards
C. (*N.*) *sumatranus* Brug
C. (*N.*) *tenuipalpis* Barraud
C. (*Mochthogenes*) *bokorensis* Klein & Sirivanakarn
C. (*M.*) *campilunati* Carter & Wijesundara
C. (*M.*) *castrensis* Edwards
C. (*M.*) *chiyutoi* Baisas
C. (*M.*) *foliatus* Brug
C. (*M.*) *hackeri* Edwards
C. (*M.*) *hinglungensis* Chu
C. (*M.*) *iphis* Barraud
C. (*M.*) *khazani* Edwards
C. (*M.*) *kiriensis* Klein & Sirivanakarn
C. (*M.*) *malayi* (Leicester)
C. (*M.*) *otachati* Klein & Sirivanakarn
C. (*M.*) *pluvialis* Barraud
C. (*M.*) *selai* Klein & Sirivanakarn
C. (*M.*) *yeageri* Baisas
C. (*Acalleomyia*) *obscurus* (Leicester)
C. (*Lophoceraomyia*) *barkerii* (Theobald)
C. (*L.*) *bernardi* (Borel)
C. (*L.*) *cinctellus* Edwards
C. (*L.*) *coerulescens* Edwards
C. (*L.*) *curtipalpis* (Edwards)

C. (*L.*) *eminentia* (Leicester)
C. (*L.*) *flavicornis* Barraud
C. (*L.*) *fraudatrix* (Theobald)
C. (*L.*) *fulleri* (Ludlow)
C. (*L.*) *harpagophallus* Wang & Feng
C. (*L.*) *hewitti* (Edwards)
C. (*L.*) *hui* Lien
C. (*L.*) *infantulus* Edwards
C. (*L.*) *jenseni* (Meijere)
C. (*L.*) *lavatae* Stone & Bohart
C. (*L.*) *mammilifer* (Leicester)
C. (*L.*) *mindanaoensis* Baisas
C. (*L.*) *minor* (Leicester)
C. (*L.*) *minutissimus* (Theobald)
C. (*L.*) *navalis* Edwards
C. (*L.*) *niger* (Leicester)
C. (*L.*) *nolledoi* Baisas
C. (*L.*) *pachecoi* Baisas
C. (*L.*) *pilifemoralis* Wang & Feng
C. (*L.*) *plantaginis* Barraud
C. (*L.*) *quadripalpis* (Edwards)
C. (*L.*) *raghavanii* Rahman, Choudhury & Kalra
C. (*L.*) *rubithoracis* (Leicester)
C. (*L.*) *seniori* Barraud
C. (*L.*) *szemaoensis* Wang & Feng
C. (*L.*) *tuberis* Bohart
C. (*L.*) *uniformis* (Theobald)
C. (*Culiciomyia*) *bahri* (Edwards)
C. (*C.*) *bailyi* Barraud
C. (*C.*) *barrinus* Bram
C. (*C.*) *fragilis* Ludlow
C. (*C.*) *javanensis* Bonne-Wepster
C. (*C.*) *nigropunctatus* Edwards
f **C. (C.) pallidothorax** Theobald
C. (*C.*) *ryukyensis* Bohart
C. (*C.*) *scanloni* Bram
C. (*C.*) *shebbearei* Barraud
C. (*C.*) *spathifurca* (Edwards)
C. (*C.*) *spiculothorax* Bram
C. (*C.*) *termi* Thurman
C. (*C.*) *thurmanorum* Bram
C. (*C.*) *viridiventer* Giles
C. (*Culex*) *afridii* Qutubuddin
C. (*C.*) *alienus* Colless
fv **C. (C.) annulirostris** Skuse
C. (*C.*) *barraudi* Edwards
fv **C. (C.) bitaeniorhynchus** Giles
C. (*C.*) *confusus* Baisas
C. (*C.*) *cornutus* Edwards
C. (*C.*) *diengensis* Brug
C. (*C.*) *epidesmus* (Theobald)
C. (*C.*) *fasyi* Baisas
C. (*C.*) *fuscifurcatus* Edwards
fv **C. (C.) fuscocephalus** Theobald
v **C. (C.) gelidus** Theobald
C. (*C.*) *geminus* Colless
C. (*C.*) *hutchinsoni* Barraud
C. (*C.*) *incognitus* Baisas
C. (*C.*) *kangi* Lien
C. (*C.*) *litoralis* Bohart
C. (*C.*) *mimeticus* Noé
C. (*C.*) *mimuloides* Barraud
C. (*C.*) *mimulus* Edwards
C. (*C.*) *murrelli* Lien
C. (*C.*) *neolitoralis* Bram
C. (*C.*) *neomimulus* Lien
C. (*C.*) *neovishnui* Lien
C. (*C.*) *nilgiricus* Edwards
C. (*C.*) *orientalis* Edwards
C. (*C.*) *palmi* Baisas

C. (*C.*) *permixtus* Hsieh & Liao
C. (*C.*) *perplexus* Leicester
fv **C. (C.) pipiens** Linnaeus
C. (*C.*) *propinquus* Colless
C. (*C.*) *pseudosinensis* Colless
v **C. (C.) pseudovishnui** Colless
C. (*C.*) *salinus* Baisas
C. (*C.*) *sinensis* Theobald
fv **C. (C.) sitiens** Wiedemann
C. (*C.*) *suborientalis* Baisas
C. (*C.*) *theileri* Theobald
fv **C. (C.) tritaeniorhynchus** Giles
C. (*C.*) *tsengi* Lien
v **C. (C.) univittatus** Theobald
fv **C. (C.) vishnui** Theobald
C. (*C.*) *whitei* Barraud
f **C. (C.) whitmorei** (Giles)

APPENDIX 8 MOSQUITOS OF THE PALAEARCTIC REGION

(*Europe, North Africa, Middle East and northern Asia*)

Anopheles (*Anopheles*) *algeriensis* Theobald
mv **A. (A.) atroparvus** Van Thiel
A. (*A.*) *chodukini* Martini
mv **A. (A.) claviger** (Meigen)
m **A. (A.) labranchiae** Falleroni
A. (*A.*) *lewisi* Ludlow
m **A. (A.) maculipennis** Meigen
A. (*A.*) *marteri* Senevet & Prunelle
m **A. (A.) melanoon** Hackett
m **A. (A.) messeae** Falleroni
A. (*A.*) *petragnani* Del Vecchio
m **A. (A.) plumbeus** Stephens
m **A. (A.) sacharovi** Favre
A. (*Cellia*) *apoci* Marsh
A. (*C.*) *atheniensis* Cardamatis
A. (*C.*) *broussesi* Edwards
A. (*C.*) *dthali* Patton
A. (*C.*) *grassei* Grjebine
m **A. (C.) hispaniola** (Theobald)
A. (*C.*) *multicolor* Cambouliu
m **A. (C.) pattoni** Christophers
fmv **A. (C.) pharoensis** Theobald
m **A. (C.) sergentii** (Theobald)
m **A. (C.) stephensi** Liston
m **A. (C.) superpictus** Grassi
A. (*C.*) *turkhudi* Liston
Toxorhynchites (*Toxorhynchites*) *christophi* (Portschinsky)
Coquillettidia (*Coquillettidia*) *buxtoni* (Edwards)
C. (*C.*) *richiardii* (Ficalbi)
Uranotaenia unguiculata Edwards
Orthopodomyia pulchripalpis (Rondani)
Heizmannia kanhsienensis Tung
Aedes (*Ochlerotatus*) *annulipes* (Meigen)
A. (*O.*) *behningi* Martini
A. (*O.*) *beklemishevi* Denisova
A. (*O.*) *berlandi* Séguy
A. (*O.*) *cantans* (Meigen)
v **A. (O.) caspius** (Pallas)
A. (*O.*) *cataphylla* Dyar
A. (*O.*) *communis* (De Geer)
A. (*O.*) *cyprius* Ludlow
A. (*O.*) *detritus* (Haliday)
A. (*O.*) *diantaeus* Howard, Dyar & Knab
v **A. (O.) dorsalis** (Meigen)

A. (O.) *duplex* Martini
A. (O.) *dzeta* Séguy
A. (O.) *excrucians* (Walker)
A. (O.) *flavescens* (Müller)
A. (O.) *grandilarva* Sazanova
A. (O.) *hakusanensis* Yamaguti & Tamaboko
A. (O.) *hungaricus* Mihályi
A. (O.) *impiger* (Walker)
A. (O.) *lepidonotus* Edwards
A. (O.) *leucomelas* (Meigen)
A. (O.) *mariae* (Sergent & Sergent)
A. (O.) *montchadskyi* Dubitsky
A. (O.) *nigrinus* (Eckstein)
A. (O.) *nigripes* (Zetterstedt)
A. (O.) *nigrocanus* Martini
A. (O.) *pulchritarsis* (Rondani)
A. (O.) *pullatus* (Coquillett)
A. (O.) *punctor* (Kirby)
A. (O.) *quasirusticus* Torres Cañamares
A. (O.) *refiki* Medschid
A. (O.) *riparius* Dyar & Knab
A. (O.) *rusticus* (Rossi)
A. (O.) *schtakelbergi* Shingarev
A. (O.) *stampari* Apfelbeck
A. (O.) *sticticus* (Meigen)
A. (O.) *subdiversus* Martini
A. (*Finlaya*) *alektorovi* Stackelberg
A. (F.) *bunanoki* Sasa & Ishimura
A. (F.) *eatoni* (Edwards)
A. (F.) *echinus* (Edwards)
A. (F.) *fengi* Edwards
v **A. (F.) geniculatus** (Olivier)
A. (F.) *hatorii* Yamada
v **A. (F.) japonicus** (Theobald)
A. (F.) *kiangsiensis* Tung
A. (F.) *kobayashii* Nakata
A. (F.) *koreicoides* Sasa, Kano & Hayashi
A. (F.) *koreicus* (Edwards)
A. (F.) *nipponicus* La Casse & Yamaguti
A. (F.) *nippononiveus* Sasa & Nakahashi
A. (F.) *peipingensis* Feng
A. (F.) *pulchriventer* (Giles)
A. (F.) *seoulensis* Yamada
fv **A. (F.) togoi** (Theobald)
A. (F.) *watasei* Yamada
A. (F.) *yunnanensis* (Gaschen)
fv **A. (Stegomyia) aegypti** (Linnaeus)
A. (S.) *cretinus* Edwards
A. (S.) *galloisi* Yamada
v **A. (Aedimorphus) vexans** (Meigen)
A. (*Indusius*) *pulverulentus* Edwards
A. (*Aedes*) *asanumai* Sasa, Kano & Takahasi
A. (A.) *aureus* Gutzevich
A. (A.) *cinereus* Meigen
A. (A.) *esoensis* Yamada
A. (A.) *nobukonis* Yamada
A. (A.) *pseudoesoensis* Yamaguti & Tamaboko
A. (A.) *rossicus* Dolbeskin, Gorickaja & Mitrofanova
A. (A.) *yamadai* Sasa, Kano & Takahasi
Culiseta (*Culiseta*) *alaskaensis* (Ludlow)
C. (C.) *annulata* (Schrank)
C. (C.) *bergrothi* (Edwards)
C. (C.) *glaphyroptera* (Schiner)
C. (C.) *kanayamensis* (Yamada)
C. (C.) *niveitaeniata* (Theobald)
C. (C.) *subochrea* (Edwards)
C. (*Culicella*) *fumipennis* (Stephens)
C. (C.) *litorea* (Shute)

C. (*C.*) *morsitans* (Theobald)
C. (*C.*) *nipponica* La Casse & Yamaguti
C. (*C.*) *ochroptera* (Peus)
C. (*C.*) *setivalva* (Monchadskii)
C. (*C.*) *silvestris* (Shingarev)
C. (*Allotheobaldia*) *longiareolata* (Macquart)
Culex (*Lasiosiphon*) *adairi* Kirkpatrick
C. (*Barraudius*) *modestus* Ficalbi
C. (*B.*) *pusillus* Macquart
C. (*Neoculex*) *deserticola* Kirkpatrick
C. (*N.*) *gamma* Séguy
C. (*N.*) *hortensis* Ficalbi
C. (*N.*) *impudicus* Ficalbi
C. (*N.*) *judaicus* Edwards
C. (*N.*) *martinii* Medschid
C. (*N.*) *rubensis* Sasa & Takahashi
C. (*N.*) *territans* Walker
C. (*Culiciomyia*) *kyotoensis* Yamaguti & La Casse
C. (*C.*) *sasai* Kano, Nitahara & Awaya
C. (*C.*) *yaoi* Tung
C. (*Culex*) *alpha* Séguy
v **C. (C.) antennatus** (Becker)
C. (*C.*) *beta* Séguy
C. (*C.*) *brumpti* Galliard
C. (*C.*) *huangae* Meng
C. (*C.*) *jacksoni* Edwards
C. (*C.*) *laticinctus* Edwards
C. (*C.*) *mauritanicus* Callot
C. (*C.*) *mimeticus* Noé
C. (*C.*) *orientalis* Edwards
fv **C. (C.) pipiens** Linnaeus
C. (*C.*) *prosecutor* Séguy
C. (*C.*) *theileri* Theobald
C. (*C.*) *torrentium* Martini
fv **C. (C.) tritaeniorhynchus** Giles
v **C. (C.) univittatus** Theobald
C. (*C.*) *vagans* Wiedemann
fv **C. (C.) vishnui** Theobald

APPENDIX 9 MOSQUITOS OF THE UNITED KINGDOM AND IRELAND

Anopheles (*Anopheles*) *algeriensis* Theobald
mv **A. (A.) atroparvus** Van Thiel
mv **A. (A.) claviger** (Meigen)
m **A. (A.) messeae** Falleroni
m **A. (A.) plumbeus** Stephens
Coquillettidia (*Coquillettidia*) *richiardii* (Ficalbi)
Orthopodomyia pulchripalpis (Rondani)
Aedes (*Ochlerotatus*) *annulipes* (Meigen)
A. (*O.*) *cantans* (Meigen)
v **A. (O.) caspius** (Pallas)
A. (*O.*) *communis* (De Geer)
A. (*O.*) *detritus* (Haliday)
v **A. (O.) dorsalis** (Meigen)
A. (*O.*) *flavescens* (Müller)
A. (*O.*) *leucomelas* (Meigen)
A. (*O.*) *punctor* (Kirby)
A. (*O.*) *rusticus* (Rossi)
A. (*O.*) *sticticus* (Meigen)
v **A. (Finlaya) geniculatus** (Olivier)
v **A. (Aedimorphus) vexans** (Meigen)
A. (*Aedes*) *cinereus* Meigen
Culiseta (*Culiseta*) *alaskaensis* (Ludlow)
C. (*C.*) *annulata* (Schrank)
C. (*C.*) *subochrea* (Edwards)
C. (*Culicella*) *fumipennis* (Stephens)
C. (*C.*) *litorea* (Shute)

C. (*C.*) *morsitans* (Theobald)
C. (*Allotheobaldia*) *longiareolata* (Macquart)
Culex (*Barraudius*) *modestus* Ficalbi
C. (*Neoculex*) *territans* Walker
fv **C. (Culex) pipiens** Linnaeus
C. (*C.*) *torrentium* Martini

GLOSSARY*

aestivation: a resting period in a state of torpor during exceptional hot or dry weather

annelid: member of the phylum Annelida, which includes the leeches and earthworms

anthropophilic: a preference for feeding on man

arachnid: class of arthropod (*q.v.*) including spiders, scorpions, ticks and mites

arthropod: member of the phylum Arthropoda, which includes insects, spiders, ticks, scorpions, crabs, etc.

axil: the space between petiole (leaf stem) and main stem in a plant

basement membrane: a membrane of modified connective tissue lying beneath an epithelium

biometry: statistical study of biological measurement

bromeliad: tropical American epiphytic plants related to and resembling pineapple plants

Carboniferous: the geological period in the Palaeozoic era during which many of the coal measures were laid down. It came between the Devonian and the Permian periods

chironomid: a harlequin-fly, or non-biting gnat, sometimes also called a midge

circadian: around, but not necessarily exactly, twenty-four hours (*circa diem*)

coelenterate: a member of the phylum Coelenterata (or Cnidaria) which includes *Hydra,* corals, sea anemones and jelly-fish

* The definitions are valid only in the context of their use in this book. Some may have wider connotations elsewhere.

coelom: the body cavity
crepuscular: at twilight
crustacean: class of arthropod (*q.v.*) including prawns, crabs and lobsters
cybernetics: study of automatic (feed-back) control in nervous, endocrine and enzyme systems and in electrical and mechanical communications
cytoplasm: the substance of a living cell external to the nucleus and bounded by the cell membrane
Devonian: the geological period in the Palaeozoic era between the Silurian and the Carboniferous (or Mississippian) periods
diapause: a period of spontaneous dormancy
diverticulum: a closed pocket leading from the main alimentary canal
dracaena: lily-like old-world shrubs and trees including the so-called dragon-palms
ecdysis: the act of moulting or shedding the skin
eclosion: the act of breaking out of the pupa: emergence
ecology: the study of living things in relation to their total environment, both living and physico-chemical
embryogenesis: the development of the embryo
encephalitis: inflammation of the brain
encephalomyelitis: concurrent inflammation of the brain and spinal cord
endocrine organs: glands secreting hormones into the blood
epidemiology: the study of the various factors controlling the presence or absence of a disease or disease-causing organism
epiphyte: a non-parasitic plant growing on another plant (usually on a tree)
ethology: the study of behaviour in relation to habitat
facultative: optional, thriving under more than one set of conditions
fatbody: fat storage organ in an insect
filiform: thread-like, filamentous
ganglion: a concentration of nerve tissue containing nerve cells outside the brain
geniculate: bent abruptly at an angle like a knee-joint
gonadotrophic: stimulating the sex cells to activity
haemocoele: the blood-filled body cavity of insects
hydra: small freshwater tentacled polyp belonging to the phylum Coelenterata, which includes corals, sea anemones and jelly-fish

labella: the sensitive tip of the proboscis

labium: the lower lip of an insect formed by the fusion of the left and right second maxillae

malpighian tubules: the excretory organs of insects analogous to the kidneys in mammals

microfilaraemia: presence of microfilaria (the larvae of filarial worms) in the circulating blood

micron: one-thousandth part of a millimetre or about $\frac{1}{25000}$ of an inch

nematocera: the suborder of flies that includes mosquitos, midges, crane-flies, etc.

nematodes: cylindrical worms, many of which are parasitic in animals and plants, for example, the filarial worms causing elephantiasis in man

neurosecretory: cells associated with nerve tissue that produce hormones and function as endocrine organs

nulliparous: not yet given birth or laid eggs

oöcyte: the unfertilised egg cell

operculum: a covering or flap

ovariole: the tube in which the oöcyte develops in the ovary

oviposition: the act of laying eggs

palaeontology: the study of fossil remains

parasitaemia: presence of parasites in the circulating blood

pathogenic: causing disease

pathogens: disease-producing agents, for example certain viruses, protozoa and bacteria

perispiracular: surrounding the spiracles or respiratory openings

phenotype: the resulting expression of the interaction of inherited and environmental influences; the visible and detectable characters of an organism

photoperiod: the relative duration of alternating periods of light and darkness as it affects living organisms

planarian: small aquatic flat-worm

polymorphism: occurring in differing forms and characters

primates: lemurs, monkeys, apes and man

precipitin: precipitate that forms when an antibody unites with its antigen

protozoa: small acellular organisms some of which are parasitic, for example malaria parasites and the trypanosomes of sleeping-sickness

rotifer: member of a class of small, usually microscopic, aquatic invertebrates

spiracles: the respiratory openings in an insect, leading directly into the branching system of tracheae and tracheoles

strain: a local wild or captive population of an organism differing in its behaviour from that of other populations of the same species

substrate: the substance acted upon or the base on which an organism lives, for example soil, mud, the inner surface of a tree-hole

tarsal: relating to the tarsus, the terminal five segments making up the insect's foot

taxonomist: a person concerned with the classification and orderly arrangement of organisms

tergites: the dorsal segments covering an insect's body

tetany: tonic spasm of muscles resulting from rapidly repeated stimulation

tetrapods: four-limbed vertebrate animals, including, amphibians, reptiles, birds and mammals

tipulid: a crane-fly or daddy-long-legs belonging to the family Tipulidae

touraco: tropical African bird belonging to the family Musophagidae

trypanosomes: protozoan parasites, including those responsible for sleeping-sickness

vascular plants: plants with conducting tubes or vessels, including all common flowering plants

vector: transmitter of disease agent, usually applied to an insect or other arthropod bringing infection to a mammal, bird or other vertebrate

viraemia: presence of virus in the circulating blood

BIBLIOGRAPHY

1 MOSQUITOS IN GENERAL

1 BATES, M. (1949) *The Natural History of Mosquitoes*. Macmillan: New York

2 CHRISTOPHERS, S. R. (1901) Anatomy and histology of the female mosquito. *Repts. Mal. comm. R. Soc.* Ser. 4

3 CHRISTOPHERS, S. R. (1960) Aedes aegypti *(L.) The Yellow Fever Mosquito. Its Life History, Bionomics and Structure*. Cambridge University Press

4 CLEMENTS, A. N. (1963) *The Physiology of Mosquitoes*. Pergamon Press: Oxford

5 DYAR, H. G. (1922) Mosquitoes of the United States. *Proc. U. S. Nat. Mus.* **62**: 1–119

6 DYAR, H. G. and KNAB, F. (1906) Mosquitoes of the United States. *J.N.Y. ent. Soc.* **14**: 188

7 EDWARDS, F. W. (1923) Oligocene mosquitoes in the British Museum, with a summary of our present knowledge concerning fossil Culicidae. *Quart. J. Geol. Soc. Lond.* **79**: 139–55

8 EDWARDS, F. W. (1941) *Mosquitoes of the Ethiopian Region. III. Culicine Adults and Pupae*. British Museum (Nat. Hist.): London

9 EVANS, A. M. (1938) *Mosquitoes of the Ethiopian Region. II. Anophelini, Adults and Early Stages*. British Museum (Nat. Hist.): London

10 GERBERG, E. (1970) *Manual for Mosquito Rearing and Experimental Techniques*. Amer. Mosq. Control Assoc. Bull. No. 5. Selma, California

11 GILLIES, M. T. and MEILLON, B. DE (1968) *The Anophelinae of Africa South of the Sahara (Ethiopian Zoogeographical Region)*. S. African Inst. Med. Res.: Johannesburg

12 HOPKINS, G. H. E. (1951) *Mosquitoes of the Ethiopian Region. I. Larval Bionomics of Mosquitoes and Taxonomy of Culicine Larvae.* British Museum (Nat. Hist.): London

13 HOWARD, L. O., DYAR, H. G. and KNAB, F. (1917) *The Mosquitoes of North and Central America and the West Indies.* Vols. I–IV. Carnegie Inst.: Washington

14 MARSHALL, J. F. (1938) *The British Mosquitoes.* British Museum (Nat. Hist.): London

15 REID, J. A. (1968) *Anopheline mosquitoes of Malaya and Borneo.* Stud. Inst. med. Res., Malaysia, No. 31

16 SNODGRASS, R. E. (1959) The anatomical life of the mosquito. *Smithson. Misc. Coll.* **139**: 87 pp.

17 STONE, A., KNIGHT, K. L. and STARCKE, H. (1959) *A Synoptic Catalog of the Mosquitoes of the World (Diptera, Culicidae).* Thomas Say Foundation, 6, Ent. Soc. Amer. (and supplements)

18 THEOBALD, F. V. (1910) *A Monograph of the Culicidae of the World.* Vols. I–V. British Museum (Nat. Hist.): London

19 WESENBERG-LUND, C. (1921) Contributions to the biology of the Danish Culicidae. *K. Danske Vidensk. Selsk.* **7**: 1–210

20 WIGGLESWORTH, V. B. (1949) *Physiology of Mosquitoes.* In Boyd's *Malariology,* **1**, W. B. Saunders: London

2 EGGS

21 ATKIN, E. E. and BACOT, A. (1917) The relation between the hatching of the eggs and the development of the larvae of *Stegomyia fasciata (Aëdes calopus)* and the presence of bacteria and yeasts. *Parasitology* **9**: 482–536

22 BACOT, A. (1916) Report of the Yellow Fever Commission. No. **3**: 1–191

23 BATES, M. (1940) Oviposition experiments with anopheline mosquitoes. *Amer. J. trop. Med.* **20**: 569–83

24 BECKEL, W. E. (1955) Oviposition site preference of *Aedes* mosquitoes (Culicidae) in the laboratory. *Mosquito News* **15**: 224–8

25 BORG, A. F. and HORSFALL, W. R. (1953) Eggs of floodwater

mosquitoes. II. Hatching stimulus. *Ann. ent. Soc. Amer.* **46**: 472–7

26 BUXTON, P. A. and HOPKINS, G. H. E. (1927) Researches in Polynesia and Melanesia. *Mem. Lond. Sch. Hyg., Trop. Med.* **1**

27 CHRISTOPHERS, S. R. (1945) Structure of the *Culex* egg and egg-raft in relation to function (Diptera). *Trans. R. ent. Soc. Lond.* **95**: 25–34

28 DUNN, L. H. (1926) Mosquitoes bred from dry material taken from holes in trees. *Bull. ent. Res.* **17**: 183

29 ELMORE, JR., C. M. and FAY, R. W. (1958) *Aedes sollicitans* and *A. taeniorhynchus* larval emergence from sod samples. *Mosquito News* **18**: 230–3

30 GILLETT, J. D. (1955*a*) Variation in the hatching response of *Aedes* eggs (Diptera: Culicidae). *Bull. ent. Res.* **46**: 241–54

31 GILLETT, J. D. (1955*b*) The inherited basis of variation in the hatching response of *Aedes* eggs (Diptera: Culicidae). *Bull. ent. Res.* **46**: 255–65

32 GILLETT, J. D. (1959) Control of hatching in the prediapause eggs of *Aedes* mosquitoes. *Nature, Lond.* **184**: 1621–3

33 GILLETT, J. D. (1961) Laboratory observations on the life-history and ethology of *Mansonia* mosquitos. *Bull. ent. Res.* **52**: 23–30

34 GJULLIN, C. M., HEGARTY, C. P. and BOLLEN, W. B. (1941) The necessity of a low oxygen concentration for the hatching of *Aedes* mosquito eggs. *J. Cell. Comp. Physiol.* **17**: 193–202

35 GJULLIN, C. M., YATES, W. W. and STAGE, H. H. (1950) Studies on *Aedes vexans* (Meig.) and *Aedes sticticus* (Meig.), flood-water mosquitoes, in the Lower Columbia River Valley. *Ann. ent. Soc. Amer.* **43**: 262–75

36 GOMA, L. K. H. (1964) Laboratory observations on the oviposition habits of *Aedes* (*Stegomyia*) *aegypti* (Linnaeus). *Ann. trop. Med. Parasit.* **58**: 347–9

37 HERMS, W. B. and FREEBORN, S. B. (1921) The egg-laying habits of Californian anophelines. *J. Parasit.* **7**: 69–79

38 HORSFALL, W. R. (1956) Eggs of floodwater mosquitoes. III. (Diptera, Culicidae). Conditioning and hatching of *Aedes vexans. Ann. ent. Soc. Amer.* **49**: 66–71

39 HORSFALL, W. R. (1961) Daily rate of oviposition of a brood of the mosquito, *Aedes vexans. J. econ. Ent.* **54**: 807–8

40 HUSBANDS, R. C. (1952) Some techniques used in the study of *Aedes* eggs in irrigated pastures in California. *Mosquito News* **12**: 145–50

41 JUDSON, C. L. (1960) The physiology of hatching of aedine mosquito eggs: hatching stimulus. *Ann. ent. Soc. Amer.* **53**: 688–91

42 JUDSON, C. L. (1963) The physiology of hatching of aedine mosquito eggs: carbon monoxide stimulation and ethyl chloride inhibition of hatching of *Aedes aegypti* (*L.*) eggs. *J. ins. Physiol.* **9**: 787–92

43 JUDSON, C. L. and HOKAMA, Y. (1964) Formation of the line of dehiscence in aedine mosquito eggs. *J. ins. Physiol.* **11**: 337–45

44 JUDSON, C. L., HOKAMA, Y. and HAYDOCK, I. (1965) The physiology of hatching of aedine mosquito eggs: some larval responses to the hatching stimulus. *J. ins. Physiol.* **11**: 1169–77

45 MACDONALD, W. W. (1960) On the systematics and ecology of *Armigeres* subgenus *Leicesteria* (Diptera, Culicidae). *Stud. Inst. Med. Res. F.M.S.* **29**: 110–53

46 MUIRHEAD-THOMSON, R. C. (1940*a*) Studies on the behaviour of *Anopheles minimus*. Part I. The selection of the breeding place and the influence of light and shade. *J. malar. Inst. India* **3**: 265–94

47 MUIRHEAD-THOMSON, R. C. (1940*b*) Studies on the behaviour of *Anopheles minimus*. Part II. The influence of water movement on the selection of the breeding place. *J. malar. Inst. India* **3**: 295–322

48 MUIRHEAD-THOMSON, R. C. (1940*c*) Studies on the behaviour of *Anopheles minimus*. Part III. The influence of water temperature on the choice and suitability of the breeding place. *J. malar. Inst. India* **3**: 323–48

49 O'GOWER, A. K. (1957*a*) The influence of the surface on oviposition by *Aedes aegypti* (Linn.) (Diptera, Culicidae). *Proc. Linn. Soc. N.S.W.* **82**: 240–44

50 O'GOWER, A. K. (1957*b*) The influence of the surface on oviposition by *Aedes albopictus* (Skuse) and *Aedes scutellaris katherinensis* Woodhill (Diptera, Culicidae). *Proc. Linn. Soc. N.S.W.* **82**: 285–8

51 O'GOWER, A. K. (1958) The oviposition behaviour of *Aedes australis* (Erickson) (Diptera, Culicidae). *Proc. Linn. Soc. N.S.W.* **83**: 245–50

52 PUTNAM, P. and SHANNON, R. C. (1934) The biology of *Stegomyia* under laboratory conditions. II. Egg-laying capacity and longevity of adults. *Proc. ent. Soc. Wash.* **36**: 217–42

53 RUSSELL, P. F. and RAO, T. R. (1942*d*) On relation of mechanical obstruction and shade to ovipositing of *Anopheles culicifacies. J. exp. Zool.* **91**: 303–29

54 TELFORD, A. D. (1957) The pasture *Aedes* of central and northern California. The egg stage, gross embryology and resistance to desiccation. *Ann. ent. Soc. Amer.* **50**: 537–43

55 TRAVIS, B. V. (1953) Laboratory studies on the hatching of marsh-mosquito eggs. *Mosquito News* **13**: 190–8

56 WALLIS, R. C. and WHITMAN, L. (1968) Oviposition of *Culiseta morsitans* (Theobald) and comments on the life cycle of the American form. *Mosquito News* **28**: 198–200

57 WOODWARD, D. B., CHAPMAN, H. C. and PETERSEN, J. J. (1968) Laboratory studies on the seasonal hatchability of egg batches of *Aedes sollicitans, A. taeniorhynchus* and *Psorophora confinnis. Mosquito News* **28**: 143–6

3 LARVAE

58 ASSEM, J. VAN DEN (1958) Some experimental evidence for the survival value of the root-piercing habits of *Mansonia* larvae (Culicidae) to predators. *Ent. exp. appl.* **1**: 125–9

59 BAR-ZEEV, M. (1958) The effect of temperature on the growth rate and survival of the immature stages of *Aedes aegypti* (L.). *Bull. ent. Res.* **49**: 157–63

60 CHRISTIE, M. (1954) A method for the numerical study of larval populations of *Anopheles gambiae* and other pool-breeding mosquitoes. *Ann. trop. Med. Parasit.* **48**: 271–6

61 FIELDING, J. W. (1919) Notes on the bionomics of *Stegomyia fasciata*, Fabr. (Part I). *Ann. trop. Med. Parasit.* **13**: 259–96

62 GIBBINS, E. G. (1932) A note on the relative size of the anal gills of mosquito larvae breeding in salt and fresh water. *Ann. trop. Med. Parasit.* **26**: 551–4

63 GIGLIOLI, M. E. C. (1964) Tides, salinity and the breeding of *Anopheles melas* (Theobald, 1903) during the dry season in the Gambia. *Riv. di Malariologia* **43**: 245–63

64 GILLETT, J. D. (1942) A larvascope for use in identification of living culicine larvae. *Bull. ent. Res.* **33**: 27–9

65 GILLETT, J. D. (1945) The larva and pupa of *Taeniorhynchus* (*Coquillettidia*) *maculipennis*, Theobald. *Bull. ent. Res.* **35**: 395–7

66 HADDOW, A. J. (1946) The mosquitoes of Bwamba County, Uganda. IV. Studies on the genus *Eretmapodites*, Theobald. *Bull. ent. Res.* **37**: 57–82

67 HAWORTH, W. E. (1922) A new breeding place for mosquitoes. *Trans. R. Soc. trop. Med. Hyg.* **16**: 201

68 HORSFALL, W. R. (1939) Habits of *Aedes thibaulti* Dyar and Knab (Diptera, Culicidae). *J. Kans. ent. Soc.* **12**: 70–1

69 KEILIN, D., TATE, P. and VINCENT, M. (1935) The perispiracular glands of mosquito larvae. *Parasitology* **27**: 257–62

70 KOCH, H. J. (1938) The absorption of chloride ions by the anal papillae of Diptera larvae. *J. exp. Biol.* **15**: 152–60

71 LAURENCE, B. R. (1960) The biology of two species of mosquito, *Mansonia africana* (Theobald) and *Mansonia uniformis* (Theobald), belonging to the subgenus *Mansonioides* (Diptera, Culicidae). *Bull. ent. Res.* **51**: 491–517

72 LEESON, H. S. (1937) The mosquitoes of the *funestus* series in East Africa. *Bull. ent. Res.* **28**: 587–603

73 LESTER, A. R. (1927) The coconut palm: its potentialities in providing breeding places for mosquitoes. *J. trop. Med.* (*Hyg.*) **30**: 137–45

74 LEWIS, D. J. (1949) Tracheal gills in some African culicine mosquito larvae. *Proc. R. ent. Soc. Lond.* (A) **24**: 60–6

75 MUIRHEAD-THOMSON, R. C. (1941) Studies on the behaviour of *Anopheles minimus*. Part IV. The composition of the water and the influence of organic pollution and silt. *J. malar. Inst. India* **4**: 63–102

76 PAL, R. (1945) On the bionomics of *Anopheles culicifacies* Giles. Part II. The ecology of the immature stages. *J. malar. Inst. India* **6**: 53–74

77 PHILIP, C. B. (1962) Breeding of *Aedes aegypti* and other mosquitoes in West African rock holes. *Ann. ent. Soc. Amer.* **55**: 706–8

78 PITTENDRIGH, C. S. (1948) The bromeliad–*Anopheles*–malaria complex in Trinidad. I. The Bromeliad flora. *Evolution* **2**: 58–89

79 RAMSAY, J. A. (1951) Osmotic regulation in mosquito larvae. *J. exp. Biol.* **27**: 145–57

80 ROZEBOOM, L. E. (1935) The relation of bacteria and bacterial filtrates to the development of mosquito larvae. *Amer. J. Hyg.* **21**: 167–79

81 ROZEBOOM, L. E. and BURGESS, R. W. (1962) Dry season survival of some plant-cavity breeding mosquitoes in Liberia. *Ann. ent. Soc. Amer.* **55**: 521–4

82 RUSSELL, P. F. and RAO, T. R. (1942) On the ecology of larvae of *Anopheles culicifacies* Giles, in borrow-pits. *Bull. ent. Res.* **32**: 341–61

83 SERVICE, M. W. (1965) The ecology of the tree-hole breeding mosquitoes in the northern Guinea Savannah of Nigeria. *J. appl. Ecol.* **2**: 1–16

84 SHANNON, R. C. and PUTNAM, P. (1934) The biology of *Stegomyia* under laboratory conditions. I. The analysis of factors which influence larval development. *Proc. ent. Soc. Wash.* **36**: 185–216

85 SHIPITSINA, N. K. (1930) On the role of the organic colloids of water in the feeding of the larva of *Anopheles maculipennis*. *Bull. Inst. Rech. Biol. Univ. Perm.* **7**: 171–94

86 SINGH, K. R. P. (1957) The effect of osmotic pressure on the growth and development of *Aedes aegypti* larvae. *J. anim. Morph. Physiol.* **4**: 48–53

87 SMITH, J. B. (1901) Some notes on the larval habits of *Culex pungens*. *Ent. News* **12**: 153–7

88 STOBBART, R. H. (1959) Studies on the exchange and regulation of sodium in the larva of *Aedes aegypti* (L.). I. The steady-state exchange. *J. exp. Biol.* **36**: 641–53

89 STOBBART, R. H. (1960) Studies on the exchange and regulation of sodium in the larva of *Aedes aegypti* (L.). II. The net transport and fluxes associated with it. *J. exp. Biol.* **37**: 594–608

90 SURTEES, G. (1959) Functional and morphological adaptations of the larval mouthparts in the sub-family Culicinae (Diptera) with a review of some related studies by Montschadsky. *Proc. R. ent. Soc. Lond.* (A) **34**: 7–16

91 THOMAS, I. M. (1950) The reactions of mosquito larvae to regular repetitions of shadows as stimuli. *Aust. J. Sci. Res.* (B) **3**: 113–23

92 WIGGLESWORTH, V. B. (1930) The formation of the peritrophic membrane in insects, with special reference to the larvae of mosquitoes. *Quart. J. micr. Sci.* **73**: 593–616

93 WIGGLESWORTH, V. B. (1933) The effects of salts on the anal gills of the mosquito larva; the function of the anal gills of the mosquito larva; the adaptation of mosquito larvae to salt water. *J. exp. Biol.* **10**: 1–37

94 WIGGLESWORTH, V. B. (1938*a*) The regulation of osmotic pressure and chloride concentration in the haemolymph of mosquito larvae. *J. exp. Biol.* **15**: 235–47

95 WIGGLESWORTH, V. B. (1938*b*) The absorption of fluid from the tracheal system of mosquito larvae at hatching and moulting. *J. exp. Biol.* **15**: 248–54

96 WIGGLESWORTH, V. B. (1942) The storage of protein, fat, glycogen and uric acid in the fatbody and other tissues of mosquito larvae. *J. exp. Biol.* **19**: 56–77

4 PUPAE

97 BRUMPT, E. (1941) Mécanisme d'éclosion des moustique. *Ann. Parasit. hum. comp.* **18**: 75–94

98 GILLETT, J. D. (1947) Notes on the subgenus *Coquillettidia* Dyar (Diptera, Culicidae). *Bull. ent. Res.* **36**: 425–38

99 HADDOW, A. J., GILLETT, J. D. and CORBET, P. S. (1959) Laboratory observations on pupation and emergence in the mosquito *Aedes* (*Stegomyia*) *aegypti* (Linnaeus). *Ann. trop. Med. Parasit.* **53**: 123–31

100 HURST, C. H. (1890) The pupal stage of *Culex*. *Stud. biol. Lab. Owens Coll.* **2**: 47–71

101 IMMS, A. D. (1907) On the larval and pupal stages of *Anopheles maculipennis* Meig. *J. Hyg.* **7**: 291–316

102 NIELSEN, E. T. and EVANS, D. G. (1960) Duration of the pupal stage of *Aedes taeniorhynchus* with a discussion of the velocity of development as a function of temperature. *Oikos* **11**: 200–22

103 NIELSEN, E. T. and HAEGER, J. S. (1954) Pupation and emergence in *Aedes taeniorhynchus* (Wied.). *Bull. ent. Res.* **45**: 757–68

104 RICHINS, C. A. (1938) The metamorphosis of the digestive tract of *Aedes dorsalis* Meigen. *Ann. ent. Soc. Amer.* **31**: 74–87

105 ROUBAUD, E. (1932) Des phénomènes d'histolyse larvaire post-

nymphale et d'alimentation imaginal autotrophe chez le moustique commun (*Culex pipiens*). *C.R. Acad. Sci., Paris* **194**: 389–91

106 WOOLLEY, T. A. (1943) The metamorphosis of the nervous system of *Aedes dorsalis* Meigen (Diptera: Culicidae). *Ann. ent. Soc. Amer.* **36**: 432–47

5 ADULTS

107 BAR-ZEEV, M. (1960) The location of hygroreceptors and moisture receptors in *Aedes aegypti* (L.). *Ent. exp. appl.* **3**: 251–6

108 BATES, M. (1947) The development and longevity of *Haemagogus* mosquitoes under laboratory conditions. *Ann. ent. Soc. Amer.* **40**: 1–12

109 BERG, M. and LANG, S. (1948) Observations of hibernating mosquitoes in Massachusetts. *Mosquito News* **8**: 70–1

110 BERTRAM, D. S. and SAMARAWICKREMA, W. A. (1958) Age determination for individual *Mansonioides* mosquitoes. *Nature, Lond.* **182**: 444–6

111 BOORMAN, J. P. T. (1960) Observations on the feeding habits of the mosquito *Aedes* (*Stegomyia*) *aegypti* (Linnaeus): the loss of fluid after a blood-meal and the amount of blood taken during feeding. *Ann. trop. Med. Parasit.* **54**: 8–14

112 BURCHAM, E. (1957) Artificial insemination of *Aedes aegypti* (L.). *Canad. Ent.* **89**: 494–5

113 BURDICK, D. J. and KARDOS, E. H. (1963) The age structure of fall, winter, and spring populations of *Culex tarsalis* in Kern County, California. *Ann. ent. Soc. Amer.* **56**: 527–36

114 CAMBOURNAC, F. J. C. and HILL, R. B. (1940) Observations on the swarming of *Anopheles maculipennis,* var. *atroparvus. Amer. J. trop. Med.* **20**: 133–40

115 DAVIDSON, G. (1955) Measurement of the ampulla of the oviduct as a means of determining the natural daily mortality of *Anopheles gambiae. Ann. trop. Med. Parasit.* **49**, 1: 24–36

116 DAY, M. F. (1954) The mechanism of food distribution to midgut or diverticula in the mosquito. *Aust. J. biol. Sci.* **7**: 515–24

117 DOWNES, J. A. (1958) Assembly and mating in the biting Nematocera. *Proc. Tenth. Int. Congr. Ent.* (1956) **2**: 425–34

118 EDWARDS, D. K. (1960) A method for continuous determination

of displacement activity in a group of flying insects. *Can. J. Zool.* **38**: 1021–5

119 ELMORE, JR., C. M. and SCHOOF, H. F. (1963) Dispersal of *Aedes taeniorhynchus* Wiedemann near Savannah, Georgia. *Mosquito News* **23**: 1–7

120 FARQUHARSON, C. O. (1918) *Harpagomyia* and other Diptera fed by *Cremastogaster* ants in southern Nigeria. *Proc. ent. Soc. Lond.* pp. 29–39

121 GILLETT, J. D. (1952) *Embryonic Diapause in the Genus Aëdes Meigen (Diptera: Culicidae)*. Thesis, University of London

122 GILLETT, J. D. (1958) Laboratory colonisation of the mosquito, *Eretmapodites chrysogaster* Grah. *Bull. ent. Res.* **49**: 287–90

123 GORDON, R. M. and CREWE, W. (1948) The mechanism by which mosquitoes and tsetse-flies obtain their blood-meal, the histology of the lesions produced, and the subsequent reactions of the mammalian host; together with some observations on the feeding of *Chrysops* and *Cimex*. *Ann. trop. Med. Parasit.* **42**: 334–56

124 GORDON, R. M. and LUMSDEN, W. H. R. (1939) A study of the behaviour of the mouthparts of mosquitoes when taking up blood from living tissue; together with some observations on the ingestion of microfilariae. *Ann. trop. Med. Parasit.* **33**: 259–78

125 GRIFFITHS, R. B. and GORDON, R. M. (1952) An apparatus which enables the process of feeding by mosquitoes to be observed in the tissues of a live rodent; together with an account of the ejection of saliva and its significance in malaria. *Ann. trop. Med. Parasit.* **46**: 311–19

126 HAEGER, J. S. (1955) The non-blood feeding habits of *Aedes taeniorhynchus* (Diptera, Culicidae) on Sanibel Island, Florida. *Mosquito News* **15**: 21–6

127 HAEGER, J. S. and PHINIZEE, J. (1959) The biology of the crabhole mosquito, *Deinocerites cancer* Theobald. *Rep. Fla. Antimosq. Assoc.* **30**: 34–7

128 HARDEN, F. W. and CHUBB, H. S. (1960) Observations of *Aedes taeniorhynchus* dispersal in extreme south Florida and the Everglades National Park. *Mosquito News* **20**: 249–55

129 HARWOOD, R. F. (1962) Trapping overwintering adults of the mosquitoes *Culex tarsalis* and *Anopheles freeborni*. *Mosquito News* **22**, 1: 28–31

130 HAYES, R. O. (1953) Studies on the artificial insemination of the

mosquito *Aedes aegypti* (Linnaeus). *Mosquito News* **13**: 145–52

131 HEY, R. C. VAN DE and CRAIG, G. B. (1958) Multiple fertilization demonstrated in *Aedes aegypti*. *Bull. ent. Soc. Amer.* **4**: 102

132 HILL, R. B. (1937) The length of life of *Anopheles maculipennis*, var. *atroparvus*. *S. med. J.* **30**: 952–3

133 HOSOI, T. (1954) Mechanism enabling the mosquito to ingest blood into the stomach and sugary fluids into the oesophageal diverticula. *Annot. Zool. Jap.* **27**: 82–90

134 HUDSON, B. N. A. (1956) The behaviour of the female mosquito in selecting water for oviposition. *J. exp. Biol.* **33**: 478–92

135 JOHNSTON, C. (1855) Auditory apparatus of the *Culex* mosquito. *Quart. J. micr. Sci.* **3**: 97–102

136 KENNEDY, J. S. (1941) On water finding and oviposition by captive mosquitoes. *Bull. ent. Res.* **32**: 279–301

137 KENNEDY, J. S. (1961) A turning point in the study of insect migration. *Nature, Lond.* **189**: 785–91

138 KIRK, H. B. (1923) Notes on the mating-habits and early life-history of the culicid *Opifex fuscus* Hutton. *Trans. Proc. N.Z. Inst.* **54**: 400–6

139 KOZHEVNIKOVA, S. M. (1953) The survival of mosquitoes of the genus *Anopheles* to an epidemiologically dangerous age under the conditions found in Stalingrad. *Med. Parasit., Moscow* **22**: 500–5

140 MARKS, E. N. (1958) Notes on *Opifex fuscus* Hutton (Diptera: Culicidae) and the scope for further research on it. *N.Z. Entom.* **2**: 20–5

141 MARSHALL, J. F. and STALEY, J. (1932) On the distribution of air in the oesophageal diverticula and intestine of mosquitoes. Its relation to emergence, feeding and hypopygial rotation. *Parasitology* **24**: 368–81

142 MELLANBY, K. (1932) The influence of atmospheric humidity on the thermal death point of a number of insects. *J. exp. Biol.* **9**: 222–31

143 NIELSEN, E. T. (1958) The initial stage of migration in salt-marsh mosquitoes. *Bull. ent. Res.* **49**: 305–13

144 NIELSEN, E. T. and GREVE, H. (1950) Studies on the swarming habits of mosquitoes and other Nematocera. *Bull. ent. Res.* **41**: 227–58

145 NIELSEN, E. T. and HAEGER, J. S. (1960) Swarming and mating in mosquitoes. *Misc. Publ. ent. Soc. Amer.* **1**: 71–95

146 NIELSEN, H. T. and NIELSEN, E. T. (1962) Swarming of mosquitoes. Laboratory experiments under controlled conditions. *Ent. exp. appl.* **5**: 14–32

147 OMORI, N. (1954) On the swarming of *Culex pipiens pallens*. *Jap. J. San. Zool.* **4**: 342–50

148 O'ROURKE, F. J. (1956) Observations on pool and capillary feeding in *Aedes aegypti* (L.) *Nature, Lond.* **177**: 1087–8

149 PAL, R. (1943) On the histological structure of the midgut of mosquitoes. *J. malar. Inst. India* **5**: 247–50

150 PEYTON, E. L. (1956) Biology of the Pacific coast tree-hole mosquito *Aedes varipalpus* (Coq.). *Mosquito News* **16**: 220–4

151 PROVOST, M. W. (1958) Mating and male swarming in *Psorophora* mosquitoes. *Proc. Tenth Int. Congr. Ent.* (1956) **2**: 553–61

152 PROVOST, M. W. (1960) The dispersal of *Aedes taeniorhynchus*. III. Study methods for migratory exodus. *Mosquito News* **20**: 148–61

153 PROVOST, M. W., LUM, P. T. M. and BRANCH, N. (1961) Rotation of male terminalia in *Aedes taeniorhynchus* (Diptera: Culicidae) as affected by temperature. *Ann. ent. Soc. Amer.* **54**: 896–900

154 RAHM, U. (1958) Die Funktion der Antennen, Palpen und Tarsen von *Aedes aegypti* L. beim Aufsuchen des Wirtes. *Rev. Suisse Zool.* **65**: 779–92

155 RAO, T. R. and RUSSELL, P. F. (1938) Some field observations on the swarming and pairing of mosquitoes, particularly *An. annularis* in south India. *J. malar. Inst. India* **1**: 395–403

156 REES, D. M. and ONISHI, K. (1951) Morphology of the terminalia and internal reproductive organs, and copulation in the mosquito, *Culiseta inornata* (Williston). *Proc. ent. Soc. Wash.* **53**: 233–46

157 RISLER, H. (1955) Das Gehörorgan der Männchen von *Culex pipiens* L. *Aedes aegypti* L. und *Anopheles stephensi* Liston (Culicidae). *Zool. Jb. Anat.* **74**: 478–90

158 ROTH, L. M. (1948) A study of mosquito behaviour. An experimental laboratory study of the sexual behavior of *Aedes aegypti* (Linnaeus). *Amer. Midl. Nat.* **40**: 265–352

159 RUDOLFS, W. (1925) Relation between temperature, humidity and activity of house mosquitoes. *J.N.Y. ent. Soc.* **33**: 163–9

160 RUSH, W. A. (1962) Observations on an overwintering population of *Culex tarsalis* with notes on other species. *Mosquito News* **22**: 176–81

161 RUSSELL, P. F. and RAO, T. R. (1942*a*) Observations on longevity of *Anopheles culicifacies* imagines. *Amer. J. trop. Med.* **22**: 517–33

162 RUSSELL, P. F. and RAO, T. R. (1942*b*) On the swarming, mating and ovipositing behaviour of *Anopheles culicifacies. Amer. J. trop. Med.* **22**: 417–27

163 SERVICE, M. W. (1964*a*) The identification of blood-meals from culicine mosquitoes from northern Nigeria. *Bull. ent. Res.* **55**: 637–43

164 SERVICE, M. W. (1964*b*) The attraction of mosquitoes by animal baits in the northern Guinea savannah of Nigeria. *J. ent. Soc. S. Africa* **27**: 29–36

165 SHUTE, G. T. (1958) The mating plug in *Anopheles gambiae. Trans. R. Soc. trop. Med. Hyg.* **52**: 305

166 SHUTE, P. G. (1935) Agglutination of the red blood-corpuscles of man, animals and birds by the salivary glands of *Anopheles maculipennis. J. trop. Med. Hyg.* **38**: 277–8

167 SIVERLY, R. E. and SCHOOF, H. F. (1962) Biology of *Culiseta melanura* (Coquillett) in southeast Georgia. *Mosquito News* **22**: 274–82

168 SNODGRASS, R. E. (1944) The feeding apparatus of biting and sucking insects affecting man and animals. *Smithson. Misc. Coll.* **104**: 113 pp.

169 SOMEREN, G. R. C. VAN (1965) Male mosquito swarms: some observations in Kenya. *Proc. R. ent. Soc. Lond.* **40**: 89–91

170 SOTOVALTA, O. (1947) The flight-tone (wing-stroke frequency) of insects. *Acta Ent. Fenn.* **4**: 1–117

171 SWELLENGREBEL, N. H. (1929) La dissociation des fonctions sexuelles et nutritives (dissociation gonotrophique) d'*Anopheles maculipennis* comme cause du paludisme dans les pays-bas et ses rapports avec 'l'infection domicilaire'. *Ann. Inst. Past. Paris* **43**: 1370–89

172 TEESDALE, C. (1950) An apparent invasion of *Aëdes* (*Banksinella*) *lineatopennis* and *Aëdes* (*B*) *albicosta* into Mombasa Island. *Proc. R. ent. Soc. Lond.* (A) **25**: 10–12

173 THOMPSON, M. T. (1905) Alimentary canal of the mosquito. *Proc. Boston Soc. Nat. Hist.* **32**: 145–202

174 VINOGRADOVA, E. B. (1960) An experimental investigation of the ecological factors inducing imaginal diapause in bloodsucking mosquitoes (Diptera, Culicidae). *Ent. Rev.* **39**: 210–19

175 WHARTON, R. H. (1953*b*) The habits of adult mosquitoes in Malaya. IV. Swarming of anophelines in nature. *Ann. trop. Med. Parasit.* **47**: 285–90

176 WHARTON, R. H. (1959) Age determination in *Mansonioides* mosquitoes. *Nature, Lond.* **811**: 830

177 WILLIAMS, C. A. (1956) Digestion of serum proteins by *Aedes aegypti*. *Proc. Fourteenth Int. Congr. Ent.*, 1953

178 WISHART, F. and RIORDAN, D. F. (1959) Flight responses to various sounds by adult males of *Aedes aegypti* (L.) (Diptera: Culicidae). *Canad. Ent.* **91**: 181–91

179 ZULUETA, J. DE (1950) Comparative oviposition experiments with caged mosquitoes. *Amer. J. Hyg.* **52**: 133–42

6 FLIGHT

180 BARR, A. R., SMITH, T. A. and BOREHAM, M. M. (1960) Light intensity and the attraction of mosquitoes to light traps. *J. econ. Ent.* **53**: 876–80

181 BROUWER, R. (1960*a*) The attraction of carbon dioxide excreted by the skin of the arm for malaria mosquitoes. *Trop. geogr. Med.* **12**: 62–6

182 BROUWER, R. (1960*b*) Variations in human body odour as a cause of individual differences of attraction for malaria mosquitoes. *Trop. geogr. Med.* **12**: 186–92

183 BROWN, A. W. A. (1951) Studies of the responses of the female *Aedes* mosquito. Part IV. Field experiments on Canadian species. *Bull. ent. Res.* **42**: 572–82

184 BROWN, A. W. A. (1954) Studies on the responses of the female *Aedes* mosquito. Part VI. The attractiveness of coloured cloths to Canadian species. *Bull. ent. Res.* **45**: 67–78

185 BROWN, A. W. A. (1958) Factors which attract *Aedes* mosquitoes to humans. *Proc. Tenth Int. Congr. Ent.* **3**: 757–63

186 BROWN, A. W. A. and CARMICHAEL, A. G. (1961) Lysine and alanine as mosquito attractants. *J. econ. Ent.* **54**: 317–24

187 CLEMENTS, A. N. (1955) The sources of energy for flight in mosquitoes. *J. exp. Biol.* **32**: 547–54

188 CURRY, D. P. (1939) A documented record of a long flight of *Aedes sollicitans. Proc. N.J. Mosq. Ext. Ass.* **26**: 36–9

189 FAY, R. W., BAKER, J. T. and JENSEN, J. A. (1959) Radioactive tagging of *Culex quinquefasciatus* (Say) with P^{32}. *Mosquito News* **19**: 139–43

190 GILLIES, M. T. (1961) Studies on the dispersion and survival of *Anopheles gambiae* Giles in East Africa by means of marking and release experiment. *Bull. ent. Res.* **52**: 99–127

191 HECHT, O. and HERNANDEZ-CORZO, J. (1963) On the visual orientation of mosquitoes in their search of resting places. *Ent. exp. appl.* **6**: 63–74

192 HOCKING, B. (1953) The intrinsic range and speed of flight of insects. *Trans. R. ent. Soc. Lond.* **104**: 223–345

193 HYLTON, A. R. (1966) Histopathological changes with age in the flight muscles of the African mosquito *Eretmapodites chrysogaster. J. invert. Path.* **8**: 75–8

194 KENNEDY, J. S. (1939) The visual responses of flying mosquitoes. *Proc. zool. Soc. Lond.* (A) **109**: 221–42

195 LAARMAN, J. J. (1958) The host-seeking behaviour of anopheline mosquitoes. *Trop. geogr. Med.* **10**: 293–305

196 MARCHAND, W. (1918) First account of thermotrophism in *Anopheles punctipennis,* with bionomic observations. *Psyche* **25**: 130–5

197 PETERSON, D. G. and BROWN, A. W. A. (1951) Studies on the responses of the female *Aëdes* mosquito. Part III. The response of *Aedes aegypti* (L.) to a warm body and its radiation. *Bull. ent. Res.* **42**: 535–41

198 PLATT, R. B., COLLINS, C. L. and WITHERSPOON, J. P. (1957) Reactions of *Anopheles quadrimaculatus* Say to moisture, temperature and light. *Ecol. Monogr.* **27**: 303–24

199 PRINGLE, J. W. S. (1948) The gyroscopic mechanism of the halteres of Diptera. *Phil. Trans.* (B) **233**: 347–84

200 RAHM, U. (1957) Zur Bedeutung des Duftes und des Schweisses bei der Attraktion von *Aedes aegypti* durch den Menschen. *Acta trop., Basel* **14**: 208–17

201 RAHM, U. (1958) Die attraktive Wirkung der vom Menschen

abgegebenen Duftstoffe auf *Aedes aegypti* L. *Z. Tropenmed. u. Parasit.* **9**: 146–56

202 RAO, T. R. (1947) Visual responses of mosquitoes artificially rendered flightless. *J. exp. Biol.* **24**: 64–78

203 REEVES, W. C. (1951) Field studies on carbon dioxide as a possible host simulant to mosquitoes. *Proc. Soc. exp. Biol. Med.* **77**: 64–6

204 REEVES, W. C. (1963) Quantitative field studies on a carbon dioxide chemotropism of mosquitoes. *Amer. J. trop. Med. Hyg.* **2**: 325–31

205 REEVES, W. C., BROOKMAN, B. and HAMMON, W. McD. (1948) Studies on the flight range of certain *Culex* mosquitoes, using a fluorescent-dye marker, with notes on *Culiseta* and *Anopheles. Mosquito News* **8**: 61–9

206 RIBBANDS, C. R. (1949) Studies on the attractiveness of human populations to anophelines. *Bull. ent. Res.* **40**: 227–38

207 ROTH, L. M. (1951) Loci of sensory end-organs used by mosquitoes (*Aedes aegypti* (L.) and *Anopheles quadrimaculatus* Say) in receiving host stimuli. *Ann. ent. Soc. Amer.* **44**: 59–74

208 SIPPELL, W. L. and BROWN, A. W. A. (1953) Studies of the responses of the female *Aedes* mosquito. Part V. The role of visual factors. *Bull. ent. Res.* **43**: 567–74

209 SMART, M. R. and BROWN, A. W. A. (1956) Studies on the responses of the female *Aedes* mosquito. Part VII. The effect of skin temperature, hue and moisture on the attractiveness of the human hand. *Bull. ent. Res.* **47**: 80–100

210 SOTAVALTA, O. (1952) The essential factor regulating the wing stroke frequency of insects in wing mutilation and loading experiments. *Ann. zool. Soc. Vanamo* **15**: 1–67

211 SUZUKI, K. (1960) Perception of movement in compound eyes studied by optomotor reaction of mosquitoes. *J. Fac. Sci. Hokkaido Univ. (Zool.)* **14**: 349–80

212 TOWNSON, H. (1967) Studies of flight in the mosquito *Aedes aegypti* and the effect on flight of infection with *Brugia pahangi.* Thesis, Univ. of Liverpool

213 WIGGLESWORTH, V. B. and GILLETT, J. D. (1934) The function of the antennae in *Rhodnius prolixus* (Hemiptera) and the mechanism of orientation to the host. *J. exp. Biol.* **11**: 120–39

214 WIGGLESWORTH, V. B. and GILLETT, J. D. (1934) The function

of the antennae in *Rhodnius prolixus:* confirmatory experiments. *J. exp. Biol.* **11**: 408

215 Willis, E. R. (1947) The olfactory responses of female mosquitoes. *J. econ. Ent.* **40**: 769–78

216 Willis, E. R. and Roth, L. M. (1952) Reactions of *Aedes aegypti* (L.) to carbon dioxide. *J. exp. Zool.* **121**: 149–79

217 Woke, P. A. (1937) Cold-blooded vertebrates as hosts for *Aedes aegypti* Linn. *J. Parasit.* **23**: 310–13

218 Wright, R. H. (1958) The olfactory guidance of flying insects. *Canad. Ent.* **90**: 81–9

219 Wright, R. H. (1968) Tunes to which mosquitoes dance. *New Scientist* **37**: 694–7

7 THE FEEDING CYCLE

220 Aitken, T. H. G. (1967) The canopy-frequenting mosquitoes of bush bush forest, Trinidad, West Indies. *Atas do Simpósio sôbre a Biota Amazoñica* **6** (*Patologia*): 65–73

221 Gillett, J. D. (1964) Studies on the cyclic behaviour of mosquitoes. *Proc. seventh Internat. Cong. trop. Med. Malaria* **5**: 39–40

222 Gillett, J. D. (1969) *Aedes simpsoni* in Chaggaland, Tanzania. *Ann. trop. Med. Parasit.* **63**: 147–56

223 Gillies, M. T. (1957) Age-groups and the biting cycle in *Anopheles gambiae*. A preliminary investigation. *Bull. ent. Res.* **48**: 553–9

224 Haddow, A. J. (1945) The mosquitoes of Bwamba County, Uganda. II. Biting activity with special reference to the influence of microclimate. *Bull. ent. Res.* **36**: 33–73

225 Haddow, A. J. (1954) Studies of the biting-habits of African mosquitoes. An appraisal of methods employed, with special reference to the twenty-four-hour catch. *Bull. ent. Res.* **45**: 199–242

226 Haddow, A. J. (1960) Studies on the biting habits and medical importance of East African mosquitoes in the genus *Aëdes*. I. Subgenera *Aëdimorphus, Banksinella* and *Dunnius*. *Bull. ent. Res.* **50**: 759–79

227 Haddow, A. J. (1961) Studies on the biting habits and medical importance of East African mosquitoes in the genus *Aëdes*. II.

Subgenera *Mucidus, Diceromyia, Finlaya* and *Stegomyia. Bull. ent. Res.* **52**: 317–51

228 HADDOW, A. J., CORBET, P. S. and GILLETT, J. D. (1960) Entomological studies from a high tower in Mpanga Forest, Uganda. *Trans. R. ent. Soc. Lond.* **113**: 249–56

229 HADDOW, A. J., GILLETT, J. D. and HIGHTON, R. B. (1947) The mosquitoes of Bwamba County, Uganda. V. The vertical distribution and biting-cycle of mosquitoes in rain-forest, with further observations on microclimate. *Bull. ent. Res.* **37**: 301–30

230 HADDOW, A. J., VAN SOMEREN, E. C. C., LUMSDEN, W. H. R., HARPER, J. O. and GILLETT, J. D. (1951) The mosquitoes of Bwamba County, Uganda. VIII. Records of occurrence, behaviour and habitat. *Bull. ent. Res.* **42**: 207–38

231 HINGSTON, R. W. G. (1932) *A Naturalist in the Guiana Forest.* Edward Arnold & Co.: London

232 LAARMAN, J. J. (1955) *The Host Seeking Behaviour of the Malaria Mosquito* Anopheles maculipennis atroparvus. Thesis. Univ. of Utrecht

233 LUMSDEN, W. H. R. (1947) Observations on the effect of microclimate on the biting of *Aëdes aegypti* (L.) (Diptera: Culicidae). *J. exp. Biol.* **24**: 361–73

234 MACDONALD, W. W. (1967) Host feeding preferences. *Bull. Wld. Hlth. Org.* **36**: 597–9

235 MATTINGLY, P. F. (1949) Studies on West African forest mosquitoes. Part I. The seasonal distribution and biting cycle and vertical distribution of four of the principal species. *Bull. ent. Res.* **40**: 149–68

236 MCCLELLAND, G. A. H. (1959) Observations on the mosquito, *Aëdes* (*Stegomyia*) *aegypti* (L.), in East Africa. I. The biting cycle in an outdoor population at Entebbe, Uganda. *Bull. ent. Res.* **50**: 227–35

237 MCCLELLAND, G. A. H. (1960) Observations on the mosquito *Aëdes* (*Stegomyia*) *aegypti* (L.) in East Africa. II. The biting cycle in a domestic population on the Kenya Coast. *Bull. ent. Res.* **50**: 687–96

238 SMITH, A. (1958) Outdoor cattle feeding and resting of *A. gambiae* Giles and *A. pharoensis* Theo. in the Pare-Taveta area of East Africa. *E. Afr. med. J.* **35**: 559–67

239 TEMPELIS, CH., REEVES, W. C., BELLAMY, R. E. and LOFY, M.

(1965) A three-year study of the feeding habits of *Culex tarsalis* in Kern County, California. *Am. J. trop. Med. Hyg.* **14**: 170–7

240 WHARTON, R. H. (1953*a*) The habits of adult mosquitoes in Malaya. III. Feeding preferences of anophelines. *Ann. trop. Med. Parasit.* **47**: 272–84

8 THE OVARIAN CYCLE

241 BEKLEMISHEV, W. N., DETINOVA, T. S. and POLOVODOVA, V. P. (1959) Determination of physiological age in anophelines and of age distribution in anopheline populations in the USSR. *Bull. World Hlth. Org.* **21**: 223–32

242 BODENSTEIN, D. (1945) The corpora allata of mosquitoes. *Connect. agric. exp. Sta. Bull.* **488**: 396–405

243 CHRISTOPHERS, S. R. (1911) The development of the egg follicle in anophelines. *Paludism* **2**: 73–88

244 CLEMENTS, A. N. (1956) Hormonal control of ovary development in mosquitoes. *J. exp. Biol.* **33**: 211–23

245 COLLESS, D. H. and CHELLAPAH, W. T. (1960) Effects of body weight and size of blood-meal upon egg production in *Aedes aegypti* (Linnaeus) (Diptera, Culicidae). *Ann. trop. Med. Parasit.* **54**: 475–82

246 CURTIN, T. J. and JONES, J. C. (1961) The mechanism of ovulation and oviposition in *Aedes aegypti*. *Ann. ent. Soc. Amer.* **54**: 298–313

247 DETINOVA, T. S. (1945) On the influence of glands of internal secretion upon the ripening of the gonads and the imaginal diapause in *Anopheles maculipennis*. *Zool. Zh.* **34**: 291–8 (In Russian)

248 DETINOVA, T. S. (1953) The age composition and epidemiological significance of populations of *Anopheles maculipennis* under the conditions found in the Province of Moscow. *Med. Parasit. Moscow* **22**: 486–95

249 DETINOVA, T. S. (1962) Age-grouping methods in Diptera of medical importance with special reference to some vectors of malaria. *Monogr. World Hlth. Org.* No. 47: 216 pp.

250 GILLETT, J. D. (1956) Initiation and promotion of ovarian development in the mosquito *Aedes* (*Stegomyia*) *aegypti* (Linnaeus). *Ann. trop. Med. Parasit.* **50**: 375–80

251 GILLETT, J. D. (1957) Variation in the time of release of the ovarian development hormone in *Aedes aegypti. Nature, Lond.* **180**: 656–7

252 GILLETT, J. D. (1958) Induced ovarian development in decapitated mosquitoes by transfusion of haemolymph. *J. exp. Biol.* **35**: 685–93

253 GILLIES, M. T. (1953) The duration of the gonotrophic cycle in *Anopheles gambiae* and *Anopheles funestus* with a note on the efficiency of hand catching. *E. Afr. med. J.* **30**: 129–35

254 GILLIES, M. T. (1954) The recognition of age groups within populations of *Anopheles gambiae* by the pre-gravid rate and the sporozoite rate. *Ann. trop. Med. Parasit.* **48**: 58–74

255 GILLIES, M. T. (1955) The pre-gravid phase of ovarian development in *Anopheles funestus. Ann. trop. Med. Parasit.* **49**: 320–5

256 GILLIES, M. T. (1956) A new character for the recognition of nulliparous females of *Anopheles gambiae. Bull. World Hlth. Org.* **15**: 451–9

257 GILLIES, M. T. (1958) A review of some recent Russian publications on the technique of age determination in *Anopheles. Trop. Dis. Bull.* **55**: 713–21

258 GOODING, R. H. (1966) Physiological aspects of digestion of the blood meal by *Aedes aegypti* (Linnaeus) and *Culex fatigans* Wiedemann. *J. med. Ent.* **3**: 53–60

259 HAMON, J., CHAUVET, G. and THÉLIN, L. (1961) Observations sur les méthodes d'évaluation de l'âge physiologique des femelles d'anophèles. *Bull. World Hlth. Org.* **24**: 437–43

260 HAMON, J., SALES, S., ADAM, J. P. and GRENIER, P. (1964) Age physiologique et cycle d'agressivite chez *Anopheles gambiae* Giles et *A. funestus* Giles dans la region de Bobo-Dioulasso (Haute-Volta) (Dipt. Culicidae). *Bull. Soc. ent. France* **69**: 110–21

261 LARSEN, J. R. (1958) Hormone-induced ovarian development in mosquitoes. *Science* **127**: 587–8

262 LARSEN, J. R. and BODENSTEIN, D. (1959) The humoral control of egg maturation in the mosquito. *J. exp. Zool.* **140**: 343–81

263 LEA, A. O. (1963) Some relationships between environment, corpora allata, and egg maturation in aedine mosquitoes. *J. ins. Physiol.* **9**: 793–809

264 LEA, A. O. (1964) Studies on the dietary and endocrine regula-

tion of autogenous reproduction in *Aedes taeniorhynchus* (Wied.). *J. med. Ent.* **1**: 40–4

265 LEA, A. O., KNIERIM, J. A., DIMOND, J. B. and DE LONG, D. M. (1955) A preliminary note on egg production from milk-fed mosquitoes. *Ohio J. Sci.* **55**: 21–2

266 MER, G. (1932) The determination of the age of *Anopheles* by differences in the size of the common oviduct. *Bull. ent. Res.* **23**: 563–6

267 POLOVODOVA, V. P. (1949) The determination of the physiological age of female *Anopheles*, by the number of gonotrophic cycles completed. *Med. Parasit. Moscow* **18**: 352–5 (In Russian)

268 ROY, D. N. (1936) On the role of blood in ovulation in *Aedes aegypti*, Linn. *Bull. ent. Res.* **27**: 423–9

269 SERVICE, M. W. (1968) Pre-gravid phase of ovarian development in *Aedes detritus* Haliday. *Nature, Lond.* **217**: 681–2

270 TEESDALE, C. (1955) Studies on the bionomics of *Aedes aegypti* (L.) in its natural habitats in a coastal region of Kenya. *Bull. ent. Res.* **46**: 711–42

271 THOMSEN, E. (1952) Functional significance of the neurosecretory brain cells and the corpus cardiacum in the female blow-fly, *Calliphora erythrocephala* Meig. *J. exp. Biol.* **29**: 137–72

272 WALLIS, R. C. (1954) A study of oviposition activity of mosquitoes. *Amer. J. Hyg.* **60**: 135–68

273 WOKE, P. A. (1937*a*) Comparative effects of the blood of man and of canary on egg-production of *Culex pipiens* Linn. *J. Parasit.* **23**: 311–13

274 WOKE, P. A. (1937*b*) Effects of various blood fractions on egg production of *Aedes aegypti* Linn. *Amer. J. Hyg.* **25**: 372–80

275 WOKE, P. A., ALLY, M. S. and ROSENBERGER, C. R. (1956) The number of eggs developed related to the quantities of human blood ingested in *Aedes aegypti* (L.) (Diptera: Culicidae). *Ann. ent. Soc. Amer.* **49**: 435–41

9 THE CIRCADIAN RHYTHM

276 ASCHOFF, J. (1960) Exogenous and endogenous components in circadian rhythms. *Cold Spr. Harb. Sym. Quant. Biol.* **25**: 11–28

277 BATES, M. (1941) Laboratory observations on the sexual behaviour of anopheline mosquitoes. *J. exp. Zool.* **86**: 153–73

278 CORBET, P. S., HADDOW, A. J. and GILLETT, J. D. (1960) Observations on the oviposition-cycle of *Aedes* (*Stegomyia*) *aegypti* (Linnaeus). IV. *Ann. trop. Med. Parasit.* **54**: 156–64

279 GILLETT, J. D. (1961) Cyclical feeding-activity in colonised mosquitoes. *Nature, Lond.* **190**: 881–3

280 GILLETT, J. D. (1962) Contribution to the oviposition-cycle by the individual mosquitoes in a population. *J. insect Physiol.* **8**: 665–81

281 GILLETT, J. D., CORBET, P. S. and HADDOW, A. J. (1959) Observations on the oviposition-cycle of *Aedes* (*Stegomyia*) *aegypti* (Linnaeus). III. *Ann. trop. Med. Parasit.* **53**: 132–6

282 GILLETT, J. D., CORBET, P. S. and HADDOW, A. J. (1961) Observations on the oviposition-cycle of *Aedes* (*Stegomyia*) *aegypti* (Linnaeus). VI. *Ann. trop. Med. Parasit.* **55**: 427–31

283 GILLETT, J. D. and HADDOW, A. J. (1957) Laboratory observations on the oviposition-cycle in the mosquito *Aedes* (*Stegomyia*) *africanus* Theobald. *Ann. trop. Med. Parasit.* **51**: 170–4

284 GILLETT, J. D., HADDOW, A. J. and CORBET, P. S. (1959) Observations on the oviposition-cycle of *Aedes* (*Stegomyia*) *aegypti* (Linnaeus). II. *Ann. trop. Med. Parasit.* **53**: 35–41

285 GILLETT, J. D., HADDOW, A. J. and CORBET, P. S. (1962) The sugar-feeding-cycle in a cage-population of mosquitoes. *Ent. exp. & appl.* **5**: 223–32

286 HADDOW, A. J., CORBET, P. S. and GILLETT, J. D. (1960) Laboratory observations on the oviposition-cycle in the mosquito *Aedes* (*Stegomyia*) *apicoargenteus* Theobald. *Ann. trop. Med. Parasit.* **54**: 392–6

287 HADDOW, A. J. and GILLETT, J. D. (1957) Observations on the oviposition-cycle of *Aedes* (*Stegomyia*) *aegypti* (Linnaeus). *Ann. trop. Med. Parasit.* **51**: 159–69

288 HADDOW, A. J. and GILLETT, J. D. (1958) Laboratory observations on the oviposition-cycle in the mosquito *Taeniorhynchus* (*Coquillettidia*) *fuscopennatus* Theobald. *Ann. trop. Med. Parasit.* **52**: 320–5

289 HADDOW, A. J., GILLETT, J. D. and CORBET, P. S. (1961) Observations on the oviposition-cycle of *Aedes* (*Stegomyia*) *aegypti* (Linnaeus). V. *Ann. trop. Med. Parasit.* **55**: 343–56

290 HARKER, J. E. (1958) Diurnal rhythms in the animal kingdom. *Biol. Rev.* **3**: 1–52

291 JOBLING, B. (1935) The effect of light and darkness on oviposition in mosquitoes. *Trans. R. Soc. trop. Med. Hyg.* **29**: 157–66

292 JONES, M. D. R. (1963) The automatic recording of mosquito activity. *J. ins. Physiol.* **10**: 343–51

293 JONES, M. D. R., FORD, M. G. and GILLETT, J. D. (1966) Light-on and light-off effects on the circadian flight activity in the mosquito *Anopheles gambiae. Nature, Lond.* **211**: 871–2

294 JONES, M. D. R., HILL, M. N. and HOPE, A. M. (1967) The circadian flight-activity of the mosquito *Anopheles gambiae:* Phase setting by the light-regime. *J. exp. Biol.* **47**: 503–11

295 KASHIN, P. (1966) Electronic recording of the mosquito bite. *J. ins. Phys.* **12**: 281–6

296 MCCLELLAND, G. A. H. (1968) Field observations on periodicity and site preference in oviposition by *Aedes aegypti* (L.) and related mosquitoes (Diptera: Culicidae) in Kenya. *Proc. R. ent. Soc. Lond.* (A) **43**

297 ROUBAUD, E. (1918) Rythmes physiologiques et vol spontané chez l'*Anopheles maculipennis C.R. Acad. Sci., Paris* **167**: 967–9

298 TAYLOR, B. and JONES, M. D. R. (1968) The circadian rhythm of flight activity in the mosquito *Aedes aegypti* (L.): the phase-setting effects of light-on and light-off. *J. exp. Biol.* **51**: 59–70

10 STRAINS AND SPECIES

299 BUCK, A. DE, SCHOUTE, E. and SWELLENGREBEL, N. H. (1932) Further investigations on the racial differences of *Anopheles maculipennis* in the Netherlands and its bearing on malaria. *Riv. Malariol.* **11**: 137–56

300 CHRISTOPHERS, S. R. (1951) Note on morphological characters differentiating *Culex pipiens* L. from *Culex molestus* Forskål and the status of these forms. *Trans. R. ent. Soc. Lond.* **102**: 376

301 DAVIDSON, G. (1967) The *Anopheles gambiae* complex. *Genetics of Insect Vectors of Disease* by Wright, J. W., and Pal, R. Elsevier Publishing Company: Amsterdam

302 FORD, E. B. (1961) The theory of genetic polymorphism. *Symp. R. ent. Soc. Lond.* No 1: 11–19

303 GILLETT, J. D. (1949) Further notes on the Ethiopian species of *Taeniorhynchus* Arribálzaga (Diptera, Culicidae) *Proc. R. ent. Soc. Lond.* (B) **18**: 97–102

304 GILLETT, J. D. (1955*a*) Behaviour differences in two strains of *Aedes aegypti*. *Nature, Lond.* **176**: 124

305 GILLETT, J. D. (1955*b*) The male of *Anopheles* (*Myzomyia*) *distinctus* var. *ugandae* Evans (Diptera: Culicidae). *Proc. R. ent. Soc. Lond.* (B) **24**: 36

306 GILLETT, J. D. (1956) Genetic differences affecting egg-laying in the mosquito *Aedes* (*Stegomyia*) *aegypti* (Linnaeus). *Ann. trop. Med. Parasit.* **50**: 362–74

307 GILLETT, J. D. (1962) Resistance to desiccation in six strains of the *Culex pipiens* complex. *Mosquito News* **22**: 342–5

308 HADDOW, A. J. (1956) Observations on the biting-cycle of African mosquitoes in the genus *Eretmapodites* Theobald. *Bull. ent. Res.* **46**: 761–72

309 HILL, G. F. (1921) Notes on some unusual breeding places of *Stegomyia fasciata* in Australia. *Ann. trop. Med. Parasit.* **15**: 91–2

310 MACDONALD, W. W. (1967) Genetics of vectorial capacity. *Bull. Wld. Hlth. Org.* **36**: 566–8

311 MACDONALD, W. W. and WHARTON, R. H. (1963) Genetic control of susceptibility to filariae in *Aedes aegypti*. *Trans. R. Soc. trop. Med. Hyg.* **57**: 4

312 MATTINGLY, P. F. (1957) Genetical aspects of the *Aedes aegypti* problem. I. Taxonomy and bionomics. *Ann. trop. Med. Parasit.* **51**: 392–408

313 MATTINGLY, P. F. (1958) Genetical aspects of the *Aedes aegypti* problem. II. Disease relationships, genetics and control. *Ann. trop. Med. Parasit.* **52**: 5–17

314 MATTINGLY, P. F., ROZEBOOM, L. E., KNIGHT, K. L., LAVEN, H., DRUMMOND, F. H., CHRISTOPHERS, S. R. and SHUTE, P. G. (1951) The *Culex pipiens* complex. *Trans. R. ent. Soc. Lond.* **102**: 331–82

315 MCCLELLAND, G. A. H. (1960) A preliminary study of the genetics of abdominal colour variations in *Aedes aegypti* (L.) (Diptera, Culicidae). *Ann. trop. Med. Parasit.* **54**: 305–20

316 MCCLELLAND, G. A. H. (1962) A contribution to the genetics of the mosquito *Aedes aegypti* with particular reference to factors determining colour. Thesis, Univ. of London, 276 pp.

317 MER, G. (1937) Variations saisonières des caractères de *Anopheles elutus* en Palestine. II. *Bull. Soc. Pat. Exot.* **30**: 38–42

318 Pittendrigh, C. S. (1950) The ecoclimatic divergence of *Anopheles bellator* and *A. homunculus*. *Evolution* **4**: 43–63

319 Roubaud, E. (1929) Cycle autogène d'attente et générations hivernales suractives inapparentes chez le moustique commun, *Culex pipiens*. *C.R. Acad. Sci., Paris* **188**: 735–8

320 Roubaud, E. (1930) Sur l'existence de races biologiques génétiquement distinctes chez le moustique commun *Culex pipiens*. *C.R. Acad. Sci., Paris* **191**: 1386–8

321 Rozeboom, L. E. (1963) Mutant forms of *Anopheles albimanus* Wiedemann (Diptera: Culicidae). *Proc. ent. Soc., Washington* **65**: 110–14

322 Rozeboom, L. E. (1963) Penetrance of the gene for dieldrin resistance in *Anopheles albimanus* Wiedemann. *Ann. J. trop. Med. Hyg.* **12**: 249–53

323 Someren, E. C. C. van, Teesdale, C. and Furlong, M. (1955) The mosquitoes of the Kenya coast: records of occurrence, behaviour and habitat. *Bull. ent. Res.* **46**: 463–93

324 Spielman, A. (1957) The inheritance of autogeny in the *Culex pipiens* complex of mosquitoes. *Amer. J. Hyg.* **65**: 404–25

325 Swellengrebel, N. H. and de Buck, A. (1938) *Malaria in the Netherlands*. Scheltema and Holkema: Amsterdam

326 Tate, P. and Vincent, M. (1936) The biology of autogenous and anautogenous races of *Culex pipiens* L. (Diptera: Culicidae). *Parasitology* **28**: 115–45

327 Thiel, P. H. van (1928) La nourriture des larves de l'*Anopheles maculipennis* en rapport avec le problème de l'existence de la variété atroparvus. *Bull. Soc. Pat. Exot.* **21**: 551–74

328 Thiel, P. H. van (1939) On zoophilism and anthropophilism of *Anopheles* biotypes and species. *Riv. Malariol.* **18**: 95–124

329 Twohy, D. W. and Rozeboom, L. E. (1957) A comparison of food reserves in autogenous and anautogenous *Culex pipiens* populations. *Amer. J. Hyg.* **65**: 316–24

330 Walker, F. R. (1848) *List of the Species of Dipterous Insects in the Collection of the British Museum:* London

331 Wright, J. W. and Pal, R. (1967) *Genetics of Insect Vectors of Disease*. Elsevier Publishing Co.: Amsterdam

11 PARASITES AND PREDATORS

332 COLLESS, D. H. (1958) Relationship between certain biological constants and trophic condition in mosquito populations. *Nature, Lond.* **182**: 951–2

333 CORBET, P. S. (1960) Recognition of nulliparous mosquitoes without dissection. *Nature, Lond.* **187**: 525–6

334 CORBET, P. S. (1963) The reliability of parasitic water-mites (Hydracarina) as indicators of physiological age in mosquitoes (Diptera: Culicidac). *Ent. exp. appl.* **6**: 215–33

335 FERGUSON, M. J. and MICKS, D. W. (1961) Microorganisms associated with mosquitoes. I. Bacteria isolated from the mid-gut of adult *Culex fatigans* Wiedemann. *J. insect. Pathol.* **3**: 112–19

336 GILLETT, J. D. (1957*a*) On the habits and life history of captive emesine bugs (Hemiptera: Reduviidae). *Proc. R. ent. Soc. Lond.* (A) **32**: 193–5

337 GILLETT, J. D. (1957*b*) Age analysis in the biting-cycle of the mosquito *Taeniorhynchus* (*Mansonioides*) *africanus* Theobald, based on the presence of parasitic mites. *Ann. trop. Med. Parasit.* **51**: 151–8

338 HINMAN, E. H. (1932) The presence of bacteria within the eggs of mosquitoes. *Science* **76**: 106–7

339 JACKSON, N. (1953) Observations on the feeding habits of a predaceous mosquito larva, *Culex* (*Lutzia*) *tigripes* Granpré and Charmoy (Diptera). *Proc. R. ent. Soc. Lond.* (A) **28**: 153–9

340 KOCH, A. (1956) The experimental elimination of symbionts and its consequences. *Exp. Parasit.* **5**: 481–518

341 LAIRD, M. (1959) Fungal parasites of mosquito larvae from the Oriental and Australian Regions, with a key to the genus *Coelomomyces* (Blastocladiales: Coelomomyces). *Can. J. Zool.* **37**: 782–91

342 LILES, J. N. and DUNN, P. H. (1959) Preliminary laboratory results on the susceptibility of *Aedes aegypti* (Linnaeus) to *Bacillus thuringiensis* Berliner. *J. ins. Path.* **1**: 309–10

343 MACAN, T. T. (1966) The influence of predation on the fauna of a moorland fishpond. *Arch. Hydrobiol.* **61**: 432–52

344 MICKS, D. W., JULIAN, S. R., FERGUSON, M. J. and DUNCAN, D. (1961) Microorganisms associated with mosquitoes. II. Location

and morphology of microorganisms in the mid-gut of *Culex fatigans* Wiedemann and certain other species. *J. insect. Path.* **3**: 120–8

345 Muspratt, J. (1952) The bionomics of an African *Megarhinus* (Dipt., Culicidae) and its possible use in biological control. *Bull. ent. Res.* **42**: 355–70

346 Sinton, J. A. (1939) Spirochaetal infections of mosquitoes. *J. trop. Med. Hyg.* **42**: 125–6

347 Steinhaus, E. A. (1946) *Insect Microbiology*. Comstock Publishing Co.: N.Y.

348 Wallace, F. G. (1943) Flagellate parasites of mosquitoes with special reference to *Crithidia fasciculata*. *J. Parasit.* **29**: 196–205

349 Wigglesworth, V. B. (1929) Delayed metamorphosis in a predaceous mosquito larva and a possible practical application. *Nature, Lond.* **123**: 17

12 MOSQUITOS AS NUISANCE

350 Corbet, P. S. (1967) Facultative autogeny in arctic mosquitoes. *Nature, Lond.* **215**: 662–3

351 Corbet, P. S. and Downe, A. E. R. (1966) Natural hosts of mosquitoes in northern Ellesmere Island. *J. Arctic Inst. N. Amer.* **19**: 153–61

352 Gillett, J. D. (1967) Natural selection and feeding speed in a blood-sucking insect. *Proc. R. Soc.* B, **167**: 316–29

353 Haufe, W. O. (1952) Observations on the biology of mosquitoes (Diptera: Culicidae) at Goose Bay, Labrador. *Canad. Entom.* **84**: 254–63

354 Haufe, W. O. and Burgess, L. (1956) Development of *Aedes* (Diptera: Culicidae) at Fort Churchill, Manitoba, and prediction of dates of emergence. *Ecology* **37**: 500–19

355 Hearle, E. (1929) The life history of *Aedes flavescens* Müller: A contribution to the biology of mosquitoes of the Canadian prairies. *Trans. R. Soc. Can.* **23**: 85–102

356 Hocking, B. (1952) Protection from northern biting flies. *Mosquito News* **12**: 91–102

357 Hudson, A., McKiel, J. A., West, A. S. and Bourns, T. K. R. (1958) Reactions to mosquito bites. *Mosquito News* **18**: 249–52

358 LAWRENCE, D. H. (1968) The mosquito, from *Poems Selected for Young People*. Macmillan, London

359 LONGSTAFF, T. G. (1932) An ecological reconnaisance in west Greenland. *J. anim. Ecol.* **1**: 119–42

360 MELLANBY, K. (1946) Man's reaction to mosquito bites. *Nature, Lond.* **158**: 554

361 MELLANBY, K. (1960) The activity of certain arctic insects at low temperatures. *J. anim. Ecol.* **9**: 296–301

362 ROCKWELL, E. M. and JOHNSON, P. (1952) The insect bite reaction. II. Evaluation of the allergic reaction. *J. invest. Dermatol.* **19**: 137–55

363 TWINN, C. R. (1955) Review of recent progress in mosquito studies in Canada. *Mosquito News* **15**: 195–203

364 TWINN, C. R., HOCKING, B., MCDUFFIE, W. C. and CROSS, H. F. (1948) A preliminary account of the biting flies at Churchill, Manitoba. *Can. J. Res.* D, **26**: 334–357

365 WHITE, J. C. (1871) On the protection acquired by the human skin and other tissues against the action of certain animal poisons after repeated inoculation. *Boston med. & surg. J.* **9**: 297–300

13 MOSQUITOS AND DISEASE

366 AGRAMONTE, A. (1908) An account of Dr Louis David Beauperthuy, a pioneer in yellow fever research. *Boston Med. Surg. J.* **158**: 927–30

367 ANDREWES, C. H. (1967) *Natural History of Viruses,* Weidenfeld and Nicolson: London

368 BARDOS, V. and CUPKOVA, E. (1962) The Calovo virus: The second virus isolated from mosquitoes in Czechoslovakia. *J. Hyg. Epidemiol. Microbiol. Immunol.* **6**: 186–92

369 BARDOS, V. and DANIELOVA, V. L. (1959) The Tahyna virus: a virus isolated from mosquitoes in Czechoslovakia. *J. Hyg. Epidemiol. Microbiol. Immunol.* **3**: 264–76

370 BEAUPERTHUY, L. D. (1854) Transmission of yellow fever and other diseases by the mosquito. *Gaz. Oficial de Cumana.* Anno **4**

371 BELLAMY, R. E., REEVES, W. C. and SCRIVANI, R. P. (1967) Experimental cyclic transmission of western equine encephalitis virus in chickens and *Culex tarsalis* through a year. *Amer. J. Epid.* **85**: 282–96

372 Boshell, J. (1955) Yellow fever in Central America; the post-war spread as a threat. *Yellow Fever, a Symposium in Commemoration of Carlos Juan Finlay*. Jefferson Med. Col., Philadelphia

373 Carey, D. E., Reuben, R. and Myers, R. M. (1968) Japanese encephalitis studies in Vellore, South India. I. Virus isolation from mosquitoes. *Ind. Jour. med. Res.* **56**: 1309–18

374 Chamberlain, R. W., Corristan, E. C. and Sikes, R. K. (1954) Studies on the North American arthropod-borne encephalitides. V. The extrinsic incubation of eastern and western equine encephalitis in mosquitoes. *Am. J. Hyg.* **60**: 269–77

375 Chamberlain, R. W., Kissling, R. E. and Sikes, R. K. (1954) Studies on the North American arthropod-borne encephalitides. VII. Estimation of amount of eastern equine encephalitis virus inoculated by infected *Aedes aegypti*. *Am. J. Hyg.* **60**: 286–91

376 Chamberlain, R. W., Rubin, H., Kissling, R. E. and Eidson, M. E. (1951) Recovery of virus of eastern equine encephalomyelitis from a mosquito, *Culiseta melanura* (Coquillett). *Proc. Soc. exp. Biol. Med.* **77**: 396–7

377 Chamberlain, R. W. and Sudia, W. D. (1955) The effects of temperature upon the extrinsic incubation of eastern equine encephalitis in mosquitoes. *Am. J. Hyg.* **62**: 295–305

378 Chamberlain, R. W. and Sudia, W. D. (1957) The North American arthropod-borne encephalitis viruses in *Culex tarsalis* Coquillett. *Am. J. Hyg.* **66**: 151–9

379 Chamberlain, R. W., Sudia, W. D. and Gillett, J. D. (1959) St Louis encephalitis virus in mosquitoes. *Amer. J. Hyg.* **70**: 221–36

380 Cheng, T. C. (1964) *The Biology of Animal Parasites*. W. B. Saunders Co.: Philadelphia and London

381 Coggleshall, L. T. (1941) Infection of *Anopheles quadrimaculatus* with *Plasmodium cynomolgi*, a monkey malaria parasite, and with *P. lophurae*, an avian malaria parasite. *Amer. J. trop. Med.* **21**: 525–30

382 Corbet, P. S., Williams, M. C. and Gillett, J. D. (1961) O'nyong-nyong fever: an epidemic virus disease in East Africa. IV. Vector studies at epidemic sites. *Trans. R. Soc. trop. Med. Hyg.* **55**: 463–80

383 Danielová, V. (1962) Multiplication dynamics of Tahyna virus

in different body parts of *Aedes vexans* mosquito. *Acta Virol.* **6**: 227–30

384 DAVIES, A. M. and YOSHPE-PURER, Y. (1954*a*) Observations on the biology of West Nile virus with special reference to its behaviour in the mosquito *Aëdes aegypti*. *Ann. trop. Med. Parasit.* **48**: 46–54

385 DAVIES, A. M. and YOSHPE-PURER, Y. (1954*b*) The transmission of Semliki Forest virus by *Aëdes aegypti*. *J. trop. Med.* (*Hyg.*) **57**: 273–5

386 DICK, G. W. A. (1953) Epidemiological notes on some viruses isolated in Uganda. *Trans. R. Soc. trop. Med. Hyg.* **47**: 13–43

387 DOW, R. P., REEVES, W. C. and BELLAMY, R. E. (1957) Field tests of avian host preference of *Culex tarsalis* Coq. *Amer. J. trop. Med. Hyg.* **6**: 294–303

388 DOWNS, W. G. (1963) General ecology of the arboviruses. *Anais Microbiol.* **11**: 57–65

389 FINLAY, C. J. (1881) El mosquito hipoteticamente considerado como agente de transmission de la fiebbre amarilla. *R. Acad. Ciencias Habana* **18**: 147–69

390 FINLAY, C. J. (1886) Yellow fever, its transmission by means of the *Culex* mosquito. *Amer. J. Med. Sci.* **92**: 395–409

391 GALINDO, P. (1958) Bionomics of *Sabethes chloropterus* Humboldt, a vector of sylvan yellow fever in middle America. *Amer. J. trop. Med. Hyg.* **7**: 429–40

392 GALINDO, P. and TRAPIDO, H. (1957) Forest mosquitoes associated with sylvan yellow fever in Nicaragua. *Amer. J. trop. Med. Hyg.* **6**: 145–52

393 GARNHAM, P. C. C., HARPER, J. A. and HIGHTON, R. B. (1946) The mosquitoes of the Kaimosi Forest, Kenya Colony, with special reference to yellow fever. *Bull. ent. Res.* **36**: 473–96

394 GILLETT, J. D. (1951) The habits of the mosquito *Aëdes* (*Stegomyia*) *simpsoni* Theobald in relation to the epidemiology of yellow fever in Uganda. *Ann. trop. Med. Parasit.* **45**: 110–21

395 GILLETT, J. D. (1955) Further studies on the biting behaviour of *Aëdes* (*Stegomyia*) *simpsoni* Theobald in Uganda. *Ann. trop. Med. Parasit.* **49**: 154–7

396 GILLETT, J. D. (1958) Laboratory tests on the maintenance of yellow fever virus in certain predatory arthropods. *Trans. R. Soc. trop. Med. Hyg.* **52**: 269–71

397 Gillett, J. D. (1968) *Trans. R. Soc. trop. Med. Hyg.* **62**: 764–5

398 Gillett, J. D. (1969) Yellow fever in East Africa today. *E. Afr. med. J.* **46**: 22–5

399 Gillett, J. D. and Ross, R. W. (1953) The laboratory transmission of yellow fever by the mosquito *Aedes* (*Stegomyia*) *strelitziae* Muspratt. *Ann. trop. Med. Parasit.* **47**: 367–70

400 Gillett, J. D. and Ross, R. W. (1955) The laboratory transmission of yellow fever by Malayan *Aedes aegypti*. *Ann. trop. Med. Parasit.* **49**: 63–5

401 Gillett, J. D., Ross, R. W., Dick, G. W. A., Haddow, A. J. and Hewitt, L. E. (1950) Experiments to test the possibility of trans-ovarial transmission of yellow fever virus in the mosquito *Aëdes* (*Stegomyia*) *africanus* Theobald. *Ann. trop. Med. Parasit.* **44**: 342–50

402 Hackett, L. W. (1945) The malaria of the Andean region of South America. *Rev. Inst. Sal. Enberm.* Trop. **6**: 239–52

403 Haddow, A. J., Smithburn, K. C., M'Ahaffy, A. F. and Bugher, J. C. (1947) Monkeys in relation to yellow fever in Bwamba County, Uganda. *Trans. R. Soc. trop. Med. Hyg.* **40**: 677–700

404 Hawking, F., Worms, M. J. and Gammage, K. (1968) 24- and 48-hour cycles of malaria parasites in the blood; their purpose, production and control. *Trans. R. Soc. trop. Med.* **62**: 731–60

405 Hughes, T. P., Jacobs, H. R. and Burke, A. W. (1941) A survey of yellow fever immunity in Uganda. *Trans. R. Soc. trop. Med. Hyg.* **35**: 131–42

406 Kershaw, W. E., Lavoipierre, M. M. J. and Chalmers, T. A. (1953) Studies on the intake of microfilariae by their insect vectors. I. *Dirofilaria immitis* and *Aëdes aegypti*. *Ann. trop. Med. Parasit.* **47**: 207–24

407 Kissling, R. E., Stamm, D. D., Chamberlain, R. W. and Sudia, W. D. (1957) Birds as winter hosts for eastern and western equine encephalomyelitis viruses. *Amer. J. Hyg.* **66**: 42–7

408 Laing, A. B. G. (1960) A review of recent research on filariasis in Malaya. *Ind. J. Malariol.* **14**: 391–408

409 Lavoipierre, M. M. J. (1958) Studies on the host-parasite relationships of filarial nematodes and their arthropod hosts. II. The arthropod as a host to the nematode: a brief appraisal of our present knowledge, based on a study of the more important

literature from 1878 to 1957. *Ann. trop. Med. Parasit.* **52**: No. 3. 326–45

410 LEWIS, D. J. (1947) Mosquitoes in relation to yellow fever in the Sudan. *Bull. ent. Res.* **37**: 543–66

411 LUMSDEN, W. H. R. (1951) The night-resting habits of monkeys in a small area on the edge of the Semliki Forest, Uganda. A study in relation to the epidemiology of sylvan yellow fever. *J. anim. Ecol.* **20**: 11–32

412 LUMSDEN, W. H. R. (1952) The crepuscular biting activity of insects in the forest canopy in Bwamba, Uganda. A study in relation to the sylvan epidemiology of yellow fever. *Bull. ent. Res.* **42**: 721–60

413 LUMSDEN, W. H. R. (1955) An epidemic of virus disease in Southern Province, Tanganyika Territory, in 1952–3. II. General description and epidemiology. *Trans. R. Soc. trop. Med. Hyg.* **49**: 33–57

414 MACDONALD, W. W., SMIGHT, C. E. G., DAWSON, P. S., GANAPATHI-PILLAI, A. and MAHADEVAN, S. (1967) Arbovirus infections in Sarawak: Further observations on mosquitoes. *J. med. Ent.* **4**: 146–57

415 MAHAFFY, A. F., SMITHBURN, K. C., JACOBS, H. R. and GILLETT, J. D. (1942) Yellow fever in western Uganda. *Trans. R. Soc. trop. Med. Hyg.* **36**: 9–20

416 MANSON, P. (1879) On the development of *Filaria sanguinis hominis* and on the mosquito considered as a nurse. *J. Linn. Soc. Lond.* **14**: 304–11

417 MOTTE, L. C. LA (1958) Japanese B Enceophalitis in bats during simulated hibernation. *Am. J. Hyg.* **67**: 101–8

418 MOTTE, L. C. LA (1960) Japanese B encephalitis virus in the organs of infected mosquitoes. *Amer. J. Hyg.* **72**: 73–87

419 MYERS, R. M. and CASEY, D. E. (1967) Concurrent isolation from patient of two arboviruses, chikungunya and dengue type 2. *Science* **157**: 1307–8

420 PAL, R. (1943) On the bionomics of *Anopheles culicifacies*. Pt. I. Longevity under controlled conditions of temperature and humidity. *J. malar. Inst. India* **5**: 77–85

421 REED, W., CARROLL, J., AGRAMONTE, A. and LAZEAR, J. W. (1900) The etiology of yellow fever. A preliminary note. *Philad. Med. J.* **6**: 790–6

422 REEVES, W. C. (1965) Ecology of mosquitoes in relation to arboviruses. *Ann. Rev. Ent.* **10**: 25–46

423 ROCA-GARCIA, M. (1944) The isolation of three neurotropic viruses from forest mosquitoes in eastern Colombia. *J. infect. Dis.* **75**: 160–9

424 ROSS, R. W. (1956) The Newala epidemic. III. The virus: Isolation, pathogenic properties and relationship to the epidemic. *J. Hyg.* **54**: 177–91

425 ROSS, R. W. and GILLETT, J. D. (1950) The cyclical transmission of yellow fever through the grivet monkey, *Cercopithecus aethiops centralis* Neumann, and the mosquito *Aëdes* (*Stegomyia*) *africanus* Theobald. *Ann. trop. Med. Parasit.* **44**: 351–6

426 ROUBAUD, E., COLAS-BELCOUR, J. and STEFANOPOULO, G. J. (1937) Transmission de la fièvre jaune par un moustique paléarctique répandu dans la region parisienne, *l'Aëdes geniculatus* Oliv. *C.R. Acad. Sci. Fr.* **205**: 182–3

427 SÉRIÉ, C., ANDRAL, L., LINDREC, A. and NERI, P. (1964) Epidémie de fièvre jaune en Ethiopie (1960–1962): observations preliminaires. *Bull. Wld. Hlth. Org.* **30**: 299–319

428 SMITHBURN, K. C. and HADDOW, A. J. (1949) The susceptibility of African wild animals to yellow fever. *Amer. J. trop. Med.* **29**: 389–405

429 SMITHBURN, K. C., HADDOW, A. J. and GILLETT, J. D. (1948) Rift Valley fever. Isolation of the virus from wild mosquitoes. *Brit. J. exp. Path.* **29**: 107–21

430 SMITHBURN, K. C., HADDOW, A. J. and LUMSDEN, W. H. R. (1949) Outbreak of sylvan yellow fever in Uganda with *Aëdes* (*Stegomyia*) *africanus* Theobald as principal vector and insect host of virus. *Ann. trop. Med. Parasit.* **43**: 74–89

431 SOPER, F. L. (1958) The 1959 status of yellow fever in the Americas. *Mosquito News* **18**: 203–16

432 STRODE, G. K. (1951) *Yellow Fever.* McGraw-Hill Publ. Co.: London

433 THEILER, M. and SMITH, H. H. (1937) The use of yellow fever virus modified by *in vitro* cultivation for human immunisation. *J. exp. Med.* **65**: 789–800

434 TRAPIDO, H. and GALINDO, P. (1957) Mosquitoes associated with sylvan yellow fever near Almirante, Panama. *Amer. J. trop. Med. Hyg.* **6**: 114–44

435 WALLIS, R. C., TAYLOR, R. M., MCCOLLUM, R. W. and RIORDAN, J. T. (1958) Study of hibernating mosquitoes in eastern equine encephalomyelitis epidemic areas in Connecticut. *Mosquito News* **18**: 1–4

436 WILLIAMS, M. C., WOODALL, J. P. and GILLETT, J. D. (1965) O'nyong-nyong fever: An epidemic virus disease in East Africa. VII. Virus isolations from man and serological studies up to July 1961. *Trans. R. Soc. trop. Med. Hyg.* **59**: 186–97

437 WILLIAMS, M. C., WOODALL, J. P., CORBET, P. S. and GILLETT, J. D. (1965) O'nyong-nyong fever: An epidemic virus disease in East Africa. VIII. Virus isolations from *Anopheles* mosquitoes. *Trans. R. Soc. trop. Med. Hyg.* **59**: 300–6

438 WOODALL, J. P., GILLETT, J. D., CORBET, P. S., WEINBREN, M. P. and WILLIAMS, M. C. (1961) The isolation of West Nile virus from the bird-biting mosquito *Mansonia metallica* in Uganda. *Ann. trop. Med. Parasit.* **55**: 398–402

14 MOSQUITOS AND HISTORY

439 ACKERKNECHT, E. H. (1965) *History and geography of the most important diseases.* Hafner Publishing Co. Inc.: New York

440 BENNETT, I. E. (1915) *History of the Panama Canal, its Construction and Builders.* Historical Publ. Co.: Washington, D.C.

441 BOYCE, R. (1909) *Mosquito or Man.* John Murray: London

442 BROWN, A. W. A. (1951) *Insect Control by Chemicals.* Chapman and Hall: London

443 BRUCE-CHWATT, L. J. (1956) Radio-isotopes for research on and control of mosquitoes. *Bull. World Hlth. Org.* **15**: 491–511

444 BUNAU-VARILLA, P. (1920) *The Great Adventure of Panama.* Doubleday, Page and Co.: Garden City, New York

445 BUSVINE, J. (1951) *Insects and Hygiene.* Methuen & Co.: London

446 BUSVINE, J. (1957) Insecticide resistance. Strains of insects of public health importance. *Trans. R. Soc. trop. Med. Hyg.* **51**: 11–31

447 CHRISTOPHERS, S. R. (1947) Mosquito repellents. *J. Hyg.* **45**: 176–231

448 CREIGHTON, C. (1965) *A history of epidemics in Britain (with additional material by Eversley, D. E. C., Underwood, E. A. and Ovenall, L.).* 2 vols. Frank Cass & Co. Ltd.: London

449 DETHIER, V. G. (1956) Repellents. In *Ann. Rev. Ent.* **1**: 181–202

450 ENOCK, R. (1914) *The Panama Canal.* Collins: London

451 FAY, R. W., MCCRAY, JR., E. M. and KILPATRICK, J. W. (1963) Mass production of sterilised male *Aedes aegypti. Mosquito News* **23**: 210–14

452 GILOTRA, S. K. (1965) Inheritance of dieldrin resistance in the larvae of *Anopheles albimanus* Wiedemann. *Am. J. trop. Med. Hyg.* **14**: 838–45

453 GILOTRA, S. K. (1966) Effects of larval environments on adult resistance in *Anopheles albimanus* Wiedemann. *Mosquito News* **26**: 26–30

454 GORGAS, W. C. (1915) *Sanitation in Panama.* D. Appleton & Co.: New York

455 HAMMOND, R. and LEWIN, C. J. (1966) *The Panama Canal.* Frederick Muller: London

456 HAMON, J. and MOUCHET, J. (1961) La resistance aux insecticides chez les insectes d'importance medicale. Methodes d'études et situation en Afrique au Sud du Sahara. *Med. Trop.* **21**: 565–96

457 HAYES, JR., G. R. and TINKER, M. E. (1958) The 1956–7 status of *Aedes aegypti* in the United States. *Mosquito News* **18**: 253–7

458 KEEVIL, J. J. (1958) *Medicine and the Navy 1200–1900.* 2 vols. E. & S. Livingstone Ltd: Edinburgh

459 MORLAN, H. B., MCCRAY, JR., E. M. and KILPATRICK, J. W. (1962) Field tests with sexually sterile males for control of *Aedes aegypti. Mosquito News* **22**: 295–300

460 NOGUEIRA, P. (1955) The early history of yellow fever. In *Yellow Fever, a Symposium in Commemoration of Carlos Juan Finlay.* Jefferson Med. Col., Philadelphia

461 ROZEBOOM, L. E. and JOHNSON, R. (1961) Inheritance of resistance to dieldrin in *Anopheles albimanus* Wiedemann. *Am. J. trop. Med. Hyg.* **10**: 775–81

462 RUSSELL, P. F. (1955) *Man's Mastery of Malaria.* Oxford Univ. Press, London

463 SHANNON, R. C. (1942) *Anopheles gambiae. Carib. med. J.* **4**: 1–7

464 SHUTE, P. G. and MARYON, M. E. (1964) Imported malaria as a serious problem during surveillance and consolidation phases of malaria eradication campaigns. *J. trop. Med. Hyg.* **67**: 19–23

465 WIGGLESWORTH, V. B. (1936*a*) The malaria epidemic in Ceylon 1934–1935. *Proc. R. Soc. Med.* **29**: 20–3

466 WIGGLESWORTH, V. B. (1936*b*) Malaria in Ceylon. *The Asiatic Rev.* **3**: London

467 WIGGLESWORTH, V. B. (1942) Some notes on the integument of insects in relation to the entry of contact insecticides. *Bull. ent. Res.* **33**: 205–18

468 WIGGLESWORTH, V. B. (1948) Mode of action of new insecticides. *Rep. 5th Commonw. Conf. Ent.* (1948), 35–7

INDEX

Note: Numbers in boldface refer to plates.

J. D. GILLETT is head of the School of Biological Sciences at Brunel University in Middlesex, England. He has served as entomologist in the Uganda Medical Department and as Assistant Director of the East African Virus Research Institute. His work with colleagues in Africa on the ecology of mosquitos has led to our present understanding of yellow fever and other mosquito-borne virus diseases.

Professor Gillett is a frequent consultant to the World Health Organization and is the author of numerous scientific papers.